D0151874

Unified Design
of Steel Structures

BICENTENNIAL

1807

⊛WILEY

2007

BICENTENNIAL

THE WILEY BICENTENNIAL–KNOWLEDGE FOR GENERATIONS

*E*ach generation has its unique needs and aspirations. When Charles Wiley first opened his small printing shop in lower Manhattan in 1807, it was a generation of boundless potential searching for an identity. And we were there, helping to define a new American literary tradition. Over half a century later, in the midst of the Second Industrial Revolution, it was a generation focused on building the future. Once again, we were there, supplying the critical scientific, technical, and engineering knowledge that helped frame the world. Throughout the 20th Century, and into the new millennium, nations began to reach out beyond their own borders and a new international community was born. Wiley was there, expanding its operations around the world to enable a global exchange of ideas, opinions, and know-how.

For 200 years, Wiley has been an integral part of each generation's journey, enabling the flow of information and understanding necessary to meet their needs and fulfill their aspirations. Today, bold new technologies are changing the way we live and learn. Wiley will be there, providing you the must-have knowledge you need to imagine new worlds, new possibilities, and new opportunities.

Generations come and go, but you can always count on Wiley to provide you the knowledge you need, when and where you need it!

WILLIAM J. PESCE
PRESIDENT AND CHIEF EXECUTIVE OFFICER

PETER BOOTH WILEY
CHAIRMAN OF THE BOARD

Unified Design of Steel Structures

Louis F. Geschwindner

Vice President of Engineering and Research
American Institute of Steel Construction
and
Professor Emeritus of Architectural Engineering
The Pennsylvania State University

WILEY

JOHN WILEY & SONS, INC.

VP & Executive Publisher	Don Fowley
Associate Publisher	Daniel Sayre
Acquisitions Editor	Jennifer Welter
Senior Production Editor	Valerie A. Vargas
Marketing Manager	Christopher Ruel
Creative Director	Harry Nolan
Senior Designer	Kevin Murphy
Production Management Services	Aptara, Inc.
Senior Illustration Editor	Anna Melhorn
Senior Editorial Assistant	Lindsay Murdock
Media Editor	Lauren Sapira
Cover Photo	Serge Drouin
Cover Design	David Levy
Bicentennial Logo Design	Richard J. Pacifico

The building on the cover is The New York Times Building, a joint venture of The New York Times Company and Forest City Ratner Companies.
This book was set in 10/12 Times Roman by Aptara, Inc. Printed and bound by Hamilton Printing. The cover was printed by Phoenix Color Inc.

This book is printed on acid free paper. ∞

Copyright © 2008 John Wiley & Sons, Inc. All rights reserved. No part of this publication may be reproduced, stored in a retrieval system or transmitted in any form or by any means, electronic, mechanical, photocopying, recording, scanning or otherwise, except as permitted under Sections 107 or 108 of the 1976 United States Copyright Act, without either the prior written permission of the Publisher, or authorization through payment of the appropriate per-copy fee to the Copyright Clearance Center, Inc., 222 Rosewood Drive, Danvers, MA 01923, website www.copyright.com. Requests to the Publisher for permission should be addressed to the Permissions Department, John Wiley & Sons, Inc., 111 River Street, Hoboken, NJ 07030-5774, (201)748-6011, fax (201)748-6008, website: http://www.wiley.com/go/permissions.

To order books or for customer service, please call 1-800-CALL WILEY (225-5945).

ISBN-13:   978-0-471-47558-3

Printed in the United States of America
10 9 8 7 6 5 4 3 2

Preface

INTENDED AUDIENCE

This book presents the design of steel building structures based on the 2005 unified specification, ANSI/AISC 360-05 *Specification for Structural Steel Buildings*. It is intended primarily as a text for a first course in steel design for civil and architectural engineers. Such a course usually occurs in the third or fourth year of an engineering program. The book can also be used in a second, building-oriented course in steel design, depending on the coverage in the first course. In addition to its use as an undergraduate text, it provides a good review for practicing engineers looking to learn the provisions of the unified specification and to convert their practice from either of the old specifications to the new specification. Users are expected to have a firm knowledge of statics and strength of materials and have easy access to the AISC *Steel Construction Manual*, 13th Edition.

UNIFIED ASD AND LRFD

A preferred approach to the design of steel structures has been elusive over the last 20 years. In 1986, the American Institute of Steel Construction (AISC) issued its first *Load and Resistance Factor Design (LRFD) Specification for Structural Steel Buildings*. This specification came after almost 50 years of publication of an Allowable Stress Design (ASD) specification. Unfortunately, LRFD was accepted by the academic community but not by the professional engineering community. Although AISC revised the format of the ASD specification in 1989, it had not updated its provisions for over 25 years. This use of two specifications was seen as an undesirable situation by the professions and in 2001 AISC began the development of a combined ASD and LRFD specification. In 2005, AISC published its first unified specification, combining the provisions of both the LRFD and ASD specifications into a single standard for the design of steel building structures. This new specification, ANSI/AISC 360-05 *Specification for Structural Steel Buildings*, reflects a major change in philosophy by AISC, one that makes the use of ASD and LRFD equally acceptable approaches for the design of steel buildings.

The reader familiar with past editions of the ASD and LRFD specifications will undoubtedly question how these two diverse design philosophies can be effectively combined into one specification. This is a reasonable question to ask. The primary answer is that this specification is not a combination of the old ASD and LRFD provisions. It is a new approach with a new ASD that uses the same strength equations as the new LRFD. A combination of the old ASD provisions with the old LRFD provisions could lead, in some cases, to a design wherein an element is treated as behaving elastically for ASD design and plastically for LRFD design. The unified specification takes a different approach. It is based on the understanding that the strength of an element or structure, called the nominal strength in the specification, can be determined independent of the design philosophy. Once that nominal strength is determined, the available strength for ASD or LRFD is determined as a function of that nominal strength. Thus, the available strength of the element is always based on the same behavior and no inconsistency in behavior results from the use of ASD or LRFD. This important aspect of the unified specification is further explained in Chapter 1.

CHANGES IN BUILDING LOADS

In addition to the provisions for steel design issued by AISC, structural engineering has seen many changes in the area of loads for which buildings must be designed. The American Society of Civil Engineers (ASCE) is continually revising ASCE-7 *Minimum Design Loads for Buildings and Other Structures*, its standard for building loads. The International Code Council (ICC) has issued its International Building Code (IBC), and the National Fire Protection Association (NFPA) has issued its model building code (NFPA 5000). The major changes brought about by these new standards are the inclusion of requirements for consideration of seismic loading, which now applies to almost the entire country. In response to the expansion of the requirements for seismic design, AISC issued ANSI/AISC 341-05 *Seismic Provisions for Structural Steel Buildings*, a standard to guide the design of steel building structures to resist seismic loads. For the calculation of loads within this text, ASCE 7-05 provisions are used. For any actual design, the designer must use the loadings established by the governing building code. The AISC seismic provisions are discussed in Chapter 13.

UNITS

ANSI/AISC 360-05 is, as much as possible, a unitless specification. In those rare instances where equations could not be written in a unitless form, two equations are given, one in U.S. customary units and one in SI units. The Manual presents all of its material in U.S. customary units. The construction industry in this country has not adopted SI units in any visible way, and it is not clear that they will in the foreseeable future. Thus, this book uses only U.S. customary units.

TOPICAL ORGANIZATION

Chapters 1 through 3 present the general material applicable to all steel structures. This is followed in Chapters 4 through 9 with a presentation of member design. Chapters 10 through 12 discuss connections and Chapter 13 provides an introduction to seismic design.

In Chapter 1, the text addresses the principles of limit states design upon which all steel design is based. It shows how these principles are incorporated into both LRFD and ASD approaches. Chapter 2 introduces the development of load factors, resistance factors, and safety factors. It discusses load combinations and compares the calculation of required strength for both LRFD and ASD. Chapter 3 discusses steel as a structural material. It describes the availability of steel in a variety of shapes and the grades of steel available for construction.

Once the foundation for steel design is established, the various member types are considered. Tension members are addressed in Chapter 4, compression members in Chapter 5, and bending members in Chapter 6. Chapter 7 covers plate girders, which are simply bending members made from individual plates. Chapter 8 treats members subjected to combined axial load and bending as well as design of bracing. Chapter 9 deals with composite members, that is, members composed of both steel and concrete working together to provide the available strength. Each of these chapters begins with a discussion of that particular member type and how it is used in buildings. This is followed by a discussion of the specification provisions and the behavior from which those provisions have been derived. The LRFD and ASD design philosophies of the 2005 specification are used throughout. Design examples that use the specification provisions directly are provided along with examples using the variety of design aids available in the AISC *Steel Construction Manual*. All examples that have an LRFD and ASD component are provided for both approaches. Throughout this book, ASD examples, or portions of examples that address the ASD approach, are presented with shaded background for ease of identification.

The member-oriented chapters are followed by chapters addressing connection design. Chapter 10 introduces the variety of potential connection types and discusses the strength of bolts, welds, and connecting elements. Chapter 11 addresses simple connections. This includes simple beam shear connections and light bracing connections. Chapter 12 deals with moment-resisting connections. As with the member-oriented chapters, the basic principles of limit states design are developed first. This is followed by the application of the provisions to simple shear connections and beam-to-column moment connections through extensive examples in both LRFD and ASD.

The text concludes in Chapter 13 with an introduction to steel systems for seismic force resistance. It discusses the variety of structural framing systems available and approved for inclusion in the seismic force resisting system.

EXAMPLES AND HOMEWORK PROBLEMS IN LRFD AND ASD

The LRFD and ASD design philosophies of the 2005 specification are used throughout. Design examples that use the specification provisions directly are provided along with examples using the variety of design aids available in the AISC *Steel Construction Manual*. All examples that have an LRFD and ASD component are provided for both approaches. Throughout this book, ASD examples, or portions of examples that address the ASD approach, are presented with shaded background for ease of identification.

GOAL: Select a double-angle tension member for use as a web member in a truss and determine the maximum area reduction that would be permitted for holes and shear lag.

GIVEN: The member must carry an ASD required strength, $P_a = 270$ kips. Use equal leg angles of A36 steel.

Step 1: Determine the minimum required gross area based on the limit state of yielding

$$A_{g\ min} = 270/(36/1.67) = 12.5 \text{ in.}^2$$

Step 2: Based on this minimum gross area, from Manual Table 1-15, select

2L6×6×^{9}/₁₆ with $A_g = 12.9$ in.^{2}

Each chapter includes homework problems at the end of the chapter. These problems are organized to follow the order of presentation of the material in the chapters. Several problems are provided for each general subject. Problems are provided for both LRFD and ASD solutions. There are also problems designed to show comparisons between ASD and LRFD solutions. These problems show that in some instances one method might give a more economical design, whereas in other instances the reverse is true.

WEBSITE

The following resources are available from the book website at www.wiley.com/college/geschwindner. Visit the Student section of the website.

- **Answers** Selected homework problem answers are available on the student section of the website.
- **Errata** We have reviewed the text to make sure that it is as error-free as possible. However, if any errors are discovered, they will be listed on the book website as a reference.
- If you encounter any errors as you are using the book, please send them directly to the author (LFG@psu.edu) so we may include them on the website, and correct these errors in future editions.

RESOURCES FOR INSTRUCTORS

All resources for instructors are available on the Instructor section of the website at www.wiley.com/college/geschwindner.

The following resources are available only to instructors who adopt the text:

- **Solutions Manual:** Solutions for all homework problems in the text.
- **Image Gallery of Text Figures**
- **Text Figures in PowerPoint format**

Visit the Instructor section of the website at www.wiley.com/college/geschwindner to register and request access to these resources.

ACKNOWLEDGEMENTS

I would like to thank all of my former students for their interactions over the years and the influence they had on the development of my approach to teaching. In particular I would like to thank Chris Crilly and Andy Kauffman for their assistance in reviewing the manuscript, checking calculations, and assistance with the figures. I would like to thank Charles Carter of AISC, a former student and valued colleague, for his authorship of Chapter 13. A special note of thanks is due Larry Kruth of Douglas Steel Fabricating Corporation for his review and assistance with figures in Chapters 10 through 12. I also want to thank those who reviewed the draft manuscripts for their valuable suggestions and those faculty members who have chosen to class test the draft of this text prior to the actual publication of the work.

REVIEWERS

Sonya Cooper, New Mexico State University
Jose Gomez, Virginia Transportation Research Council
Jeffery A. Laman, Penn State University
Dr. Craig C. Menzemer, The University of Akron
Levon Minnetyan, Clarkson University
Candace S. Sulzbach, Colorado School of Mines

CLASS TESTERS

Dr. Chris Tuan, University of Nebraska at Omaha; Catherine Frend, University of Minnesota; David G Pollock, Washington State University; Kelly Salyards, Bucknell University; P. K. Saha, Alabama A&M University; Marc Leviton, Louisiana State University; Chai H. Yoo, Auburn University; Dr. Anil Patnaik, South Dakota School of Mines and Technology; Bozidar Stjadinovic, University of California; Dimitris C. Rizos, University of South Carolina; Chia-Ming Uang, University of California, San Diego.

Finally, I want to thank my wife, Judy, for her understanding and that not-so-subtle nudge when it was really needed. Her continued support has permitted me to complete this project.

Louis F. Geschwindner
State College, Pennsylvania

Contents

Chapter 1

Hearst Tower, New York City.
Photo courtesy WSP Cantor Seinuk.

Introduction

1.1 SCOPE

A wide variety of designs can be characterized as *structural steel design.* This book deals with the design of steel structures for buildings as governed by *ANSI/AISC 360-05 Specification for Structural Steel Buildings*, published by the American Institute of Steel Construction (AISC) in 2005, and referred to as the Specification in this book. The areas of application given throughout this book specifically focus on the design of steel building structures. The treatment of subjects associated with bridges and industrial structures, if addressed at all, is kept relatively brief.

The book addresses the concepts and design criteria for the two design approaches detailed by the Specification: Load and Resistance Factor Design (LRFD) and Allowable Strength Design (ASD). Both methods are discussed later in this chapter.

In addition to the Specification, the primary reference for this book is the 13th edition of the *AISC Steel Construction Manual.* This reference handbook contains tables of the basic values needed for structural steel design, design tables to simplify actual design, and the complete Specification. Throughout this book, this is referred to as the Manual.

1.2 PRINCIPLES OF STRUCTURAL DESIGN

From the time an owner determines a need to build a building, through the development of conceptual and detailed plans, to completion and occupancy, a building project is a multi-faceted task that involves many professionals. The owner and the financial analysis team evaluate the basic economic criteria for the building. The architects and engineers form the design team and prepare the initial proposals for the building, demonstrating how the users' needs will be met. This teamwork continues through the final planning and design stages, where the detailed plans, specifications, and contract documents are readied for the construction phase. During this process, input may also be provided by the individuals who will transform the plans into a real life structure. Thus, those responsible for the construction phase of the project often help improve the design by taking into account the actual on-site requirements for efficient construction.

Once a project is completed and turned over to the owner, the work of the design teams is normally over. The operation and maintenance of the building, although major activities in the life of the structure, are not usually within the scope of the designer's responsibilities, except when significant changes in building use are anticipated. In such cases, a design team should verify that the proposed changes can be accommodated.

The basic goals of the design team can be summarized by the words *safety*, *function*, and *economy*. The building must be safe for its occupants and all others who may come in contact with it. It must neither fail locally nor overall, nor exhibit behavioral characteristics that test the confidence of rational human beings. To help achieve that level of safety, building codes and design specifications are published that outline the minimum criteria that any structure must meet.

The building must also serve its owner in the best possible way to ensure that the functional criteria are met. Although structural safety and integrity are of paramount importance, a building that does not serve its intended purpose will not have met the goals of the owner.

Last, but not least, the design, construction, and long-term use of the building should be economical. The degree of financial success of any structure will depend on a wide range of factors. Some are established prior to the work of the design team, whereas others are determined after the building is in operation. Nevertheless, the final design should, within all reasonable constraints, produce the lowest combined short- and long-term expenditures.

The AISC Specification follows the same principles. The mission of the AISC Committee on Specifications is to: "Develop the practice-oriented specification for structural steel buildings that provide for life safety, economical building systems, predictable behavior and response, and efficient use." Thus, this book emphasizes the practical orientation of this specification.

1.3   PARTS OF THE STEEL STRUCTURE

All structures incorporate some or all of the following basic types of structural components:

1. Tension members
2. Compression members
3. Bending members
4. Combined force members
5. Connections

The first four items represent structural members. The fifth, connections, provides the contact regions between the structural members, and thus assures that all components work together as a structure.

Detailed evaluations of the strength, behavior, and design criteria for these members are presented in the following chapters:

Tension members:	Chapter 4
Compression members:	Chapter 5
Bending members:	Chapters 6 and 7
Combined force members:	Chapter 8
Connections:	Chapters 10, 11, and 12

The strength and behavior of structural frames composed of a combination of these elements are covered in Chapters 8 and 13, and the special considerations that apply to composite (steel and concrete working together) construction are presented in Chapter 9. An introduction to the design of steel structures for earthquake loading is presented in Chapter 13. The properties of structural steel and the various shapes commonly used are discussed in Chapter 3, and a brief discussion of the types of loads and load combinations is presented in Chapter 2.

Tension members are typically found as web and chord members in trusses and open-web steel joists, as diagonal members in structural bracing systems, and as hangers for balconies, mezzanine floors and pedestrian walkways. They are also used as sag rods for purlins and girts in many building types, as well as to support platforms for mechanical equipment and pipelines. Figures 1.1 and 1.2 illustrate typical applications of tension members in actual structures.

In the idealized case, tension members transmit concentric tensile forces only. In certain structures, reversals of the overall load may change the tension member force from tension to compression. Some members will be designed for this action; others will have been designed with the assumption that they carry tension only.

The idealized tension member is analyzed with the assumption that its end connections are pins, which prevent any moment or shear force from being transmitted to the member. However, in an actual structure, the type of connection normally dictates that some bending may be introduced to the tension member. This is also the case when the tension member is directly exposed to some form of transverse load. Moments will also be introduced if the element is not perfectly straight, or if the axial load is not applied along the centroidal axis of the member.

The primary load effect in the tension member is a concentric axial force, with bending and shear considered as secondary effects.

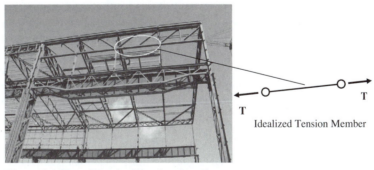

Idealized Tension Member

Figure 1.1 Use of Tension Members in a Truss.
Photo courtesy Ruby & Associates.

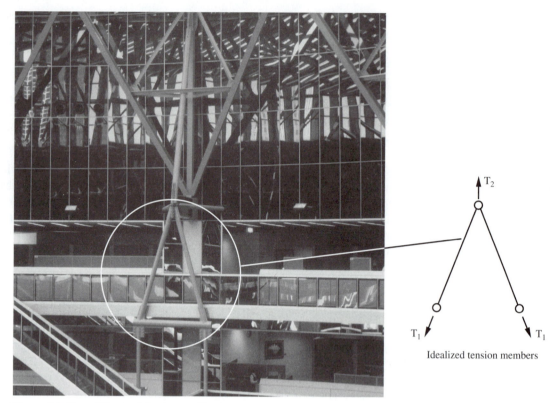

Figure 1.2 Use of Tension Members as Hangers.

Compression members are also known as columns, struts, or posts. They are used as web and chord members in trusses and joists and as vertical members in all types of building structures. Figure 1.3 shows a typical use of structural compression members.

The idealized compression member carries only a concentric, compressive force. Its strength is heavily influenced by the distance between the supports, as well as by the support conditions. The basic column is therefore defined as an axially loaded member with pinned ends. Historically, design rules for compression members have been based on the behavior and strength of this idealized compression member.

The basic column is practically nonexistent in real structures. Realistic end supports rarely resemble perfect pins; the axial load is normally not concentric, due to the way the surrounding structure transmits its load to the member; and beams and similar components are likely to be connected to the column in such a way that moments are introduced. All of these conditions produce bending effects in the member, making it a combined force member or beam-column, rather than the idealized column.

The primary load effect in the pinned-end column is therefore a concentric axial compressive force accompanied by the secondary effects of bending and shear.

Bending members are known as beams, girders, joists, spandrels, purlins, lintels, and girts. Although all of these are bending members, each name implies a certain structural application within a building:

1. Beams, girders, and joists form part of common floor systems. The beams are most often considered as the members that are directly supported by girders, which in turn are usually supported by columns. Joists are beams with fairly close spacing.

Idealized
compression
member

Figure 1.3 Use of Columns in a Building Frame.

A girder may generally be considered a higher-order bending member than a beam or joist. However, variations to this basic scheme are common.

2. The bending members that form the perimeter of a floor or roof plan in a building are known as spandrels or spandrel beams. Their design may be different from other beams and girders because the load comes primarily from only one side of the member.

3. Bending members in roof systems that span between other bending members are usually referred to as purlins.

4. Lintels are bending members that span across the top of openings in walls, usually carrying the weight of the wall above the opening as well as any other load brought into that area. They typically are seen spanning across the openings for doors and windows.

5. Girts are used in exterior wall systems. They transfer the lateral load from the wall surface to the exterior columns. They may also assist in supporting the weight of the wall.

Figure 1.4 shows the use of a variety of bending members in an actual structure.

The basic bending member carries transverse loads that lie in a plane that contains the longitudinal centroidal axis of the member. The primary load effects are bending moment and shear force. Axial forces and torsion may occur as secondary effects.

Idealized spandral

Idealized beam

Figure 1.4 Building Structure Showing Bending Members.
Photo courtesy Greg Grieco.

The most common *combined force member* is known as a beam-column, implying that this structural element is simultaneously subjected to bending and axial compression. Although bending and axial tension represents a potential loading case for the combined force member, this case is not as critical or common as the beam-column.

Figure 1.5a is a schematic illustration of a multi-story steel frame where the beams and columns are joined with rigid connections. Because of the geometric configuration, the types of connections, and the loading pattern, the vertical members are subjected to axial loads and bending moments. This is a typical case of practical beam-columns; other examples are the members of the gable frame shown in Figure 1.5b and the vertical components of a single story portal frame shown in Figure 1.5c.

The beam-column may be regarded as the general structural element, where axial forces, shear forces, and bending moments act simultaneously. Thus, the basic column may be thought of as a special case, representing a beam-column with no moments or transverse loads. Similarly, the basic bending member may be thought of as a beam-column with no axial load. Therefore, the considerations that must be accounted for in the design of both columns and beams must also apply to beam-columns.

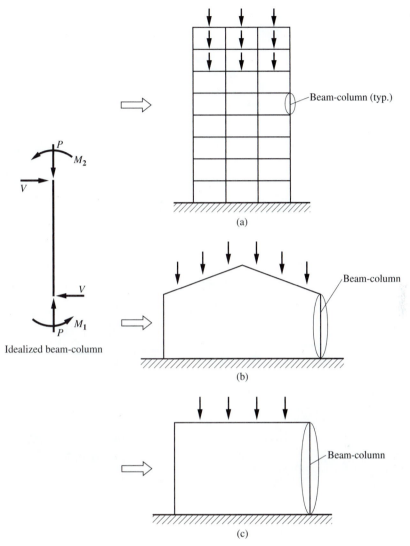

Figure 1.5 Schematic Representation of Steel Frames in Which the Vertical Members are Subjected to Axial Loads and Bending Moments.

Because of the generalized nature of the combined force element, all load effects are considered primary. However, when the ratio of axial load to axial load capacity in a beam-column becomes high, column behavior will overshadow other influences. Similarly, when the ratio of applied moment to moment capacity is high, beam behavior will outweigh other effects. The beam-column is an element in which a variety of different force types interact. Thus, practical design approaches are normally based on interaction equations.

Connections are the collection of elements that join the members of a steel structure together. Whether they connect the axially loaded members in a truss or the beams and columns of a multi-story frame, connections must ensure that the structural members function together as a unit.

The fasteners used in structural steel connections today are almost entirely limited to bolts and welds. The load effects that the various elements of the connection must resist

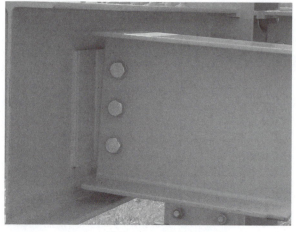

(a) Tee connection

(b) Shear end plate

(c) Shear tab

(d) Moment end plate

Figure 1.6 Building Connections.

Photos courtesy American Institute of Steel Construction.

are a function of the specific connection type being considered. They include all of the possible forces and moments. Figure 1.6 illustrates a variety of connections. The idealized representations for connections are presented in Chapters 10, 11, and 12.

1.4   TYPES OF STEEL STRUCTURES

It is difficult to classify steel structures into neat categories, due to the wide variety of systems available to the designer. The elements of the structure, as defined in Section 1.3, are combined to form the total structure of a building which must safely and economically carry all imposed loads. This combination of members is usually referred to as the framing system.

Steel framed buildings come in a wide variety of shapes and sizes and in combinations with other structural materials. A few examples are given in the following paragraphs, to set the stage for the application of structural design presented in subsequent chapters.

1.4.1   Bearing Wall Construction

This is primarily used for one- or two-story buildings, such as storage warehouses, shopping centers, office buildings, and schools. This system normally uses brick or concrete block masonry walls, on which are placed the ends of the flexural members supporting the floor or roof. The flexural members are usually hot-rolled structural steel shapes, alone or in combination with open web steel joists or cold-formed steel shapes.

1.4.2   Beam-and-Column Construction

This is the most commonly used system for steel structures today. It is suitable for large-area buildings such as schools and shopping centers, which often have no more than two stories, but may have a large number of spans. It is also suitable for buildings with many stories. Columns are placed according to a regular, repetitive grid that supports the beams, girders, and joists, which are used for the floor and roof systems. The regularity of the floor plan lends itself to economy in fabrication and erection, because most of the members will be of the same size. Further economy may be gained by using continuous beams or drop-in spans with cantilever beams, as illustrated schematically in Figure 1.7.

For multi-story buildings, the use of composite steel and concrete flexural members affords additional savings. Further advances can be expected as designers become more familiar with the use of composite columns and other elements of mixed construction systems.

Beam-and-column structures rely on either their connections or a separate bracing system to resist lateral loads. A frame in which all connections are moment resistant provides resistance against the action of lateral loads, such as wind and earthquakes, and overall structural stability, through the bending stiffness of the overall frame.

A frame without member-end restraint needs a separate lateral load resisting system, often afforded by having the elements along one or more of the column lines act as braced frames. One of the most common types of bracing is the vertical truss, which is designed to take the loads imposed by wind and seismic action. Other bracing schemes involve shear walls and reinforced concrete cores. The latter type may also be referred to as a braced core system, and can be highly efficient because of the rigidity of the box-shaped cross section of the core. The core serves a dual purpose in this case: In addition to providing the bracing system for the building, it serves as the vertical conduit in the completed structure for all of the necessary services including elevators, staircases, electricity, and other utilities.

Combinations of these types of construction are also common. For example, frames may have been designed as moment resistant in one direction of the building and as truss braced in the other. Of course, such a choice recognizes the three-dimensional nature of the structure.

Figure 1.8 shows an idealized representation of several types of beam-and-column framed structures.

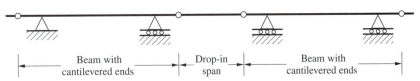

Figure 1.7 Use of Cantilever Beams with Drop-In Spans.

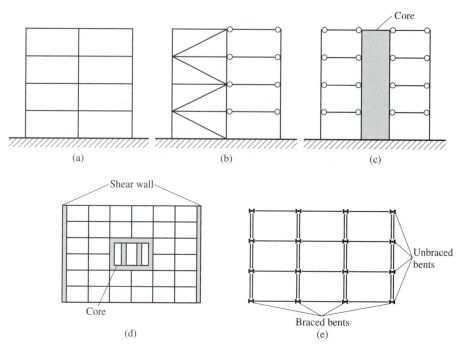

Figure 1.8 An Idealized Illustration of Several Types of Beam-and-Column Framed Structures: (a) moment-resistant frame; (b) truss-braced frame; (c) core-braced frame; (d) floor plan of shear wall and core-braced building; (e) floor plan of building with a combination of braced and unbraced bents.

1.4.3   Long-Span Construction

This type of construction encompasses steel-framed structures with long spans between the vertical load-carrying elements, such as covered arenas. The long distances may be spanned by one-way trusses, two-way space trusses, or plate and box girders. Arches or cables could also be used, although they are not considered here.

Long-span construction is also used in buildings that require large, column-free interiors. In such cases the building may be a core- or otherwise braced structure, where the long span is the distance from the exterior wall to the core.

Many designers would also characterize single-story rigid frames as part of the long-span construction systems. Depending on the geometry of the frame, such structures can span substantial distances, often with excellent economy.

1.4.4   High-Rise Construction

High-rise construction refers to multistory buildings. The large heights and unique problems encountered in the design of such structures warrant treating them independently from typical beam-and-column construction. In addition, over the past 30 years several designers have developed a number of new concepts in multi-story frame design, such as the super composite column and the steel plate shear wall.

Particular care must be exercised in the choice and design of the lateral load resisting system in high-rise construction. It is not just a matter of extrapolating from the principles used in the analysis of lower rise structures, because many effects play a significant role in the design of high-rise buildings, but have significantly less impact on frames of smaller height. These effects are crucial to the proper design of the high-rise structure.

Some of these effects may be referred to as second-order effects, because they cannot be quantified through a normal, linearly elastic analysis of the frame. Although they may be present in lower-height structures, they may be more significant in high-rise structures. An example of second-order effects is the additional moment induced in a column due to the eccentricity of the column loads that develops when a structure is displaced laterally. When added to the moments and shears produced by gravity and wind loads, the resulting effects may be significantly larger than those computed without considering the second-order effects. A designer who does not incorporate both will be making a serious and perhaps unconservative error.

Framing systems for high-rise buildings reflect the increased importance of lateral load resistance. Thus, attempts at making the perimeter of a building act as a unit or tube have proven quite successful. This tube may be in the form of a truss, such as the John Hancock Building in Chicago, Illinois, shown in Figure 1.9a or a frame as in the former World Trade Center in New York City, shown in Figure 1.9b; a solid wall tube with cutouts for windows, such as the Aon Center in Chicago, shown in Figure 1.9c; or several interconnected or bundled tubes, such as the Sears Tower in Chicago, shown in Figure 1.9d.

1.4.5   Single-Story Construction

Many designers include the single story frame as part of the long-span construction category. These structures lend themselves particularly well to fully welded construction. The pre-engineered building industry has capitalized on the use of this system through fine-tuned designs of frames for storage warehouses, industrial buildings, temporary and permanent office buildings, and similar types of structures.

1.5   DESIGN PHILOSOPHIES

A successful structural design results in a structure that is safe for its occupants, can carry the design loads without overstressing any components, does not deform or vibrate excessively, and is economical to build and operate for its intended life span. Although economy may appear to be the primary concern of an owner, safety must be the primary concern of the engineer. Costs of labor and materials will vary from one geographic location to another, making it almost impossible to design a structure that is equally economical in all locations. Because the foremost task of the designer is to produce a safe and serviceable structure, design criteria such as those published by the American Institute of Steel Construction are based on technical models and considerations that predict structural behavior and material response. The use of these provisions by the designer will dictate the economy of a particular solution in a particular location and business climate.

To perform a structural design, it is necessary to quantify the causes and effects of the loads that will be exerted on each element throughout the life of the structure. This is generally termed the *load effect* or the *required strength*. It is also necessary to account for the behavior of the material and the shapes that compose these elements. This is referred to as the *nominal strength* or *capacity of the element*.

In its simplest form, structural design is the determination of member sizes and their corresponding connections, so that the strength of the structure is greater than the load effect. The degree to which this is accomplished is often termed the *margin of safety*. Numerous approaches for accomplishing this goal have been used over the years.

Figure 1.9 High-Rise Buildings: (a) the John Hancock Center; (b) the World Trade Center; (Photo courtesy Leslie E. Robertson, RLLP.) (c) the Aon Center; (d) the Sears Tower.

Although past experience might seem to indicate that the structural designer knows the exact magnitude of the loads that will be applied to the structure, and the exact strength of all of the structural elements, this is usually not the case. Design loads are provided by many codes and standards and, although the values that are given are specific, significant uncertainty is associated with those magnitudes. Loads, load factors, and load combinations are discussed in Chapter 2.

As is the case for loading, significant uncertainty is associated with the determination of behavior and strength of structural members. The true indication of load-carrying capacity is given by the magnitude of the load that causes the failure of a component or the structure as a whole. Failure may occur either as a physical collapse of part of the building, or considered to have occurred if deflections, for instance, are larger than certain predetermined values. Whether the failure is the result of a lack of strength (collapse) or stiffness (deflection), these phenomena reflect the limits of acceptable behavior of the structure. Based on these criteria, the structure is said to have reached a specific limit state. A strength failure is termed an ultimate limit state whereas a failure to meet operational requirements, such as deflection, is termed a serviceability limit state.

Regardless of the approach to the design problem, the goal of the designer is to ensure that the load on the structure and its resulting load effect, such as bending moment, shear force, and axial force, in all cases, are sufficiently below each of the applicable limit states. This assures that the structure meets the required level of safety or reliability.

Three approaches for the design of steel structures are permitted by the AISC Specification:

1. Allowable Strength Design (ASD)
2. Load and Resistance Factor Design (LRFD)
3. Inelastic Design

Each design approach represents an alternate way of formulating the same problem, and each has the same goal. All three are based on the nominal strength of the element or structure. The nominal strength, most generally defined as R_n, is determined in exactly the same way, from the exact same equations, whether used in ASD or LRFD. Some formulations of Inelastic Design, such as Plastic Design, also use these same nominal strength equations whereas other approaches to inelastic design model in detail every aspect of the structural behavior and do not rely on the equations provided through the Specification. The use of a single nominal strength for both ASD and LRFD permits the unification of these two design approaches. It will become clear throughout this book how this approach has simplified steel design for those who have struggled in the past with comparing the two available philosophies. The following sections describe these three design approaches.

1.6   FUNDAMENTALS OF ALLOWABLE STRENGTH DESIGN (ASD)

Allowable Strength Design was formerly referred to as allowable stress design. It is the oldest approach to structural design in use today and has been the foundation of AISC Specifications since the original provisions of 1923. Allowable stress design was based on the assumption that under actual load, stresses in all members and elements would remain elastic. To meet this requirement, a safety factor was established for each potential stress-producing state. Although historically ASD was thought of as a stress-based design approach, the allowable strength was always obtained by the proper combination of the allowable stress and the corresponding section property, such as area or elastic section modulus.

The current Allowable Strength Design approach is based on the concept that the required strength of a component is not to exceed a certain permitted or allowable strength under normal in-service conditions. The required strength is determined on the basis of specific ASD load combinations and an elastic analysis of the structure. The allowable strength incorporates a factor of safety, Ω, and uses the nominal strength of the element under consideration. This strength could be presented in the form of a stress if the appropriate section property were used. In doing this, the resulting stresses will most likely again be within the elastic range, although this is not a preset requirement of the Specification.

The magnitude of the factor of safety and the resulting allowable strength depend on the particular governing limit state against which the design must produce a certain margin of safety. Safety factors are obtained from the Specification. This requirement for ASD is provided in Section B3.4 of the Specification as

$$R_a \leq \frac{R_n}{\Omega} \tag{1.1}$$

which can be stated as:

$$\text{Required Strength (ASD)} \leq \frac{\text{Nominal Strength}}{\text{Safety Factor}} = \text{Allowable Strength}$$

The governing strength depends on the type of structural element and the limit states being considered. Any single element can have multiple limit states that must be assessed. The safety factor specified for each limit state is a function of material behavior and the limit state being considered. Thus, it is possible for each limit state to have its own unique safety factor.

Design by ASD requires that the allowable stress load combinations of the building code be used. Loads and load combinations are discussed in detail in Chapter 2.

1.7   FUNDAMENTALS OF LOAD AND RESISTANCE FACTOR DESIGN (LRFD)

Load and Resistance Factor Design explicitly incorporates the effects of the random variability of both strength and load. Because the method includes the effects of these random variations and formulates the safety criteria on that basis, it is expected that a more uniform level of reliability, thus safety, for the structure and all of its components, will be attained.

LRFD is based on the concept that the required strength of a component under LRFD load combinations is not to exceed the design strength. The required strength is obtained by increasing the load magnitude by load factors that account for load variability and load combinations. The design strength is obtained by reducing the nominal strength by a resistance factor that accounts for the many variables that impact the determination of member strength. Load factors for LRFD are obtained from the building codes for strength design and will be discussed in Chapter 2. As for ASD safety factors, the resistance factors are obtained from the Specification.

The basic LRFD provision is provided in Section B3.3 of the Specification as:

$$R_u \leq \phi R_n \tag{1.2}$$

which can be stated as

$$\text{Required Strength (LRFD)} \leq \text{Resistance Factor} \times \text{Nominal Strength} = \text{Design Strength}$$

LRFD has been a part of the AISC Specifications since it was first issued in 1986.

1.8   INELASTIC DESIGN

The Specification permits a wide variety of formulations for the inelastic analysis of steel structures through the use of Appendix 1. Any inelastic analysis method will require that the structure and its elements are modeled in sufficient detail to account for all types of behavior. An analysis of this type must be able to track the structure's behavior from the unloaded condition through every load increment to complete structural failure. The only inelastic design approach that will be discussed in this book is plastic design (PD).

Plastic design is an approach that has been available as an optional method for steel design since 1961, when it was introduced as Part 2 of the then current Specification. The limiting condition for the structure and its members is the attainment of the load that would cause the structure to collapse. This load would usually be called the ultimate strength or the plastic collapse load. For an individual structural member this means that its plastic moment capacity has been reached. In most cases, due to the ductility of the material and the member, the ultimate strength of the entire structure will normally not have been reached at this stage. The less-highly stressed members can take additional load until a sufficient number of members have exhausted their individual capacities so that no further redistribution or load sharing is possible. At the point where the structure can take no additional load, the structure is said to have collapsed. This load magnitude is called the *collapse load* and is associated with a particular collapse mechanism.

The collapse load for plastic design is the service load times a certain load factor. The limit state for a structure that is designed according to the principles of plastic design is therefore the attainment of a mechanism. For this to occur, all of the structural members must be able to develop the full-yield stress in all fibers at the most highly loaded locations.

There is a fine line of distinction between the load factor of PD and the safety factor of ASD. The former is the ratio between the plastic collapse load and the service or specified load for the structure as a whole, whereas the latter is an empirically developed, experience-based term that represents the relationship between the elastic strength of the elements of the structure and the various limiting conditions for those components. Although numerically close, the load factor of plastic design and the factor of safety of allowable stress design are not the same parameter.

1.9   STRUCTURAL SAFETY

The preceding discussions of design philosophies indicate that although the basic goal of any design process is to assure that the end product will be a safe and reliable structure, the ways in which this is achieved may vary substantially.

In the past, the primary goal for safety was to ensure an adequate margin against the consequences of overload. Load factor design and its offshoots were developed to take these considerations into account. In real life, however, many other factors also play a role. These include, but are not limited to the following:

1. Variations of material strength
2. Variations of cross-sectional size and shape
3. Accuracy of method of analysis
4. Influence of workmanship in shop and field
5. Presence and variation of residual stresses
6. Lack of member straightness
7. Variations of locations of load application points

These factors consider only some of the sources of variation of the strength for a structure and its components. An even greater source of variation is the loading, which is further complicated by the fact that different types of load have different variational characteristics.

Thus, a method of design that does not attempt to incorporate the effects of strength and load variability will be burdened with sources of uncertainty that are unaccounted for. The realistic solution, therefore, is to deal with safety as a probabilistic concept. This is the foundation of load and resistance factor design, where the probabilistic characteristics of load and strength are evaluated, and the resulting safety margins determined statistically. Each load type is given its own specific factor in each combination and each material limit state is also given its own factor. This method recognizes that there is always a finite, though very small, chance that structural failure will actually occur. However, this method does not attempt to attach specific values to this probability. No specific level of probability of failure is given or implied by the Specification.

In ASD, the variability of load and strength are not treated explicitly as separate issues. They are lumped together through the use of a single factor of safety. The factor of safety varies with each strength limit state but does not vary with load source. ASD can be thought of as LRFD with a single load factor. LRFD designs are generally expected to have a more uniform level of reliability than ASD designs. That is, the probability of failure of each element in an LRFD design will be the same, regardless of the type of load or load combination. However, a detailed analysis of reliability under the LRFD provisions shows that reliability varies under various load combinations. In ASD there is no attempt to attain uniform reliability; rather the goal is to simply have a safe structure, though some elements will be safer than others.

For the development of LRFD, load effect (member force), Q, and resistance (strength), R, are assumed to have a variability that can be described by the normal distributions shown with the bell-shaped curves in Figure 1.10. Structures can be considered safe as long as the resistance is always greater than the load effect, $R > Q$. If it were appropriate to concentrate solely on the mean values, Q_m and R_m, it would be relatively easy to assure a safe structure. However, the full representation of the data shows an area where the two curves overlap. This area represents cases where the load effect exceeds the resistance and would therefore define occurrences of failure. Safety of the structure is a function of the size of this region of overlap. The smaller the region of overlap, the smaller the probability of failure.

Another approach to presenting the data is to look at the difference between resistance and load effect. Figure 1.11 shows the same data as that in Figure 1.10 but presented as $(R - Q)$. For all cases where $(R - Q) < 0$, the structure is said to have failed and for all cases where this difference is positive, the structure is considered safe. In this presentation

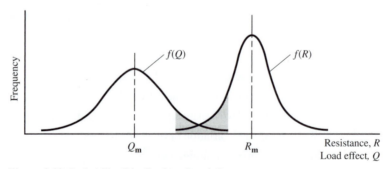

Figure 1.10 Probability Distribution, R and Q.

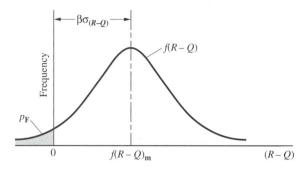

Figure 1.11 Probability Distribution, $(R - Q)$

of the data, the shaded area to the left of the origin represents the probability of failure. To limit that probability of failure, the mean value, $(R - Q)_m$, must be maintained at an appropriate distance from the origin. This distance is shown in Figure 1.11 as $\beta\sigma_{(R-Q)}$, where β is the reliability index and $\sigma_{(R-Q)}$ is the standard deviation of $(R - Q)$.

A third representation of the data is shown in Figure 1.12. In this case, the data is presented as $\ln(R/Q)$. The logarithmic form of the data is a well-conditioned representation and is more useful in the derivation of the factors required in LRFD. If we know the exact distribution of the resistance and load effect data, the probability of failure can be directly related to the reliability index β. Unfortunately, we know the actual distributions for relatively few resistance and load effect components. Thus, we must rely on other characteristics of the data, such as means and standard deviations.

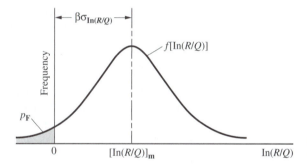

Figure 1.12 Probability Distribution, $\ln(R/Q)$

The statistical analyses required to establish an appropriate level of reliability have been carried out by the appropriate specification committees, and the resulting load factors, resistance factors, and safety factors have been established. Load factors are presented in the building codes whereas resistance factors and safety factors for each limit state are given in the Specification. A more detailed discussion of the statistical basis of steel design is available in *Load and Resistance Factor Design of Steel Structures*.[1]

1.10   LIMIT STATES

Regardless of the design approach, ASD or LRFD, or the period in history when a design is carried out, 1923 or 2007, all design is based on the ability of a structure or its elements to resist load. This ability is directly related to how an element carries that load and how it might be expected to fail, which is referred to as the element's limit state. Each structural

---

[1]Geschwindner, L. F., Disque, R. O., and Bjorhovde, R. *Load and Resistance Factor Design of Steel Structures*. Englewood Cliffs, NJ: Prentice Hall, 1994.

element can have multiple limit states and the designer is required to determine which of these limit states will actually limit the structure's strength.

There are two types of limit states to be considered: strength limit states and serviceability limit states. Strength limit states are those limiting conditions that, if exceeded, will lead to collapse of the structure or a portion of the structure, or such large deformations that the structure can no longer be expected to resist the applied load. Strength limit states are identified by the Specification, and guidance is provided for determination of the nominal strength, R_n, the safety factor, Ω, and the resistance factor, ϕ. Examples of the more common strength limit states found in the Specification are yielding, rupture, and buckling.

Serviceability limit states are less well defined than strength limit states. If a serviceability limit state is exceeded, it usually means that the structure has reached some performance level that someone would find objectionable. The Specification defines serviceability in Section L1 as "a state in which the function of a building, its appearance, maintainability, durability, and comfort of its occupants are preserved under normal usage." Chapter L of the Specification, which treats design for serviceability, lists camber, deflections, drift, vibration, wind-induced motion, expansion and contraction, and connection slip as items to be considered although no specific limits are set on any of these limit states.

Strength and serviceability limit states will be addressed throughout this book as appropriate for the elements or systems being considered.

1.11   BUILDING CODES AND DESIGN SPECIFICATIONS

The design of building structures is regulated by a number of official, legal documents that are known by their common name as *building codes*. These cover all aspects of the design, construction, and operation of buildings, and are not limited to just the structural design aspects.

Two model codes are currently in use in the United States: the ICC International Building Code and the NFPA 5000 Building Construction and Safety Code. These have been published by private organizations and are adopted, in whole or in part, by state and local governments as the legal requirements for buildings that are to be built within their area of jurisdiction. In addition to the model codes, cities or other governmental entities have written their own local building codes.

To the structural engineer, the most important sections of a building code deal with the loads that must be used in the design, and the requirements pertaining to the use of specific structural materials. The load magnitudes are normally taken from *Minimum Design Loads for Buildings and Other Structures*, a national standard published by the American Society of Civil Engineers as ASCE-7. Alteration of the loads presented in ASCE-7 may be made by the model code authority or the local building authority upon adoption, although this practice adds complexity for designers who may be called upon to design structures in numerous locations under different political entities.

The AISC Specification is incorporated into the two model building codes by reference. The Specification, therefore, becomes part of the code, and thus part of the legal requirements of any locality where the model code is adopted. Locally written building codes also continue to exist and the AISC Specifications are normally adopted within those codes by reference also. Through these adoptions the AISC Specification becomes the legally binding standard by which all structural steel buildings must be designed. However, regardless of the specification rules, it is always the engineers' responsibility to be satisfied that their structure can carry the intended loads safely, without endangering the occupants.

1.12   PROBLEMS

1. List and define the three basic goals of a design team for the design of any building.

2. All structures are composed of some or all of five basic structural types. List these five basic structural components and provide an example of each.

3. Provide an example of each of the following types of construction. To the extent possible, identify specific buildings in your own locale.

   a. Bearing wall

   b. Beam-and-column

   c. Long-span

   d. High-rise

   e. Single-story

4. What type of structural system uses the combined properties of two or more different types of materials to resist the applied loads?

5. List and describe two types of lateral bracing systems commonly used in high-rise buildings.

6. In designing a steel structure, what must be the primary concern of the design engineer?

7. Provide a simple definition of structural design.

8. Describe the difference between the ultimate limit state of a structure and a serviceability limit state.

9. Give a description of both the LRFD and ASD design approaches.

10. Provide an example of three strength limit states.

11. Provide an example of three serviceability limit states.

Chapter 2

Puerto Rico Convention Center.
Photo courtesy Walter P Moore.

Loads, Load Factors, and Load Combinations

2.1 INTRODUCTION

Material design specifications, like the AISC Specification, do not normally prescribe the magnitudes of loads that are to be used as the basis for design. These loads vary based on the usage or type of occupancy of the building, and their magnitudes are dictated by the applicable local, regional, or state laws, as prescribed through the relevant building code.

Building code loads are given as nominal values. These values are to be used in design, even though it is well known that the actual load magnitude will differ from these specified values. This is a common usage of the term *nominal*, the same as will be used for the nominal depth of a steel member to be discussed later. These nominal values are determined on the basis of material properties for dead load, load surveys for live loads, weather data for snow and wind load, and geological data for earthquake or seismic loads. These loads are further described in Section 2.2. To be reasonably certain that these loads are not exceeded in a given structure, code load values have tended to be higher or conservative, compared to the loads on a random structure at an arbitrary point in time. This somewhat higher load level also accounts for the fact that all structural loads will exhibit some random variations as a function of time and load type.

To properly address this random variation of load, an analysis reflecting time and space interdependence should be used. This is called a *stochastic analysis*. Many studies have dealt with this highly complex phenomenon, especially as it pertains to live load in buildings. However, the use of time-dependent loads is cumbersome and does not add significantly to the safety or economy of the final design. For most design situations the building code will specify the magnitude of the loads as if they were constant or unchanging. Their time and space variations are accounted for through the use of the maximum load occurring over a certain reference or return period. As an example, the American live load criteria are based on a reference period of 50 years, whereas the Canadian criteria use a 30-year interval.

The geographical location of a structure plays an important role for several load types, such as those from snow, wind, or earthquake.

2.2   BUILDING LOAD SOURCES

Many types of loads may act on a building structure at one time or another and detailed data for each is given later. Loads of primary concern to the building designer include:

1. Dead load
2. Live load
3. Snow load
4. Wind load
5. Seismic load
6. Special loads

Each of these primary load types are characterized as to their magnitude and variability by the building code, and are described in the ensuing paragraphs.

2.2.1   Dead Load

Theoretically, the dead load of a structure remains constant throughout its lifespan. The dead load includes the self-weight of the structure, as well as the weight of any permanent construction materials such as stay-in-place formwork, partitions, floor and ceiling materials, machinery, and other equipment. The dead load may potentially vary from the magnitude used in the design, even in cases where actual element weights are accurately calculated.

The weight of all dead load elements can be exactly determined only by actually weighing and/or measuring the various pieces that compose the structure. This is almost always an impractical solution and the designer therefore usually relies on published data of building material properties to obtain the nominal dead loads to be used in design. These data can be found in such publications as ASCE 7, the model building codes, and product literature. Some variation will thus likely occur in the real structure. Similarly, differences are bound to occur between the weight of otherwise identical structures, representing another source of dead load variability. However, compared to other structural loads, dead load variations are relatively small and the actual mean values are quite close to the published data.

2.2.2   Live Load

Live load is the load on the structure that occurs from all of the non-permanent installations. It includes the weight of the occupants, the furniture and moveable equipment, plus anything else that the designer could possibly anticipate might occur in the structure. The fluctuations in live load are potentially quite substantial. They vary from being essentially

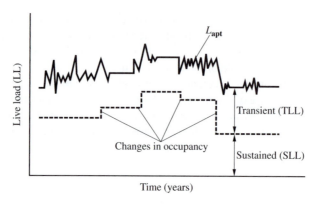

Figure 2.1 Variation of Live Load with Time.

zero immediately before the occupants take possession to a maximum value at some arbitrary point in time during the life of the structure. The magnitude of the live load to be used in a design is obtained from the appropriate building code. The actual live load on the structure at any given time may differ significantly from that specified by the building code. This is one reason why numerous attempts have been made to model live load and its variation and why measurements in actual buildings continue to be made. Although the nominal live loads found in modern building codes have not changed much over the years, the actual use of buildings has, and load surveys continue to show that the specified load levels are still an adequate representation of the loads the structure should be designed to resist.

The actual live load on a structure at any given point in time is called the *arbitrary point-in-time live load*. Figure 2.1 shows the variation of the live load on a structure as might be obtained from a live load survey. The load specified by the building code is always higher than the actual load found in the building survey. In addition, a portion of the live load remains constant. This load comes from the relatively permanent fixtures and furnishings and can be referred to as the *sustained live load (SSL)*. The occupants who enter and leave the space form another part of the live load, raising and lowering the overall live load magnitude with time. This varying live load is called a *transient live load (TLL)*.

2.2.3   Snow Loads

Although snow might be considered a form of live load, unique conditions govern its magnitude and distribution. It is the primary roof load in many geographical areas and heavily depends on local climate, building exposure, and building geometry.

Snow load data are normally based on surveys that result in isoline maps showing areas of equal depth of ground snowfall. Using this method, annual extreme snowfalls have been determined over a period of many years. These data have been analyzed through statistical models and the expected lifetime maximum snow loads estimated. The reference period is again the 50-year anticipated life of the structure.

A major difficulty is encountered in translating the ground snow load into a roof snow load. This is accomplished through a semiempirical relationship whereby the ground snow load is multiplied by factors to account for such things as roof geometry and the thermal characteristics of the roof. Work continues to be done to improve the method of snow load computation and to collect snowfall data.

2.2.4   Wind Load

By its very nature, wind is a highly dynamic natural phenomenon. For this reason it is also a complex problem from a structural perspective. Wind forces fluctuate significantly, and are also influenced by the geometry of the structure, including the height, width, depth, plan and elevation shape, and the surrounding landscape. The basic building code approach to wind load analysis is to treat wind as a static load problem, using the Bernoulli equation to translate wind speed into wind pressure. In an approach similar to that used for snow load, a semi-empirical equation is used to give the wind load at certain levels as a function of a number of factors representing such effects as wind gusts, topography, and structural geometry.

The data used for determining wind loads are based on measured wind speeds. Meteorological data for 3-second wind speed gusts have been accumulated over the contiguous United States and corrected to a standard height of 33 ft. These data are then used to model the long-term characteristics for a mean recurrence interval of 50 years. ASCE 7 and the model codes provide maps to be used as the foundation of wind force calculations. Because local site characteristics often dictate wind behavior, there are locations for which special attention must be given to wind load calculation. In addition, some buildings require special attention in determining wind load magnitude. In these cases it might be valuable to conduct wind tunnel tests before the structural design for wind is carried out.

2.2.5   Seismic Load

The treatment of seismic load effects is extremely complicated because of the highly variable nature of this natural phenomenon and the many factors that influence the impact of an earthquake on any particular structure. In addition, because the force the building feels is the result of the ground moving, inertia effects must be considered.

For most buildings it is sufficient to treat seismic effects through the use of an equivalent static load, provided that the magnitude of this equivalent static load properly reflects the dynamic characteristics of the seismic event. Many characteristics of the problem must be quantified in order to establish the correct magnitude of this static load. These include such factors as the ground motion and response spectra for the seismic event and the structural and site characteristics for the specific project. At the present time, there are many more approaches to earthquake load determination than there are current model building codes, because many jurisdictions still use out-of-date model codes. In addition, the extension of seismic design requirements to all areas of the country through the current model building codes is making seismic design a requirement for many more structures than had been the case just a few years ago.

2.2.6   Special Loads

Several other loads will become important for particular structures in particular situations. These include impact, blast, and thermal effects.

Impact

Most building loads are static or essentially so, meaning that their rate of application is so slow that the kinetic energy associated with their motion is insignificant. For example, a person entering a room is actually exerting a dynamic load on the structure by virtue of their motion. However, because of the small mass and slow movement of the individual, their kinetic energy is essentially zero.

When loads are large and/or their rate of application is very high, the influence of the energy brought to bear on the structure as the movement of the load is suddenly restrained must be taken into account. This phenomenon, known as *impact*, occurs as the kinetic energy of the moving mass is translated into a load on the structure. Depending on the rate of application, the effect of the impact is that the structure experiences a load that may be as large as twice the static value of the same mass.

Impact is of particular importance for structures where machinery and similar actions occur. Cranes, elevators, and equipment such as printing presses could all produce impact loads that would need to be considered in a design. In addition, vibrations may be induced into a structure either by these high magnitude impact loads or the normally occurring occupancy loads. Although normal live load occupancy, such as walking, is not likely to produce increased design load magnitudes, the potential for vibration from these activities should be addressed in any design.

Blast

Blast effects on buildings have become a more important design consideration during the first years of the twenty-first century. Prior to that time, when blast effects were considered they were normally the accidental kind. These types of blast do not occur as often as impact for normal structures, but should be considered under certain circumstances. Many structures designed for industrial installations, where products of a volatile nature are manufactured, are designed with resistance to blast as a design consideration. When the structure is called upon to resist the effects of blast, a great deal of effort must be placed on determining the magnitude of the blast to be resisted.

The threat of terrorism has been increasingly recognized since the attacks on the World Trade Center and Pentagon on September 11, 2001. In order to take that threat into account, owners must determine the level of threat to be designed for and design engineers must establish the extent to which a particular threat will influence a particular structure. Generally speaking, analysis and design data for blast effects is somewhat limited. Work is currently being done to establish design guidelines that help determine blast effects and member strength in response to blast.

Thermal Effects

Steel expands or contracts under changing temperatures, and in so doing may exert considerable forces on the structure if the members are restrained from moving. For most building structures, the thermal effects are less significant than other loads for structural strength. Because the movement of the structure results from the total temperature change and is directly proportional to the length of the member experiencing the change, the use of expansion joints becomes important. When expansion or contraction is not permitted, the resulting forces must be accommodated in the members.

The AISC Specification includes guidance on the design of steel structures exposed to fire. Appendix 4 provides criteria for the design and evaluation of structural steel components, systems, and frames for fire conditions. In the current building design environment, design for fire is usually accomplished by a prescriptive approach defined in the Specification as design by qualification testing. If the actual thermal effects of a fire are to be addressed, the Specification permits design by engineering analysis.

2.3   BUILDING LOAD DETERMINATION

Once the appropriate building load sources are identified, their magnitudes must be determined. Methods to determine these magnitudes are set by the applicable building code for each load source. The following sections provide general guidance to determine the building load magnitudes but for specific details, the applicable building code must be consulted.

2.3.1   Dead Load

Building dead load determination can be either quite straightforward or very complex. If the sizes of all elements of the structural system are known before an analysis is conducted, actual material weights may be determined and applied in the structural analysis. Selected unit weights of typical building materials are given in Table 2.1. Manual Table 17.13 provides the weights of building materials and product catalogs provide weights of things such as building mechanical equipment.

If the final sizes are not known, as would normally be the case in the early stages of design, assumptions need to be made to estimate the self-weight of the structure. This then necessitates an iterative process of refinement as the design and its corresponding weight are brought together.

2.3.2   Live Load

As discussed earlier, live load magnitudes are established by the applicable building code. Table 2.2 provides values for the minimum uniformly distributed live loads for buildings for selected occupancies.

In design, a simplified approach for determining the load on a particular element is through the use of the tributary area, A_T. The tributary area method is a way to visualize the load on a structural element without performing the actual equilibrium calculations. It does, however, provide the same result because it is fundamentally based on an equilibrium analysis. Simplified tributary areas for some structural members are given in Figure 2.2.

Although the concept of the tributary area can be used to determine the load on a member, an equally important concept is the *influence area*, A_I. The influence area is significant because it reflects the area over which any applied load would have an influence on the member of interest. The member under consideration would feel no portion of the load applied outside of the influence area. Table 2.3 provides the relationship between tributary area and influence area for several specific structural elements where $A_I = K_{LL} A_T$. Several of the values in this table are simplified from the actual relationship and it is always permissible to calculate the actual influence area.

Table 2.1   Unit Weights of Typical Building Materials

Material	(lb/ft^3)
Aluminum	165
Brick	120
Concrete	
Reinforced, with stone aggregate	150
Block, 60 percent void	87
Steel, rolled	490
Wood	
Fir	32–44
Plywood	36

Table 2.2 Minimum Uniformly Distributed Live Loads for Building Design[a]

Occupancy or use	(lb/ft^2)
Residential dwellings, apartments, hotel rooms, school classrooms	40
Offices	50
Auditoriums (fixed seats)	60
Retail stores	73–100
Bleachers	100
Library stacks	150
Heavy manufacturing and warehouses	250

[a]Data are taken from ASCE 7.

As the influence area increases for a particular member, the likelihood of the full code specified nominal live load actually occurring on the structure decreases. Because the code does not know ahead of time the likelihood of that full area being loaded, the magnitude of the specified load is set without consideration of loaded area. Thus, the tabulated values are referred to as the unreduced nominal live load. To account for the size of the influence area and thereby provide a more realistic predictor of the actual live load on the structure, a live load reduction factor is introduced. For influence areas greater than 400 ft^2 the live load may be reduced according to the live load reduction equation:

$$L = L_o\left(0.25 + \frac{15}{\sqrt{A_I}}\right) \tag{2.1}$$

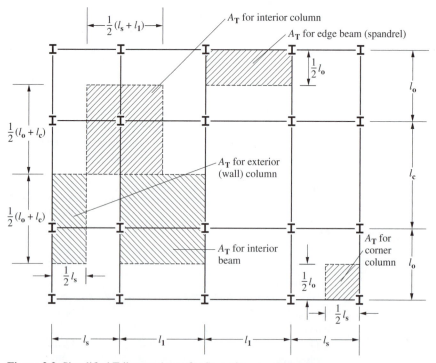

Figure 2.2 Simplified Tributary Areas for Some Structural Members.

Table 2.3   Live Load Element Factor, $K_{LL}{}^a$

Element	K_{LL} (note 1)
Interior columns	4
Exterior columns without cantilever slabs	4
Edge columns with cantilever slabs	3
Corner columns with cantilever slabs	2
Edge beams without cantilever slabs	2
Interior beams	2
All other members not identified above, including:	1
Edge beams with cantilever slabs	
Cantilever beams	
One-way slabs	
Two-way slabs	
Members without provisions for continuous shear	
transfer normal to their span	

Note 1. In lieu of the values above, K_{LL} is permitted to be calculated.
[a]Data are taken from ASCE 7.

where

$$L = \text{reduced live load}$$
$$L_o = \text{code specified design live load}$$
$$A_I = \text{influence area} = K_{LL}A_T$$

Limitations on the use of this live load reduction are spelled out in ASCE 7.

2.3.3   Snow Load

Roof snow load calculations start with determination of the ground snow load for the building site. Table 2.4 provides typical ground snow load values for selected locations. The complete picture of ground snow load is provided in the appropriate building code. In many locations, however, the snowfall depth is a very localized phenomena and the variability is such that it is not appropriate to map those values. In these situations, local building officials should be consulted to determine what the local requirements are.

Table 2.4   Typical Ground Snow Loads, $p_g{}^a$

Location	(lb/ft^2)
Portland, Maine	60
Minneapolis, Minnesota	50
Hartford, Connecticut	30
Chicago, Illinois	25
St. Louis, Missouri	20
Raleigh, North Carolina	15
Atlanta, Georgia	5

[a]Data are taken from ASCE 7.

The determination of roof snow load is complex and there are many acceptable approaches. Roof snow load on an unobstructed flat roof, as given in ASCE 7, is

$$p_f = 0.7 C_e C_t I p_g \tag{2.2}$$

where p_g is the ground snow load determined from the appropriate map, C_e is the exposure factor, C_t is the thermal factor, and I is the importance factor. Numerous other factors enter into the determination of roof snow load, including roof slope, roof configuration, snowdrift, and additional load due to rain on the snow. The applicable building code or ASCE 7 should be referred to for the complete provisions regarding snow load determination.

2.3.4   Wind Load

As with snow load and other geographically linked environmental loads, the starting point for wind load calculation is the map of 3-second gusts provided in the building code. Table 2.5 provides the wind speed data for several selected locations with varying wind velocities. These data must be transformed into wind pressure on a given building to determine the appropriate design wind loads. This transformation must take into account such factors as the importance of the building, height above the ground, relative sheltering of the site, topography, and the direction of the dominate winds. ASCE 7 provides the following equation to convert the mapped data to velocity pressure:

$$q_z = 0.00256 K_z K_{zt} K_d V^2 I \tag{2.3}$$

where

$$q_z = \text{velocity pressure at a specific height above ground}$$
$$K_z = \text{exposure coefficient}$$
$$K_{zt} = \text{topography factor}$$
$$K_d = \text{directionality coefficient}$$
$$V = \text{wind speed}$$
$$I = \text{importance factor}$$

Once the velocity pressure is determined through Equation 2.3, it must be converted to the external design wind pressure. For the main wind force resisting system, this is given by

$$p = q G C_p - q_h (G C_{pi}) \tag{2.4}$$

Table 2.5   Representative Wind Velocities and Resulting Dynamic Pressures[a]

Location	Wind velocity (mph)	Velocity pressure (lb/ft^2)
Miami, Florida	145	53.8
Houston, Texas	120	36.9
New York, New York	105	28.2
Chicago, Illinois	90	20.7
San Francisco, California	85	18.5

[a]Data are taken from ASCE 7.

where

$p =$ design wind pressure

$q =$ velocity pressure from Equation 2.3

$G =$ gust factor

$C_p =$ pressure coefficient

$GC_{pi} =$ internal pressure coefficient

The actual forces applied to the structure are then determined by multiplying the design wind pressure by the tributary area. Because each building code has potentially different requirements for wind load determination, the designer must review the provisions specified in the controlling code. If there is no building code, ASCE 7 should be used.

2.3.5 Seismic Loads

Perhaps the most rapidly fluctuating area of building load determination is that for seismic design. Although there have been many advances in the use of dynamic analysis for earthquake response, common practice is still to model the phenomenon using a static load. For those cases where it applies, ASCE 7 permits the determination of the building base shear through the expression

$$V = C_s W \tag{2.5}$$

where

$V =$ base shear

$C_s =$ seismic response coefficient

$W =$ total building weight

The seismic response coefficient need not be greater than

$$C_s = \frac{S_{D1}}{T(R/I)} \tag{2.6}$$

where

$C_s =$ seismic response coefficient

$S_{D1} =$ design spectral response acceleration

$T =$ building period

$R =$ response modification factor

$I =$ Importance factor

For the design of steel structures to resist seismic forces, the designer must select an appropriate value for the response modification factor, R. In cases where appropriate, the selection of $R = 3$ permits the structure to be designed according to the AISC Specification without using the seismic provisions. If a value of R greater than 3 is used, the design must proceed according to the additional provisions of *ANSI/AISC 341-05 Seismic Provisions for Structural Steel Buildings*. This is discussed further in Chapter 13.

As with the other environmental loads discussed here, the details of load determination for seismic response must be found in the appropriate building code.

2.4   LOAD COMBINATIONS FOR ASD AND LRFD

In addition to specifying the load magnitudes for which building structures must be designed, building codes specify how the individually defined loads should be combined to obtain the maximum load effect. Care must be exercised in combining loads to determine the most critical combination because all loads are not likely to be at their maximum magnitude at the same time. For instance, it is unlikely that the maximum snow load and maximum wind load would occur simultaneously because the wind would undoubtedly blow some of the snow off the structure. Another unlikely occurrence would be a design earthquake occurring at the same time as the maximum design wind. Thus, building codes specify which loads are to be combined and at what magnitude they should be considered. The designer must exercise judgment when combining loads in situations where the normal expectations of the building code might not be satisfied or where some particular combination would result in a greater demand than previously identified.

The two design philosophies addressed in the AISC Specification are the direct result of the two approaches to load combinations presented in current building codes. ASD uses load combinations defined in ASCE 7 as being for allowable stress design, and LRFD uses load combinations defined as being for strength design. The provisions in ASCE 7 for allowable stress design combine loads normally at their nominal or serviceability levels, the load magnitudes discussed in Section 2.3. These load combinations were historically used to determine the load effect under elastic stress distributions and those stresses were compared to the allowable stresses established at some arbitrary level below failure, indicated by either the yield stress or the ultimate stress. Considering load combinations that include only dead, live, wind, snow, and seismic loads, the load combinations presented in ASCE 7-05 Section 2.4 for ASD are:

1. Dead
2. Dead + Live
3. Dead + Roof Live
4. Dead + 0.75 Live + 0.75 Roof Live
5. Dead + Wind
6. Dead + 0.7 Earthquake
7. Dead + 0.75 (Wind or 0.7 Earthquake) + 0.75 Live + 0.75 Roof Live
8. 0.6 Dead + Wind
9. 0.6 Dead + 0.7 Earthquake

As used with the current AISC Specification, these load combinations are not restricted to an elastic stress distribution as done in the past. The current Specification is a strength-based specification, not a stress-based one, and the requirement for elastic stress distribution is no longer applicable. This has no impact on the use of these load combinations but may have some historical significance to those who were educated primarily with this former interpretation.

The second approach available in ASCE 7 combines loads at an amplified level. These combinations, referred to as strength load combinations, permits one to investigate the ability of the structure to resist loads at its ultimate strength. In this approach, loads are multiplied by a load factor that incorporates both the likelihood of the loads occurring simultaneously at their maximum level and the margin against which failure of the structure is measured. Again, considering load combinations that include only dead, live, wind, snow, and seismic loads, the load combinations presented in ASCE 7 Section 2.3, if the live load is not greater than 100 psf, for LRFD are:

1. 1.4 Dead

2. 1.2 Dead + 1.6 Live + 0.5 Roof Live

3. 1.2 Dead + 1.6 Roof Live + 0.5 Live

4. 1.2 Dead + 0.5 Live + 0.5 Roof Live + 1.6 Wind

5. 1.2 Dead + 0.5 Live + 1.0 Earthquake + 0.2 Snow

6. 0.9 Dead + 1.6 Wind

7. 0.9 Dead + 1.0 Earthquake

The design method to be used, and thus the load combinations, are at the discretion of the designer. All current building codes permit either ASD or LRFD, and the AISC Specification provisions address all limit states for each approach. As discussed in Chapter 1, the resulting design may differ for each design philosophy, because the approach taken to assure safety is different, but safety is assured when following the appropriate building code and the AISC Specification regardless of the design approach.

2.5   LOAD CALCULATIONS

In order to understand the impact of these two approaches on analysis, it is helpful to compute the load effect for a variety of structural members according to both ASD and LRFD load combinations. The floor plan of a moderate-height multistory building is given in Figure 2.3. Load case 2 for dead plus live load is considered for several beams and columns. The building is an office building with a nominal live load of 50 pounds per square foot (psf) and a calculated dead load of 70 psf.

1. *Girder AB on line 2-2 if the floor deck spans from line 1-1 to 2-2 to 3-3*:

Tributary area: $A_T = (40)(20) = 800 \text{ ft}^2$

Influence area: $A_I = 2A_T = 1600 \text{ ft}^2$

Live load reduction:

$$0.25 + \frac{15}{\sqrt{1600}} = 0.625$$

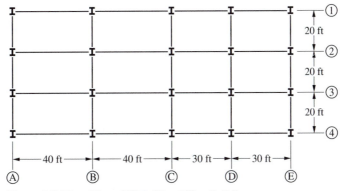

Figure 2.3 Floor Plan of High-Rise Office Building.

LRFD

Amplified loads per lineal foot:

Dead load $= 1.2$ (70 psf) (20 ft) $= 1680$ plf

Live load $= 1.6$ (0.625) (50 psf) (20 ft) $= 1000$ plf

1.2 Dead $+ 1.6$ Live $= 1680 + 1000 = 2680$ lbs/ft $= 2.68$ kips/ft

Required Moment (LRFD), $M_u = \dfrac{wl^2}{8} = \dfrac{2.68(40)^2}{8} = 536$ ft-kips

ASD

Nominal loads per lineal foot:

Dead load $= 70$ psf (20 ft) $= 1400$ pounds per lineal foot (plf)

Live load $= 0.625$ (50 psf) (20 ft) $= 625$ plf

Dead $+$ Live $= 1400 + 625 = 2030$ lbs/ft $= 2.03$ kips/ft

Required Moment (ASD), $M_a = \dfrac{wl^2}{8} \dfrac{2.03(40)^2}{8} = 406$ ft-kips

2. *Floor beam 2-3 on line D-D if the floor deck spans from line C-C to D-D to E-E*:

Tributary area: $A_T = (20)(30) = 600\ \text{ft}^2$

Influence area: $A_I = 2A_T = 1200\ \text{ft}^2$

Live load reduction:

$$0.25 + \frac{15}{\sqrt{1200}} = 0.683$$

LRFD

Amplified loads per lineal foot:

Dead load $= 1.2$ (70 psf) (30 ft) $= 2520$ plf

Live load $= 1.6$ (0.683) (50 psf) (30 ft) $= 1640$ plf

1.2 Dead $+ 1.6$ Live $= 2520 + 1640 = 4160$ plf $= 4.16$ kips/ft

Required Moment (LRFD), $M_u = \dfrac{wl^2}{8} = \dfrac{4.16(20)^2}{8} 208$ ft-kips.

ASD

Nominal loads per lineal foot:

Dead load $= 70$ psf (30 ft) $= 2100$ plf

Live load $= 0.683$ (50 psf) (30 ft) $= 1020$ plf

Dead $+$ Live $= 2100 + 1020 = 3120$ plf $= 3.12$ kips/ft

Required Moment (ASD), $M_a = \dfrac{wl^2}{8} = \dfrac{3.12(20)^2}{8} = 156$ ft-kips

3. *Interior column D-2 regardless of deck span direction*:

   Tributary area: $A_T = (30)(20) = 600 \text{ ft}^2$

   Influence area: $A_I = 4A_T = 2400 \text{ ft}^2$

   Live load reduction:

$$0.25 + \frac{15}{\sqrt{2400}} = 0.556$$

LRFD

Amplified load entering column at this level:

Dead load $= 1.2 \, (70 \text{ psf}) \, (600 \text{ ft}^2) = 50{,}400 \text{ lbs}$

Live load $= 1.6 \, (0.556) \, (50 \text{ psf}) \, (600 \text{ ft}^2) = 26{,}700 \text{ lbs}$

Dead $+$ Live $= 50{,}400 + 26{,}700 = 77{,}100 \text{ lbs} = 77.1 \text{ kips}$

Required axial force (LRFD), $P_u = 77.1 \text{ kips}$

ASD

Nominal load entering column at this level:

Dead load $= 70 \text{ psf} \, (600 \text{ ft}^2) = 42{,}000 \text{ lbs}$

Live load $= 0.556 \, (50 \text{ psf}) \, (600 \text{ ft}^2) = 16{,}700 \text{ lbs}$

Dead $+$ Live $= 42{,}000 + 16{,}700 = 58{,}700 \text{ lbs} = 58.7 \text{ kips}$

Required axial force (ASD), $P_a = 58.7 \text{ kips}$

---

4. *Exterior column D-4 regardless of deck span direction*:

   Tributary area: $A_T = (30)(10) = 300 \text{ ft}^2$

   Influence area: $A_I = 4A_T = 1200 \text{ ft}^2$

   Live load reduction:

$$0.25 + \frac{15}{\sqrt{1200}} = 0.683$$

LRFD

Amplified load entering column at this level:

Dead load $= 1.2 \, (70 \text{ psf}) \, (300 \text{ ft}^2) = 25{,}200 \text{ lbs}$

Live load $= 1.6 \, (0.683) \, (50 \text{ psf}) \, (300 \text{ ft}^2) = 16{,}400 \text{ lbs}$

1.2 Dead $+ 1.6$ Live $= 25{,}200 + 16{,}400 = 41{,}600 \text{ lbs} = 41.6 \text{ kips}$

Required axial force (LRFD), $P_u = 41.6 \text{ kips}$

ASD

Nominal load entering column at this level:

Dead load $= 70 \text{ psf} \, (300 \text{ ft}^2) = 21{,}000 \text{ lbs}$

Live load $= 0.683 \, (50 \text{ psf}) \, (300 \text{ ft}^2) = 10{,}200 \text{ lbs}$

Dead $+$ Live $= 21{,}000 + 10{,}200 = 31{,}200 \text{ lbs} = 31.2 \text{ kips}$

Required axial force (ASD), $P_a = 31.2 \text{ kips}$

2.6 CALIBRATION

The basic requirements of the ASD and LRFD design philosophies were presented in Sections 1.6 and 1.7 and Equations 1.1 and 1.2. The required load combinations for ASD and LRFD, as found in ASCE 7, have been presented earlier in this chapter. This section establishes the relationship between the resistance factor, ϕ, and the safety factor, Ω.

Early development of the LRFD approach to design concentrated on the determination of resistance factors and load factors that would result in a level of structural reliability consistent with previous practice but more uniform for different load combinations. Because the design of steel structures before that time had no particular safety-related concerns, the LRFD approach was calibrated to the then-current ASD approach. This calibration was carried out for the live load plus dead load combination at a live-to-dead load ratio, $L/D = 3.0$. It was well known that for any other load combination or live load-to-dead load ratio, the two methods could give different answers for the same design situation.

The current Specification has been developed with this same calibration, which results in a direct relationship between the resistance factor of LRFD and the safety factor of ASD. For the live load plus dead load combination in ASD, using Equation 1.1, and representing the load effect simply in terms of L and D,

$$(D + L) \leq \frac{R_n}{\Omega}$$

This same combination in LRFD, using Equation 1.2, yields

$$(1.2D + 1.6L) \leq \phi R_n$$

If it is assumed that the load effect is equal to the available strength and each equation is solved for the nominal strength, the results for ASD are:

$$\Omega(D + L) = R_n$$

and for LRFD:

$$\frac{(1.2D + 1.6L)}{\phi} = R_n$$

With L/D taken as 3, the above equations are set equal. They are then solved for the safety factor, which gives:

$$\Omega = \frac{1.5}{\phi}$$

The resistance factors in the Specification were developed through a stochastic analysis to be consistent with the specified load factors and result in the desired reliability for each limit state. More detail on the development of these resistance factors can be found in Section B3.3 of the Commentary to the Specification. Once the resistance factors were established, the corresponding safety factors were determined. This relationship has been used throughout the Specification to set the safety factor for each limit state.

Although the relationship is simple, there is actually no reason to use it to determine safety factors, because the Specification explicitly defines resistance factors and safety factors for every limit state.

2.7   PROBLEMS

1. Name and describe five basic types/sources of building loads.

2. Categorize the following loads as dead load, live load, snow load, wind load, seismic load, or special load.

   a. Load on an office floor due to filing cabinets, desks, and computers.

   b. Load on a roof from a permanent air handling unit.

   c. Load on stadium bleachers from students jumping up and down during a college football game.

   d. Load on a building caused by an explosion.

   e. Weight on a steel beam from a concrete slab that it is supporting.

   f. Load experienced by an office building in California as it shakes during an earthquake.

   g. Load on a skyscraper in Chicago on a blustery day causing the building to sway back and forth.

3. What is one source you can consult to find the snow load data for a particular region as well as maps showing wind gust data to calculate wind loads?

4. Where in the AISC Manual can you find a table of selected unit weights of common building materials?

5. What analysis method allows the designer to visualize the load on a particular structural element without performing an actual equilibrium calculation?

6. In determining the snow load on a structure, what value that can be obtained from the applicable building code is multiplied by a series of factors to obtain the actual snow load?

7. Name four factors that must be taken into account when converting wind speed data referenced by the building code into wind pressure on a given building.

8. If a response modification factor of 3 is chosen in designing a steel building to resist seismic loads, what design specification should be consulted?

9. Which design approach combines loads that are normally at their nominal or serviceability level?

10. Strength load combinations that are incorporated by the LRFD method take into account what two factors?

11. Using ASCE 7-05, determine the minimum uniformly distributed live load for a hospital operating room.

12. Using ASCE 7-05, determine the minimum uniformly distributed live load for library stacks.

13. Using ASCE 7-05, determine the minimum uniformly distributed live load for an apartment building.

14. Determine the nominal uniformly distributed self-weight of a 6-in. thick reinforced concrete slab.

15. A building has a column layout as shown in the next column with 30-ft bays in each direction. It must support a uniform dead load of 90 psf and a uniform live load of 80 psf. Determine the required strength of the members noted below for design by (a) LRFD and (b) ASD.

   i. The beam on column line 3 between column lines A and B if the deck spans from line 2-2 to 3-3 to 4-4.

   ii. The girder on column line C between column lines 3 and 4 if the deck spans from line B-B to C-C to D-D.

   iii. The column at the corner on lines 4 and A.

   iv. The column on the edge at the intersection of lines C and 4.

   v. The interior column at the intersection of column lines D and 3.

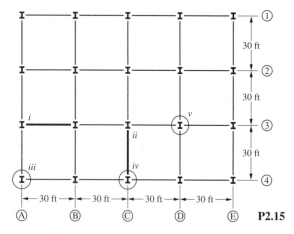

P2.15

16. If the framing plan shown below were for the roof of a structure that carried a dead load of 55 psf and a roof live load of 30 psf, determine the required strength of the members noted below for (a) design by LRFD and (b) design by ASD.

   i. The girder on column line A between column lines 1 and 2 if the deck spans from line A-A to B-B.

   ii. The beam on column line 3 between column lines B and C if the deck spans from line 2-2 to 3-3 to 4-4.

   iii. The column at the corner on lines 1 and E.

   iv. The column on the edge at the intersection of lines 1 and B.

   v. The interior column at the intersection of column lines C and 2.

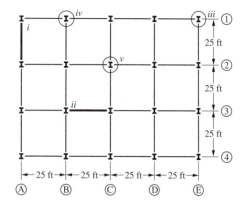

P2.16

Chapter 3

The Palazzo, Las Vegas.
Photo courtesy Walter P Moore.

Steel Building Materials

3.1 INTRODUCTION

Steel has been produced in the United States since the 1800s. Its first use in a bridge was a railroad bridge across the Mississippi River in St. Louis, built in 1874 by James B. Eads. The bridge, known as the Eads Bridge, is still an inspiring steel structure crossing the river in the shadow of the St. Louis Gateway Arch. The first skyscraper is generally considered to be the Home Insurance Building, designed by William LeBaron Jenney and erected at 135 South La Salle Street, Chicago. The building was started on May 1, 1884 and completed in the fall of 1885. It was originally a 10-story building but later had 2 additional stories added. The original structural design called for wrought iron beams bolted through angle-iron brackets to cast iron columns. As the framework reached the 6th floor, the Carnegie-Phipps Steel Company of Pittsburgh, Pennsylvania indicated that they were now rolling "Bessemer Steel" and requested permission to substitute steel members for the wrought iron beams on all remaining floors. Thus, this was the first use of steel beams in a building. The Home Insurance Building was demolished in 1929.

The first all-steel skyscraper was the Rand-McNally Building at 165 West Adams Street in Chicago designed by Daniel Burnham and John Root. This 10-story building was built from 1888–1890 and was constructed of built-up members made from standard rolled steel bridge shapes that were riveted together. It began a continuous evolution in steel building structures that continues today as new ideas are brought into play by architects and engineers who continue to build with steel.

This evolution in steel buildings has occurred in the materials used, the applications of innovative designers, and the specifications that direct their designs.

3.2   APPLICABILITY OF THE AISC SPECIFICATION

The specification that guides the design of our modern steel buildings was first published by AISC in 1923. At that time its purpose was to promote uniform practice in the design of steel buildings. Up to then, numerous approaches were being used across the industry. Steel producers each had their own standard for design whereas the larger cities also required that their own standards be used. This multiplicity of standards was no standard at all. It lead to a confusion of approaches whereby designers were continually called upon to change how they designed, depending on where their current building project was to be located.

The 1923 specification defined "the practice adopted by the American Institute of Steel Construction for the design, fabrication, and erection of structural steel buildings." It went on to provide direction on how to obtain a satisfactory structure. The following requirements were to be fulfilled:

1. The material used must be suitable, of uniform quality, and without defects affecting the strength or service of the structure.
2. Proper loads and conditions must be assumed in the design.
3. The unit stresses must be suitable for the material used.
4. The workmanship must be good, so that defects or injuries are not produced in the manufacture.
5. The computations and design must be properly made so that the unit stresses specified shall not be exceeded, and the structure and its details shall possess the requisite strength and rigidity.

The specification also provided guidance on the material to be used, stating "Structural steel shall conform to the Standard Specifications of the American Society for Testing Materials for Structural Steel for Buildings, Serial Designation A 9-21, as amended to date." These principles from 1923 are still important to steel construction almost a century later.

The 2005 AISC Specification for Structural Steel Buildings supersedes all previous AISC Specifications and thus brings together, into one document, the necessary provisions for the design of steel building structures. Over the years, the specification has lost the terms fabrication and erection from its scope, because the development of standard practice of building design and construction has changed responsibilities of the various parties. In addition, the AISC Specification has regularly been used to guide the design of structures other than building structures. In recognition of this practice, and to ensure that the specification is properly applied, the scope of this edition has been revised to state "This specification sets forth criteria for the design of structural steel buildings and other structures, where other structures are defined as those structures designed, fabricated, and erected in a manner similar to buildings, with building-like vertical and lateral load-resisting-elements."

Additionally, the specification indicates that it "shall apply to the design of the structural steel system, where the steel elements are defined in the AISC Code of Standard Practice for Steel Buildings and Bridges Section 2.1. In that document, structural steel is defined as those elements of the structural frame that are shown and sized in the structural Design Drawings, essential to support the design loads . . ." and are given here in Table 3.1. Examples of many

Table 3.1 Definitions of Structural Steel[a]

Anchor rods that will receive structural steel.

Base plates.

Beams, including built-up beams, if made from standard structural shapes and/or plates.

Bearing plates.

Bearings of steel for girders, trusses, or bridges.

Bracing, if permanent.

Canopy framing, if made from standard structural shapes and/or plates.

Columns, including built-up columns, if made from standard structural shapes and/or plates.

Connection materials for framing structural steel to structural steel.

Crane stops, if made from standard structural shapes and/or plates.

Door frames, if made from standard structural shapes and/or plates and if part of the structural steel frame.

Edge angles and plates, if attached to the structural steel frame or steel (open-web) joists.

Embedded structural steel parts, other than bearing plates, that will receive structural steel.

Expansion joints, if attached to the structural steel frame.

Fasteners for connecting structural steel items: permanent shop bolts, nuts, and washers; shop bolts, nuts, and washers for shipment; field bolts, nuts, and washers for permanent connections; and permanent pins.

Floor-opening frames, if made from standard structural shapes and/or plates and attached to the structural steel frame or steel (open-web) joists.

Floor plates (checkered or plain), if attached to the structural steel frame.

Girders, including built-up girders, if made from standard structural shapes and/or plates.

Girts, if made from standard structural shapes.

Grillage beams and girders.

Hangers, if made from standard structural shapes, plates, and/or rods and framing structural steel to structural steel.

Leveling nuts and washers.

Leveling plates.

Leveling screws.

Lintels, if attached to the structural steel frame.

Machinery supports, if made from standard structural shapes and/or plates and attached to the structural steel frame.

Marquee framing, if made from standard structural shapes and/or plates.

Monorail elements, if made from standard structural shapes and/or plates and attached to the structural steel frame.

Posts, if part of the structural steel frame.

Purlins, if made from standard structural shapes.

Relieving angles, if attached to the structural steel frame.

Roof-opening frames, if made from standard structural shapes and/or plates and attached to the structural steel frame or steel (open-web) joists.

Roof-screen support frames, if made from standard structural shapes.

Sag rods, if part of the structural steel frame and connecting structural steel to structural steel.

Shear stud connectors, if specified to be shop attached.

Shims, if permanent.

Struts, if permanent and part of the structural steel frame.

Tie rods, if part of the structural steel frame.

Trusses, if made from standard structural shapes and/or built-up members.

Wall-opening frames, if made from standard structural shapes and/or plates and attached to the structural steel frame.

Wedges, if permanent.

[a]From Code of Standard Practice for Steel Buildings and Bridges, AISC 2005.

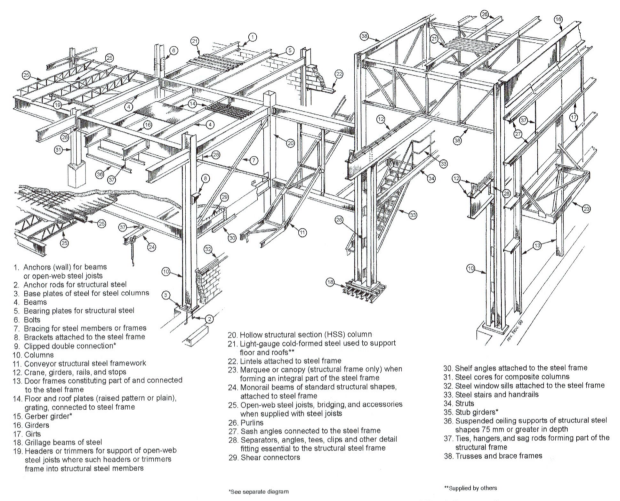

1. Anchors (wall) for beams
   or open-web steel joists
2. Anchor rods for structural steel
3. Base plates of steel for steel columns
4. Beams
5. Bearing plates for structural steel
6. Bolts
7. Bracing for steel members or frames
8. Brackets attached to the steel frame
9. Clipped double connection*
10. Columns
11. Conveyor structural steel framework
12. Crane, girders, rails, and stops
13. Door frames constituting part of and connected
    to the steel frame
14. Floor and roof plates (raised pattern or plain),
    grating, connected to steel frame
15. Gerber girder*
16. Girders
17. Girts
18. Grillage beams of steel
19. Headers or trimmers for support of open-web
    steel joists where such headers or trimmers
    frame into structural steel members

20. Hollow structural section (HSS) column
21. Light-gauge cold-formed steel used to support
    floor and roofs**
22. Lintels attached to steel frame
23. Marquee or canopy (structural frame only) when
    forming an integral part of the steel frame
24. Monorail beams of standard structural shapes,
    attached to steel frame
25. Open-web steel joists, bridging, and accessories
    when supplied with steel joists
26. Purlins
27. Sash angles connected to the steel frame
28. Separators, angles, tees, clips and other detail
    fitting essential to the structural steel frame
29. Shear connectors

30. Shelf angles attached to the steel frame
31. Steel cores for composite columns
32. Steel window sills attached to the steel frame
33. Steel stairs and handrails
34. Struts
35. Stub girders*
36. Suspended ceiling supports of structural steel
    shapes 75 mm or greater in depth
37. Ties, hangers, and sag rods forming part of the
    structural frame
38. Trusses and brace frames

*See separate diagram   **Supplied by others

Figure 3.1 Steel Elements. Copyright, Code of Standard Practice, Canadian Institute of Steel Construction.

of these elements are shown in Figure 3.1. All elements discussed within this text will meet the above definition.

3.3   STEEL FOR CONSTRUCTION

Since the introduction of the first AISC Specification, a variety of steels have been approved for use in steel construction. Which steels were specifically approved at any time has changed with the changing techniques of manufacture and steel chemistry. Steels available for use in construction have increased in strength as manufacturing has become more refined. One important aspect of all steel is that it generally behaves in a uniform and consistent manner. Thus, although the strength might be different for different grades of steel, the steel can be expected to behave the same, regardless of grade, up to its various strength limits.

The characteristics of steel that are important to the structural engineer can be determined through a simple uniaxial tension test. This standard test is conducted according to the requirements of ASTM A370 Standard Test Methods and Definitions for Mechanical Testing of Steel Products. A specimen of a specific dimension is subjected to a tensile force

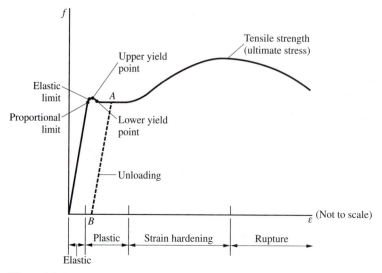

Figure 3.2 Typical Stress-Strain Plot for Mild Carbon Steel.

and the resulting stress and strain are plotted for the duration of the test. The stress, f, and strain, ε, are shown plotted in Figure 3.2 and defined as follows:

$$f = \frac{P}{A} \quad \text{and} \quad \varepsilon = \frac{\Delta L}{L}$$

where

$A =$ cross-sectional area at start of test

$L =$ length of specimen at start of test

$P =$ tensile force

$f =$ axial tensile stress

$\Delta L =$ change in length of specimen

$\varepsilon =$ axial strain

The curve shown in Figure 3.2 is typical of mild carbon steel. Several characteristics of this stress-strain curve are worth noting. First, the initial portion of the curve, which indicates the response that would be expected under most normal or service loading conditions, follows a straight line up to a point called the *proportional limit*. For structural steel with yield stresses at 65 ksi or less, this proportional limit is the point where the curve first deviates from linear and is called the *yield point*. The ratio of stress to strain in this region is constant and called Young's Modulus, or the Modulus of Elasticity, E. All structural steels exhibit the same initial stress-strain behavior and thus have the same E. The value of E obtained through a wide number of tests is consistently between 29,000 ksi and 30,000 ksi. For all calculations according to the AISC Specification, $E = 29,000$ ksi has historically been used. Within the straight-line portion of the curve, the material is said to behave elastically. A load can be applied and then removed with the structure returning to its original configuration, showing no permanent deformation.

After reaching the yield stress, the stress-strain curve for mild carbon steel exhibits a long plateau where the stress remains essentially constant while the strain increases. This region is called the *plastic region*. Any structure that is loaded into this region exhibits a permanent plastic deformation as shown by the unloading line in Figure 3.2. The length of this plastic region depends on the particular type of steel but typically is 15 to 20 times the strain at yield.

At the end of the plastic region, the curve again rises with increasing stress and strain. This increase is called *strain hardening* and continues until the specimen reaches its tensile strength, or ultimate stress, F_u, at the peak of the stress-strain curve. Once the tensile strength is reached, the specimen rapidly sheds load and increases strain until the specimen completely ruptures.

Yield stress, tensile strength, and modulus of elasticity are the engineering data used throughout design to fully describe the material and to determine the strength of the structural elements. The ratio of the tensile strength to the yield stress is also an important characteristic of steel. It is used to control the basic material behavior so that at various limit states, the expected behavior can be assured.

Figure 3.3 shows the lower strain region of the stress-strain curves for three steels with different yield stresses, 36 ksi, 50 ksi, and 100 ksi. Elastic behavior for the higher strength steels is the same as for lower strength steels as seen in the figure. As already noted, $E = 29,000$ ksi for all steel. The differences occur after the proportional limit is reached. For steels with a yield stress less than or equal to 65 ksi, the plateau defining the plastic region can be expected to occur. However, for steels with a yield stress greater than 65 ksi, it is expected that no well-defined yield point will exist and no well-defined plastic plateau will occur. For these steels it is necessary to define yield strength by some other means. ASTM A370 provides for yield strength determination by the 0.2% offset method or the 0.5% elongation method. In either case, the stress-strain curve must be obtained and the

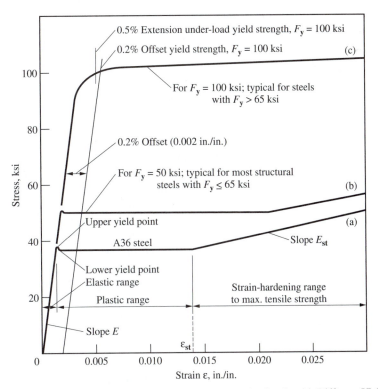

Figure 3.3 Enlarged Typical Stress-Strain Curves for Steels with Different Yield Stresses.

specified offset or elongation used to determine the appropriate stress value. The results of these two approaches are shown in Figure 3.3, and the two methods would yield different yield strength values.

3.4 STRUCTURAL STEEL SHAPES

Structural steel design serves to determine the appropriate shape and quantity of steel needed to carry a given applied load. This is normally accomplished by selecting, from a predetermined list of available shapes, the lightest-weight member. However, it could also result from the combination of steel elements into some particular desired form. The early days of steel construction had very little standardization of available shapes. Although each mill would produce its own shapes, the variety of available shapes was limited and most structural members were composed of these available shapes riveted together. One of AISC's original goals was to standardize the shapes being produced. Over the years, shapes became standardized and more shapes, designed specifically for the needs of building construction, became available. Modern production practices now make a wide variety of shapes available to the designer so that design can almost always be accomplished by selecting one of these standard shapes. In situations where these standard shapes do not meet the needs of a project, members composed of plate material can be produced to carry the imposed loading.

3.4.1 ASTM A6 Standard Shapes

The first standard shapes to be discussed are those defined by ASTM A6: W-shapes, S-shapes, HP-shapes, M-shapes, C-shapes, MC-shapes, and L-shapes. Cross-sections of these shapes are shown in Figure 3.4 where it can be seen that W-, M-, S-, and HP-shapes all take the form of an I. C- and MC-shapes are channels and L-shapes are called angles. Part 1 of the Manual contains tables of properties for all of the standard shapes.

W-Shapes

W-shapes are usually referred to as wide flange shapes and are the most commonly used shapes in buildings. They have two flanges with essentially parallel inner and outer faces and a single web located midway on the flanges. The overall shape of the wide flange may vary from being a fairly deep and narrow section, as shown in Figure 3.4a, to an almost square section, as shown in Figure 3.4b. These shapes have two axes of symmetry; the *x*-axis is the strong axis and the *y*-axis is the weak axis. Wide flange shapes can be as deep as 44 in. and as shallow as 4 in. A typical wide flange shape would be called out as a W16×26 where the W indicates it is a W-shape, the 16 indicates it has a nominal depth of 16 in., and the 26 indicates its weight is 26 pounds per foot. The nominal depth is part of the name of the shape and indicates an approximate member depth but does not indicate its actual depth. The production of wide flange shapes results in shapes being grouped in a family according to the size of the rolls that produce the shape. All shapes in a family have the same dimension between the inner faces of the flanges. The different weights are accomplished by increasing the actual depth of the member. Manual Table 1-1 provides the dimensions and section properties needed for design for all W-shapes.

HP-Shapes

HP-shapes are wide flange shapes normally used as bearing piles. These shapes have parallel face flanges like the wide flange shapes but unlike the W-shapes, their webs and flanges are of

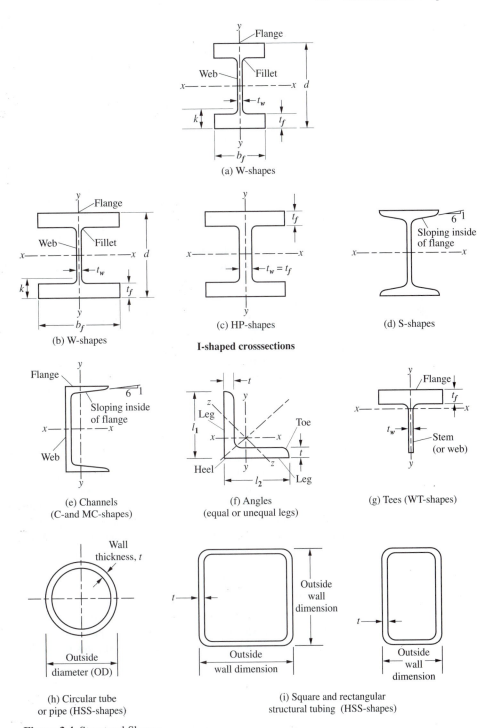

Figure 3.4 Structural Shapes.

the same nominal thickness and they are all close to being square, as shown in Figure 3.4c. An HP14×117 would be an HP-shape with a nominal depth of 14 in. and a weight of 117 pounds per foot. Manual Table 1-4 provides the dimensions and section properties needed for the design for all HP-shapes.

S-Shapes

S-shapes are American Standard Beams and were previously referred to as I-beams. They were the standard shapes used in construction prior to the development of the rolling process that permitted the introduction of the wide flange shapes. Although these shapes are still available, their use is infrequent and their availability should be confirmed prior to specifying them. These shapes have relatively narrow flanges compared to their depth and the flanges have a sloping interior face, as shown in Figure 3.4d. The Manual lists 28 S-shapes and their properties are found in Table 1-3. As with the shapes previously discussed, the numbers in the name refer to the nominal depth and the weight per foot. In all cases except the S24×121 and S24×106, the nominal depth and the actual depth are the same.

M-Shapes

M-shapes are miscellaneous shapes that do not fit into the definitions of W-, HP-, and S-shapes. The Manual lists 18 miscellaneous shapes. They are not particularly common and should be used in design only after confirmation that they are economically available. A typical designation would be M12×11.8. As with the other shapes, the 12 indicates the nominal depth and the 11.8 indicates the weight per foot. Dimensions and properties for these M-shapes are found in Manual Table 1-2.

C-Shapes

C-shapes are American Standard Channels and are produced by essentially the same process as S-shapes. They have two flanges and a single web located at the end of the flanges, as shown in Figure 3.4e. These shapes have only one axis of symmetry and, like the W-shapes, the *x*-axis is the strong axis and the *y*-axis is the weak axis. As with the S-shapes, the flanges have sloping inner faces. One of the 31 C-shapes found in Manual Table 1-5 is a C8×18.7. All C-shapes have an actual depth equal to the nominal depth.

MC-Shapes

MC-shapes are miscellaneous channels that cannot be classified as C-shapes. Their designations follow the same rules as the previous shapes with a typical shape being an MC6×18. Manual Table 1-6 lists 39 MC-shapes, and their sizes fit into the same overall range as the C-shapes.

L-Shapes

L-shapes are angles that can have equal or unequal legs. The largest angle legs are 8 in. and the smallest are 2 in., with the dimension taken from heel to toe of the angle. A typical angle designation would be L6×4×$\frac{7}{8}$ where the first two numbers are the dimensions of the legs and the third is the leg thickness. Leg dimensions are actual dimensions and the leg thickness is the same for both legs. For unequal leg angles, the longest leg is given first. Equal leg angles have one axis of symmetry whereas unequal leg angles have no axis of symmetry. All angles have three axes of interest to the designer: the geometric axes are the *x*-axis, parallel to the short leg; the *y*-axis is parallel to the long leg; and the minor principal axis, which for equal leg angles is perpendicular to the axis of symmetry, is the *z*-axis. Manual Table 1-7 provides the dimensions and section properties needed for the design for all angles.

WT-Shapes

WT-shapes are tees that have been cut from W-shapes. They are also called split tees. These shapes are designated as WT5×56 where both numbers are one-half of what would indicate the parent W-shape that they were cut from. Dimensions and properties for WT-shapes are given in Manual Table 1-8.

MT-Shapes and ST-Shapes

MT-shapes and *ST-shapes* are tees that have been cut from the parent M- and S-shapes. The properties and dimensions for these shapes are found in Manual Tables 1-9 and 1-10.

3.4.2   Hollow Shapes

Another group of shapes commonly found in building construction are the hollow shapes referred to as tubes or pipes. These shapes are produced by bending and welding flat plates or by hot rolling to form a seamless section. For all hollow structural shapes (HSS), ASTM specifications set the requirements for both the material and the sizes.

Round HSS

Round hollow structural shapes are round hollow structural sections. They are manufactured through a process called Formed-From-Round which takes a flat strip of steel and gradually bends it around its longitudinal axis and joins it by welding. Once the weld has cooled, the round shape is passed through additional shaping and sizing rolls to fix the final diameter. A round HSS would be indicated as HSS5.563×0.258 where the first number is the diameter and the second is the nominal thickness. These shapes are found in Manual Table 1-13.

Square and Rectangular HSS

Square and Rectangular HSS may be formed as Round HSS with the final sizing used to change the shape into a rectangle, or formed from a flat plate through a Formed-Square Weld-Square process wherein the plate is gradually bent into its near final size. Another process starts with two flat pieces that are each bent and then the two half sections are joined to form the final shape. A typical rectangular HSS would be HSS12×8×$\frac{1}{2}$. The first number indicates the actual height of the section, the second the actual width, and the third the nominal thickness of the section wall. Manual Tables 1-11 and 1-12 provide the dimensions and section properties needed for the design of rectangular and square HSS-shapes, respectively.

Steel Pipes

Steel pipes are another hollow round section used in building construction. They are produced to different material standards than the round HSS. Pipes are available as standard

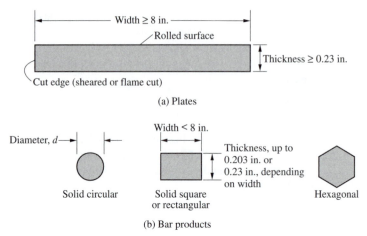

(a) Plates

(b) Bar products

Figure 3.5 Plate and Bar Products.

weight (Std.), extra strong (x-Strong), and double-extra strong (xx-Strong), which refer to the wall thickness for a given outside diameter. The standard designation for a pipe section would be Pipe 5 x-Strong, indicating that it was to meet the pipe material standards, have a nominal 5-in. outside diameter, and a thickness corresponding to the extra strong designation. This particular pipe would have an actual outside diameter of 5.56 in. and a nominal wall thickness of 0.375 in. Manual Table 1-14 provides the properties for steel pipes.

3.4.3 Plates and Bars

In addition to the shapes already discussed, steel is available as plates and bars, as shown in Figure 3.5. These elements are rarely used alone as shapes but are combined to form built-up shapes or used alone as connecting elements to join other shapes.

Plates

Plates are flat rectangular elements hot rolled to a given thickness and sheared to the appropriate width. At one time, plates were also available that were rolled to a given width as well as thickness. These plates were called universal mill plates. Because of the manufacturing process, these plates had different patterns of residual stresses than the sheared plates that resulted in lower strength. Current manufacturing practice is to produce all plates as sheared plates. By industry definition, plates are a minimum of 8 in. in width and may vary in thickness from $3/16$ in. The designation for a plate is PL$1/2$×10×2 ft–4 in. where the first number is the thickness, the second the plate width, and the third the length. Table 3.2 gives the preferred standard practice for plate thickness increments.

Bars

Bars are available in rectangular, circular, and hexagonal shapes with the rectangular bar the most commonly used shape in building construction. The only difference between rectangular bars and plates is the width. Any rectangular solid element less than 8 in. in width is technically referred to as a bar. Because the distinction between bars and plates is not significant to the designer, the designation for these narrow elements is the same as for a plate. Thus, PL$1/2$×6×2 ft–4 in. is a 6-in. wide bar.

Table 3.2 Preferred Dimensions for Plates and Bars

Product	Range of Thicknesses/Diameters		
	$t \leq \frac{3}{8}$ in.	$\frac{3}{8} < t \leq 1$ in.	1 in. $< t$
Plates	$\frac{1}{16}$	$\frac{1}{8}$	$\frac{1}{4}$
Square and rectangular bars	$\frac{1}{8}$	$\frac{1}{8}$	$\frac{1}{8}$
Circular bars	$\frac{1}{8}$	$\frac{1}{8}$	$\frac{1}{8}$

Note. Table gives increments in thickness or diameter.

3.4.4 Built-up Shapes

Other shapes are available and may be found in the Manual but are of limited application in building construction. The manual also contains tables for combinations of standard shapes that have, over the years, been found to be useful to the designer. Figure 3.6 shows a variety of built-up shapes formed from combining plates and shapes.

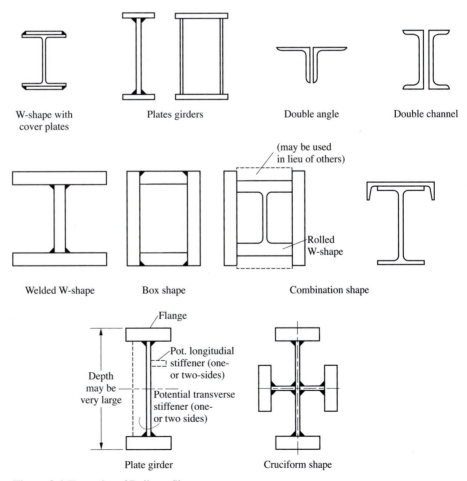

Figure 3.6 Examples of Built-up Shapes.

3.5 CHEMICAL COMPONENTS OF STRUCTURAL STEEL

The basic mechanical properties of structural steel were presented in Section 3.3 with a limited discussion of the types of steel available for use by the building industry. Essentially three types of steel are used for shapes in the construction industry: carbon steel, high-strength low-alloy steel, and corrosion-resistant high-strength low-alloy steel. For plates and bars, quenched and tempered steels are also available.

The chemical composition of steel significantly influences the properties that are of ultimate importance to the engineer. Steel is primarily made of iron but also contains such other elements as carbon, silicon, nickel, manganese, and copper. The primary element, in addition to iron, is carbon. The addition of carbon increases steel strength but decreases ductility and weldability. Even though carbon is the most significant component of steel, after iron, it is still a very small percent of the final product. Steels generally have a carbon content of up to 0.3% by weight.

Although the formula for any specific steel might be different from any other steel, certain elements are required in order to meet a specific set of criteria. These specifications come from the ASTM standards for each steel type and are discussed in Section 3.6. However, the chemical elements that may be found in the most dominant steel for wide flange shapes are reviewed here. The specific percentage requirements for ASTM A992 steel are given in Table 3.3.

Carbon

Carbon (C) is the most common element found in all steel. It is the most economical element used to increase strength. However, it also decreases ductility. Carbon content usually ranges from about 0.15% to 0.30%. Anything lower than 0.15% would produce steel with too low a strength, and anything higher than 0.30% would yield steel with poor characteristics for use in construction.

Manganese

Manganese (Mn) has an effect on strength similar to that of carbon. It is a necessary component because of the way it combines with oxygen and sulfur and its impact on the rolling process. In addition, manganese improves the notch toughness of steel. It is added

Table 3.3 Chemical Requirements for A992 Steel

Element	Composition, %
Carbon, max	0.23
Manganese	0.50 to 1.50
Silicon, max	0.40
Vanadium, max	0.11
Columbium, max	0.05
Phosphorus, max	0.035
Sulfur, max	0.045
Copper, max	0.60
Nickel, max	0.45
Chromium, max	0.35
Molybdenum, max	0.15
Nitrogen, max	0.015

to steel to offset reductions in notch toughness due to the presence of other elements. It has a negative effect on material weldability.

Silicon

Silicon (*Si*) is an important element for removing oxygen from hot steel.

Phosphorus

Phosphorus (*P*) increases strength and decreases ductility. It improves resistance to atmospheric corrosion, particularly when used in combination with copper. It has a negative impact on weldability that is more severe than that of manganese. It is generally an undesirable element but is permitted in very limited quantities in all steel.

Sulfur

Sulfur (*S*) is also permitted in very limited quantities in all steel. It has a similar negative impact to that of phosphorus.

Copper

Copper (*Cu*) in limited quantities is beneficial to the strength of steel. It increases strength with only a limited negative impact on ductility. If its content is held relatively low, it will have little effect on weldability. It is the most significant contributing element in the production of corrosion-resistant steel.

Vanadium

Vanadium (*V*) is another strengthening element. It refines the grain size and thus increases strength. Its biggest advantage is that while increasing strength, it does not negatively impact weldability or notch toughness.

Columbium

Columbium (*Cb*) is a strengthening element that, in small quantities, can increase the yield point and, to a lesser extent, the tensile strength. However, it has a significant negative impact on notch toughness.

Nitrogen

Nitrogen (*N*) is normally found in very low quantities but does provide some increase in strength. When used in combination with vanadium it can improve weldability.

Nickel

Nickel (*Ni*) can provide a moderate improvement in strength and enhances corrosion resistance. It can also improve resistance to corrosion for steel subjected to seawater when in combination with copper or phosphorous. It generally produces a slight improvement in notch toughness.

Chromium

Chromium (Cr) is typically used in combination with copper to improve corrosion resistance. It also provides some strengthening in steels containing copper and vanadium. Chromium is an integral component of stainless steel.

Molybdenum

Molybdenum (Mo) increases strength but significantly decreases notch toughness, although its negative impact can be lessened through appropriate processing or balancing with other elements.

3.6   GRADES OF STRUCTURAL STEEL

3.6.1   Steel for Shapes

More grades of steel are produced than approved by AISC for use in structures. An ASTM number designates each approved steel. The steels approved for structural shapes are grouped as carbon (A36, A53, A500, A501, and A529); high-strength low-alloy (A572, A618, A913, and A992); and corrosion-resistant high-strength low-alloy (A242, A588, and A847). Figure 3.7 shows these approved steels, their minimum yield and tensile stresses, and the shapes for which they are applicable.

A36 Steel

A36 steel was the most commonly available structural steel for many years. It was first introduced in the 1961 AISC Specification and until the late 1990s was the steel of choice for most steel shapes except for HSS, pipe, and plates. It is a mild carbon steel so it is well suited for bolted or welded construction and, even if higher strength steels were being used for members, this steel was the norm for connecting elements. It continues to be the preferred steel for M-, S-, C-, MC-, and L-shapes. It has a minimum yield stress, $F_y = 36$ ksi, and a tensile stress, $F_u = 58$ to 80 ksi, although $F_u = 58$ ksi is used for calculations throughout the specification.

A53 Steel

A53 steel is the single standard for steel pipe approved for construction. This standard provides for three types and two grades. These pipes are generally intended for mechanical and pressure applications and the only grade approved for construction is Grade B. This grade comes as Type E, which calls out electric-resistance welding of the seam, or Type S, which is a seamless pipe. A53 Grade B has a minimum yield stress, $F_y = 35$ ksi, and a minimum tensile stress, $F_u = 60$ ksi.

A500 Steel

A500 steel is a carbon steel used for structural tubing in rounds and shapes, otherwise known as HSS. It comes in two grades approved by AISC for construction: grade B, which is the preferred grade, and Grade C. The standard permits either welded or seamless manufacture. Round HSS Grade B has a minimum yield stress, $F_y = 42$ ksi, and a minimum tensile stress, $F_u = 58$ ksi, whereas rectangular HSS Grade B has a minimum yield stress, $F_y = 46$ ksi, with a minimum tensile stress, $F_u = 58$ ksi.

Table 2–3
Applicable ASTM Specifications for Various Structural Shapes

Legend: ■ = Preferred material specification · ▒ = Other applicable material specification · (blank) = Material specification does not apply.

Steel Type	ASTM Designation		F_y Min. Yield Stress (ksi)	F_u Tensile Stress[a] (ksi)	W	M	S	HP	C	MC	L	HSS Rect.	HSS Round	Pipe
Carbon	A36		36	58-80[b]	▒	■	■	▒	■	■	■			
	A53 Gr. B		35	60										■
	A500	Gr. B	42	58									■	
		Gr. B	46	58								■		
		Gr. C	46	62								▒		
		Gr. C	50	62								▒		
	A501		36	58								▒	▒	▒
	A529[c]	Gr. 50	50	65-100	▒	▒	▒	▒	▒	▒	▒			
		Gr. 55	55	70-100	▒	▒	▒	▒	▒	▒	▒			
High-Strength Low-Alloy	A572	Gr. 42	42	60	▒	▒	▒	▒	▒	▒	▒			
		Gr. 50	50	65[d]	▒	▒	▒	■	▒	▒	▒			
		Gr. 55	55	70	▒	▒	▒	▒	▒	▒	▒			
		Gr. 60[e]	60	75	▒	▒	▒	▒	▒	▒	▒			
		Gr. 65[e]	65	80	▒	▒	▒	▒	▒	▒	▒			
	A618[f]	Gr. I & II	50[g]	70[g]								▒	▒	
		Gr. III	50	65								▒	▒	
	A913	50	50[h]	60[h]	▒	▒	▒	▒						
		60	60	75	▒	▒	▒	▒						
		65	65	80	▒	▒	▒	▒						
		70	70	90	▒	▒	▒	▒						
	A992		50-65[i]	65[i]	■									
Corrosion Resistant High-Strength Low-Alloy	A242		42[j]	63[j]	▒	▒	▒	▒	▒	▒	▒			
			46[k]	67[k]	▒	▒	▒	▒	▒	▒	▒			
			50[l]	70[l]	▒	▒	▒	▒	▒	▒	▒			
	A588		50	70	▒	▒	▒	▒	▒	▒	▒			
	A847		50	70								▒	▒	

■ = Preferred material specification.
▒ = Other applicable material specification, the availability of which should be confirmed prior to specification.
☐ = Material specification does not apply.

[a] Minimum unless a range is shown.
[b] For shapes over 426 lb/ft, only the minimum of 58 ksi applies.
[c] For shapes with a flange thickness less than or equal to $1\frac{1}{2}$ in. only. To improve weldability a maximum carbon equivalent can be specified (per ASTM Supplementary Requirement S78). If desired, maximum tensile stress of 90 ksi can be specified (per ASTM Supplementary Requirement S79).
[d] If desired, maximum tensile stress of 70 ksi can be specified (per ASTM Supplementary Requirement S91).
[e] For shapes with a flange thickness less than or equal to 2 in. only.
[f] ASTM A618 can also be specified as corrosion-resistant; see ASTM A618.
[g] Minimum applies for walls nominally $\frac{3}{4}$-in. thick and under. For wall thicknesses over $\frac{3}{4}$ in., $F_y = 46$ ksi and $F_u = 67$ ksi.
[h] If desired, maximum yield stress of 65 ksi and maximum yield-to-tensile strength ratio of 0.85 can be specified (per ASTM Supplementary Requirement S75).
[i] A maximum yield-to-tensile strength ratio of 0.85 and carbon equivalent formula are included as mandatory in ASTM A992.
[j] For shapes with a flange thickness greater than 2 in. only.
[k] For shapes with a flange thickness greater than $1\frac{1}{2}$ in. and less than or equal to 2 in. only.
[l] For shapes with a flange thickness less than or equal to $1\frac{1}{2}$ in. only.

Figure 3.7 Applicable ASTM Specifications for Various Structural Shapes. Copyright © American Institute of Steel Construction, Inc. Reprinted with Permission. All rights reserved.

A501 Steel

A501 steel is a carbon steel similar to A36 but used for round and rectangular HSS. It has a minimum yield stress, $F_y = 36$ ksi, and a minimum tensile stress, $F_u = 58$ ksi.

A529 Steel

A529 steel is a carbon-manganese steel available in Grades 50 and 55. It is approved for the smaller shapes with flange thickness no greater that 1.5 in. A529 Grade 50 has a minimum yield stress, $F_y = 50$ ksi, and a tensile stress, $F_u = 65$ to 100 ksi, whereas Grade 55 has a minimum yield stress, $F_y = 55$ ksi, and a tensile stress, $F_u = 70$ to 100 ksi.

A572 Steel

A572 is a high-strength low-alloy steel, also referred to as columbium-vanadium structural steel, available in five grades. It is a versatile high-strength steel with good weldability. Availability of shapes and plates is a function of grade, generally depending on element thickness. It is available in all shapes other than HSS and pipe. The full range of minimum yield stress is 42 ksi to 65 ksi, depending on grade, and the minimum tensile stress ranges from 60 ksi to 80 ksi, again depending on grade. A572 is the preferred steel for HP-shapes.

A618 Steel

A618 is a high-strength low-alloy steel used for HSS. Grades I, II, and III are approved for use in structures by AISC. It is the only high-strength low-alloy steel available in HSS. Grade II has limited atmospheric corrosion resistance and Grade III can be produced with increased corrosion resistance if required. The minimum yield stress depends on the particular product and may vary from 46 to 50 ksi. The minimum tensile stress varies from 65 to 70 ksi, again depending on grade and product wall thickness.

A913 Steel

A913 is a high-strength low-alloy steel produced by quenching and self-tempering. It is available in Grades 50, 60, 65, and 70. This steel is currently not produced domestically but can be obtained from one foreign producer. The minimum yield stress ranges from 50 to 70 ksi and the minimum tensile stress ranges from 60 to 90 ksi.

A992 Steel

A992 steel has become the steel of choice for wide flange shapes. It was first approved for use in 1998 as a replacement for A572 Grade 50. This standard was developed partly as a result of an improved understanding of the impact of material property variations on structural behavior and partly as a result of the changes occurring in properties resulting from the use of scrap as the main resource for steel production. The chemical components for A992 steel were given in Table 3.3 and discussed in Section 3.5. It has a minimum yield stress, $F_y = 50$ ksi, and a minimum tensile stress, $F_u = 65$ ksi. An additional requirement is that the yield-to-tensile ratio can not exceed 0.85.

A242 Steel

A242 is a high-strength low-alloy corrosion-resistant steel also called *weathering steel*. It was one of the first corrosion-resistant steels and has a corrosion resistance approximately four times that of normal carbon steel. It is available in three grades but is now less common than the newer A588. The minimum yield stress ranges from 42 to 50 ksi and the minimum tensile stress ranges from 63 to 70 ksi.

A588 Steel

A588 is a high-strength low-alloy corrosion-resistant steel with substantially better corrosion resistance than carbon steel with or without copper. It is available for all shapes, except HSS and pipe, as well as plate. For all shapes, and for plates up to 4 in., it has a minimum yield stress, $F_y = 50$ ksi, and a minimum tensile stress, $F_u = 70$ ksi. Plates up to 8 in. are available at reduced stress values.

A847 Steel

A847 is the high-strength low-alloy corrosion-resistant steel used for HSS. It has the same minimum yield and tensile stresses as A588.

3.6.2   Steel for Plates and Bars

Many of the steels already discussed for shapes are also available for plates and bars. Figure 3.8 shows the ASTM designation, the corresponding yield and tensile stresses, and the plate thickness for which they apply. The only steels available for plates and bars that are not also available for shapes are the two quenched and tempered steels, A514 and A852.

A514 Steel

A514 is a high-yield strength-quenched and tempered alloy steel suitable for welding. It is available only as plate material up to 6 in. There are 14 different grades, which vary according to the chemical content and maximum thickness. The minimum yield stress is either 90 or 100 ksi and the ultimate tensile stress ranges from 100 to 130 ksi. This is the highest yield stress steel approved for use according to the AISC Specification.

A852 Steel

A852 is a quenched and tempered high-strength low-alloy corrosion-resistant steel. It is intended primarily for use in welded construction where durability and notch toughness are important. It is available as plate only up to 4 in. It has a minimum yield stress, $F_y = 70$ ksi, and a tensile strength, $F_u = 90$ to 110 ksi.

3.6.3   Steel for Fasteners

Fasteners for steel construction today include high-strength bolts, common bolts, threaded rods, and anchor rods. In addition, nuts, washers, and direct-tension-indicators must be specified. The ASTM steels approved for these elements are shown in Figure 3.9. Many grades of steel are appropriate for the variety of mechanical fasteners used in steel construction but only the three steels commonly specified for bolts are discussed here.

Table 2–4
Applicable ASTM Specifications for Plates and Bars

Steel Type	ASTM Designation		F_y Min. Yield Stress (ksi)	F_u Tensile Stress[a] (ksi)	Plates and Bars									
					to 0.75 incl.	over 0.75 to 1.25	over 1.25 to 1.5	over 1.5 to 2 incl.	over 2 to 2.5 incl.	over 2.5 to 4 incl.	over 4 to 5 incl.	over 5 to 6 incl.	over 6 to 8 incl.	over 8
Carbon	A36		32	58-80										■
	A36		36	58-80	■	■	■	■	■	■	■	■	■	
	A529	Gr. 50	50	70-100	▨	b	b	b	b					
	A529	Gr. 55	55	70-100	▨	b	b							
High-Strength Low-Alloy	A572	Gr. 42	42	60	▨	▨								
	A572	Gr. 50	50	65	▨	▨								
	A572	Gr. 55	55	70	▨	▨								
	A572	Gr. 60	60	75	▨	▨								
	A572	Gr. 65	65	80	▨	▨								
Corrosion Resistant High-Strength Low-Alloy	A242		42	63	▨			▨	▨					
	A242		46	67	▨									
	A242		50	70	▨									
	A588		42	63	▨						▨	▨		
	A588		46	67	▨									
	A588		50	70	▨									
Quenched and Tempered Alloy	A514[c]		90	100-130	▨	▨	▨	▨	▨	▨				
	A514[c]		100	110-130	▨	▨	▨	▨	▨					
Quenched and Tempered Low-Alloy	A852[c]		70	90-110	▨	▨	▨	▨						

■ = Preferred material specification.
▨ = Other applicable material specification, the availability of which should be confirmed prior to specification.
☐ = Material specification does not apply.

a Minimum unless a range is shown.
b Applicable to bars only above 1-in. thickness.
c Available as plates only.

Figure 3.8 Applicable ASTM Specifications for Plates and Bars. Copyright © American Institute of Steel Construction, Inc. Reprinted with Permission. All rights reserved.

A307 Bolts

A307 bolts are also called *common bolts* or *black bolts*. Although the ASTM standard specifies three grades, only Grade A is approved for use as bolts in general applications. These bolts have an ultimate tensile strength of 60 ksi and are thus at a strength level similar to A36 steel. Although these bolts continue to be listed by AISC, they are rarely used in steel-to-steel structural connections.

A325 Bolts

A325 bolts are a quenched and tempered steel-heavy hex structural bolt. They are the dominant high-strength bolts used in construction. Two types are available: type 1, the

Table 2–5
Applicable ASTM Specifications for Various Types of Structural Fasteners

ASTM Designation		F_y Min. Yield Stress (ksi)	F_u Tensile Stress[a] (ksi)	Diameter Range (in.)	High-Strength Bolts		Common Bolts	Nuts	Washers	Direct-Tension-Indicator Washers	Threaded Rods	Shear Stud Connectors	Anchor Rods		
					Conventional	Twist-Off-Type Tension-Control[d]							Hooked	Headed	Threaded & Nutted
A108		—	65	0.375 to 0.75, incl.								■			
A325[d]		—	105	over 1 to 1.5 incl.	■										
		—	120	0.5 to 1, incl.	■										
A490		—	150	0.5 to 1.5	■										
F1852		—	105	1.125		■									
		—	120	0.5 to 1, incl.		■									
A194 Gr. 2H		—	—	0.25 to 4				░							
A563		—	—	0.25 to 4				░							
F436[b]		—	—	0.25 to 4					░						
F959		—	—	0.5 to 1.5						░					
A36		36	58-80	to 10							■		░	░	░
A193 Gr. B7[e]		—	100	over 4 to 7							░				
		—	115	over 2.5 to 4							░				
		—	125	2.5 and under							░				
A307	Gr. A	—	60	0.25 to 4			■						░	░	░
	Gr. C	—	58-80	0.25 to 4									░	░	░
A354 Gr. BD		—	140	2.5 to 4 incl.							░		░	░	░
		—	150	0.25 to 2.5, incl.							░		░	░	░
A449		—	90	1.75 to 3 incl.	c						░		░	░	░
		—	105	1.125 to 1.5, incl.	c						░		░	░	░
		—	120	0.25 to 1, incl.	c						░		░	░	░
A572	Gr. 42	42	60	to 6							░		░	░	░
	Gr. 50	50	65	to 4							░		░	░	░
	Gr. 55	55	70	to 2							░		░	░	░
	Gr. 60	60	75	to 1.25							░		░	░	░
	Gr. 65	65	80	to 1.25							░		░	░	░
A588		42	63	Over 5 to 8, incl.							░		░	░	░
		46	67	Over 4 to 5, incl.							░		░	░	░
		50	70	4 and under							░		░	░	░
A687		105	150 max.	0.625 to 3							░		░	░	░
F1554	Gr. 36	36	58-80	0.25 to 4									■	■	■
	Gr. 55	55	75-95	0.25 to 4									░	░	░
	Gr. 105	105	125-150	0.25 to 3									░	░	░

■ = Preferred material specification.
░ = Other applicable material specification, the availability of which should be confirmed prior to specification.
☐ = Material specification does not apply.

— Indicates that a value is not specified in the material specification.
a Minimum unless a range is shown or maximum (max.) is indicated.
b Special washer requirements may apply per RCSC Specification Table 6.1 for some steel-to-steel bolting applications and per Part 14 for anchor-rod applications.
c See AISC Specification Section A3.3 for limitations on use of ASTM A449 bolts.
d When atmospheric corrosion resistance is desired, Type 3 can be specified.
e For anchor rods with temperature and corrosion resistance characteristics.

Figure 3.9 Applicable ASTM Specifications for Various Types of Structural Fasteners. Copyright © American Institute of Steel Construction, Inc. Reprinted with Permission. All rights reserved.

normal medium carbon bolt; and type 3, which is the same bolt, provided in a weathering steel. The tensile strength of these bolts is 120 ksi for bolts up to 1 in. diameter and 105 ksi for bolts with a $1\frac{1}{8}$ to $1\frac{1}{2}$ in. diameter.

A490 Bolts

A490 bolts are also a quenched and tempered steel-heavy hex structural bolt. These fasteners are used when a higher tensile strength is required. As with the A325 bolts, they are available as type 1 or type 3 with the same distinction. The minimum tensile strength for bolts with diameters from $\frac{1}{2}$ to $1\frac{1}{2}$ in. diameter is 150 ksi.

F1852 Bolts

F1852 provides the standard specification for "twist-off" tension control bolt-nut-washer assemblies. These structural fasteners are unique in that they do not have a hex head but rather a splined shank that permits installation through the use of a special torque wrench. These connectors are essentially A325 bolts but must be manufactured to a separate standard because their geometric characteristics differ from normal bolts. The tensile strength of these fasteners is 120 ksi for diameters of $\frac{1}{2}$ to 1 in. inclusive and 105 ksi for $1\frac{1}{8}$ in.

3.6.4   Steel for Welding

Steel used for welding is callled *filler metal* because it essentially fills in the gap between the base metal pieces that it is joining. The most critical aspect of selecting filler metal, actually the welding electrode, is matching the welding electrode with the base metal. In all cases, the weld must not be the weak part of the joint. The American Welding Society provides the specification for appropriate matching of the base metal and electrode in Table 3.1 of their standard, ANSI/AWS D1.1. The most commonly used weld strength is 70 ksi. A discussion of welding processes and material matching is presented in Chapter 10.

3.6.5   Steel for Shear Studs

Shear studs are mechanical fasteners welded to structural shapes and embedded in concrete that permit steel and concrete to work together. This is called *composite construction*. Because these studs are welded to the steel shape, their properties are specified jointly between AWS and ASTM. ANSI/AWS D1.1 specifies that Type B shear stud connectors made from ASTM A108 material be used. These studs have a tensile strength of 65 ksi.

3.7   AVAILABILITY OF STRUCTURAL STEEL

Structural engineers normally use the list of shapes found in the Manual as the basis for design. Unfortunately, all shapes are not equally available in the marketplace and the selection of difficult shapes to obtain could negatively impact the overall cost of a project. For instance, some shapes are available from a wide variety of producers, such as a W10×30, which can be obtained from eight different foreign and domestic mills as of January 2006. However, the largest shapes like the W44×335 are available only from one offshore mill. Also, several of the smaller M-shapes are not rolled by any mill. In order to judge availability of shapes from the 12 mills identified, the annual January issue of *Modern Steel Construction* should be consulted. Shape availability data is also maintained by the mills on the AISC Web site at www.asic.org/steelavailability.

Another important source of steel are the service centers. These organizations obtain steel directly from the mills and stock the full range of shapes. Although obtaining the steel needed for any given project falls to the steel fabricator, it is always beneficial to the engineer to have some knowledge of availability.

3.8 PROBLEMS

1. When was the first AISC Specification published and what was its purpose?

2. In addition to buildings, what other types of structures are included in the scope of the 2005 AISC Specification?

3. Sketch and label a typical stress-strain curve for steel subjected to a simple uniaxial tension test.

4. What is the value of the Modulus of Elasticity used for calculations according to the AISC Specification, and what does this value represent in relation to the graph of stress versus strain for steel?

5. What happens to a steel element when it is loaded beyond the elastic limit and then unloaded?

6. Describe the difference between the yield stress and ultimate stress of a steel element.

7. Sketch and label 10 different structural shape cross sections whose properties are given in the AISC Specification.

8. What is the nominal and actual depth of a W36×135 wide flange member? What is the weight of this member per linear foot? (Hint: use your AISC Manual.)

9. What is the weight per linear foot of a L5×3×$\frac{1}{2}$ member? (Hint: Use your AISC Manual.)

10. What is the outside diameter and wall thickness of a Pipe 4 xx-Strong? (Hint: Use your AISC Manual.)

11. What is the difference between a rectangular bar section and a plate?

12. What are the three types of steel used for shapes in the construction industry?

13. What effects does the addition of carbon have on steel?

14. Name three elements that help to improve the corrosion resistance of steel.

15. What grade of steel is most commonly used today in the production of W-shapes and what is its yield stress and tensile stress?

16. What grade of steel is preferred for the fabrication of most structural shapes other than W-shapes, and what is its yield stress and tensile stress? (Hint: see Figure 3.7.)

17. What grade of steel is typically used for high-strength bolts used in construction?

18. What resources can be consulted to determine the availability of a particular steel structural shape?

Chapter 4

Temporary PATH Station at the World Trade Center Site, New York City.
Photo courtesy John Bartelstone.

Tension Members

4.1 INTRODUCTION

The most efficient way to carry a force in steel is through tension. Because tension forces result in a fairly uniform stress distribution in the member cross section, all of the material is able to work to its fullest capacity. The normal assumption that tensile forces are applied to a member through the centroid of the cross section means that other structural actions, such as buckling or bending, are not normally present to reduce the material's ability to carry load. Thus, tension members are perhaps the simplest to design and a good starting point for studying structural steel design.

Tension members are fairly common elements in building structures, although they may not be found in every structure. The structural members considered in this chapter are those subjected to a concentric tensile force as their primary force. Secondary effects, such as load misalignment and the influence of connections, will be addressed; however, the interaction of tension and bending is saved for later treatment.

Table 4.1 lists the sections of the Specification and parts of the Manual discussed in this chapter.

Table 4.1   Sections of Specification and Parts of Manual Found in this Chapter

	Specification
D1	Slenderness Limitations
D2	Tensile Strength
D3	Area Determination
D4	Built-Up Members
D5	Pin-Connected Members
D6	Eyebars
J3.2	Minimum Spacing
J3.5	Maximum Spacing and Edge Distance
J3.6	Tension and Shear Strength of Bolts and Threaded Parts
J4.1	Strength of Elements in Tension
J4.3	Block Shear Strength
	Manual
Part 1	Dimensions and Properties
Part 5	Design of Tension Members

4.2   TENSION MEMBERS IN STRUCTURES

A wide variety of tension members can be found in building structures. Among the more important are members of trusses, bracing members, hangers, and sag rods.

Tension members are found in trusses as chords, diagonals, and verticals. Figure 4.1 illustrates a typical simply supported truss, with the tension members indicated. Tension members used as bracing for structures are normally long and slender, as seen in Figure 4.2. Because these slender members are relatively flexible, they must be carefully designed and erected, particularly if there is any chance of them experiencing load reversal causing them to be called upon to carry a compression load. Even the smallest compressive force in a member that has been designed as a tension-only member can cause significant serviceability problems in the final structure.

Other examples of tension members are hangers that connect lower floors to some support above, as seen in Figure 4.3, and sag rods that support purlins in the roof structure or the girts in the walls of a steel-framed building. It is easy to see the importance of most tension members because they normally carry an obvious, direct load. Sag rods, as seen in Figure 4.4, however, may not be as easy to understand. The load they carry is not as obvious. Their failure can produce unsightly displacements in the walls, and could cause stability

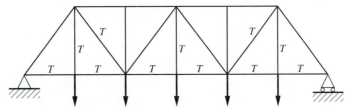

Figure 4.1 A Simply-Supported Truss with Tension Members Indicated.

Figure 4.2 Tension Bracing Members.

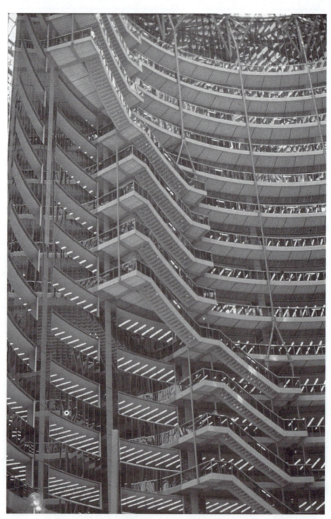

Figure 4.3 Tension Hangers.

Figure 4.4 Sag Rods in a Roof System.

problems for the purlins or girts, but they will unlikely be seen as carrying significant direct loading.

4.3   CROSS-SECTIONAL SHAPES FOR TENSION MEMBERS

Tension members can be structural steel shapes, plates, and combinations of shapes and plates; eyebars and pin-connected plates; rods and bars; or wire rope and steel cables. Wire rope and steel cables are not covered by the Specification nor considered here, although they are important elements in the special structures where they occur.

Eyebars are not in practical use today, but can be found in older applications, particularly in trusses and similar applications, as seen in Figure 4.5. Although rarely used, they are still

Figure 4.5 Eyebars in an Historic Building Roof Structure.

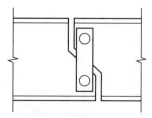

Figure 4.6 A Pin-Connected Member as Part of a Connection.

covered in the Specification. The pin-connected plate shown in Figure 4.6 is actually a part of a connection. This configuration is used in industrial structures and most commonly in bridge girders.

Several common shapes used for tension members are shown in Figure 4.7, and some typical built-up shapes are given in Figure 4.8. The solid round bar is frequently used, either as a threaded rod or welded to other members. The threaded end provides a simple connection to the structure, but the design must take into account the reduction of the cross-sectional area caused by the threads. Upset rods are occasionally used instead of the normal rods; the enlarged end permits threading without reducing the cross-sectional area below the main portion of the rod. The differences between these two types of rods can be seen in Figure 4.9.

Square, rectangular, and circular HSS have become more common as tension members over the past few years, largely due to their attractive appearance and ease of maintenance. However, the end connections may become complicated and expensive, depending on the particular application. HSS are especially useful for longer tension components, when slenderness and related serviceability considerations may be important.

Single angles, as shown in Figure 4.8a, are used extensively in towers, such as those supporting cellular telephone communications and high-voltage power lines. Double angles and double channels, as shown in Figures 4.8b and d, are probably the most popular tension members for planar trusses due to the fact that gusset plates can be conveniently placed in the space between the individual shapes. The end connections for these members are therefore straightforward to design and fabricate, and allow for symmetry in the vertical plane.

Large tensile forces usually require cross sections that may dictate that the member be made from a wide-flange shape, a tee, double channels, or built-up shapes, such as those given in Figures 4.8e and f. Built-up cross sections were more commonly used in the past,

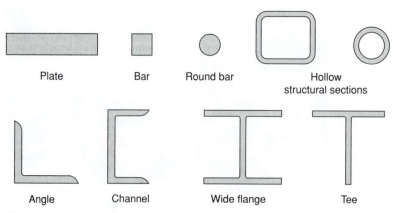

Figure 4.7 Common Shapes for Tension Members.

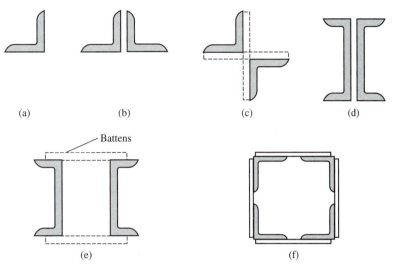

Figure 4.8 Typical Built-Up Shapes Used as Tension Members.

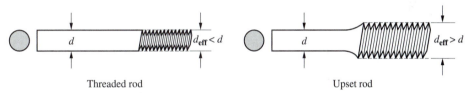

Figure 4.9 Threaded and Upset Rods.

when the cost of labor was lower; today, large-force tension members would probably be made from rolled shapes. Current structural applications of such elements are found in long-span roof trusses, bridge trusses, and bracing members in large industrial structures.

4.4   BEHAVIOR AND STRENGTH OF TENSION MEMBERS

Tension members are covered in Chapter D of the Specification. Two possible limit states are defined in Section D2 for tension members: yielding and rupture. The controlling limit state depends on the ability of the member to undergo plastic deformation. Both of these failure modes represent limit states of strength that must be taken into account in the design of the tension member. The design basis for ASD and LRFD were presented in Sections 1.6 and 1.7, respectively. Equations 1.1 and 1.2 are repeated here in order to reinforce the relationship between the nominal strength, resistance factor, and safety factor presented throughout the Specification.

For ASD, the allowable strength is

$$R_a \leq \frac{R_n}{\Omega} \tag{1.1}$$

For LRFD, the design strength is

$$R_u \leq \phi R_n \tag{1.2}$$

As indicated earlier, the Specification provides the relationship for the determination of the nominal strength and the corresponding resistance factor and safety factor for each limit state to be considered. The provisions for tension members are:

4.4.1 Yielding

Yielding occurs when the uniformly distributed stress throughout the cross section reaches the yield stress over the length of the member. Although the member will continue to resist the load that caused yielding to occur, it will undergo excessive stretching and this elongation will make the member unusable. The longer the member, the greater the elongation. Because the limit state of yielding on the gross section of the member is accompanied by this large deformation, it will readily warn of any impending failure.

The yield limit state is defined as

$$P_n = F_y A_g \tag{4.1}$$

where

    $P_n =$ nominal tensile yield strength

    $F_y =$ yield stress

    $A_g =$ gross area of the member

The design strength and allowable strength are to be determined using

$$\phi_t = 0.90\,(\text{LRFD}) \qquad \Omega_t = 1.67\,(\text{ASD})$$

4.4.2 Rupture

Holes in a member will cause stress concentrations to develop under the service load, as shown in Figure 4.10. Elastic theory shows that the stress concentration results in a peak stress approximately three times the average stress. As the peak stress reaches yield, the member will continue to strain and load can continue to increase. With increasing load, the strain in the region of the hole increases into the strain hardening region, and the member ruptures once the stress in this area exceeds the ultimate strength. Although the material in the region of the holes yields initially, it yields over a very short length, resulting in a small total elongation. Thus, the material can reach its ultimate strength through strain hardening, without excessive elongation, and failure occurs through rupture. The limit state of rupture on the effective net area of the cross section is accompanied by small deformations from yielding, giving little or no warning of the impending sudden failure, and offering limited opportunities to take corrective action before the rupture.

The rupture limit state is defined as

$$P_n = F_u A_e \tag{4.2}$$

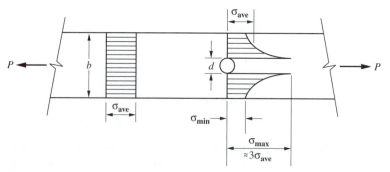

Figure 4.10 Stress Concentration Due to Hole in Member.

where

P_n = nominal tensile rupture strength

F_u = ultimate stress

A_e = effective net area of the member

The design strength and allowable strength are to be determined using

$$\phi_t = 0.75\,(\text{LRFD}) \qquad \Omega_t = 2.00\,(\text{ASD})$$

If the two limit states were to result in the same available strength, using the LRFD formulation

$$0.75 F_u A_e = 0.9 F_y A_g \tag{4.3}$$

or

$$A_e/A_g = 0.9 F_y/0.75 F_u \tag{4.4}$$

The limit state of yielding on the gross section governs when the right-hand side of Equation 4.3 is less than the left-hand side. Using Equation 4.4, yielding on the gross section governs if

$$A_e/A_g > 0.9 F_y/0.75 F_u \tag{4.5}$$

and rupture on the effective net section governs if

$$A_e/A_g < 0.9 F_y/0.75 F_u \tag{4.6}$$

Steel with a small (F_y/F_u) value, such as ASTM A36, with $0.9 F_y/0.75 F_u = 0.9(36)/(0.75(58)) = 0.74$, will allow more of the cross section to be removed in the form of bolt holes before the rupture limit state will govern than steels with a higher (F_y/F_u) value, such as ASTM A992, with $0.9 F_y/0.75 F_u = 0.9(50)/(0.75(65)) = 0.92$.

The comparisons discussed above are applicable only for normal bolted connections and their corresponding areas. Equations 4.1 through 4.6 are not intended to cover tension members with large cutouts. These require special design considerations, and are beyond the scope of this book because they are not common in most building structures.

Although welded connections do not normally require the removal of material from the cross section, the placement of the welds and the type of cross section may require a reduction from the gross area to determine the effective net area.

4.5 COMPUTATION OF AREAS

The design of tension members uses the following cross-sectional area definitions:

1. gross area, A_g
2. net area, A_n
3. effective net area, A_e

The criteria governing the computation of the various areas required for tension member analysis and design are given in Section D3 of the Specification. They are discussed in further detail here.

4.5.1 Gross Area

The gross area of a member might also be thought of as the full cross-section area. A section is made perpendicular to the longitudinal axis of the element, along which the tensile force is acting, and the gross area, A_g, is the area of that cross section. No holes or other area reductions can be present where the section is taken.

In the case of plates, bars, and solid circular shapes, the value of A_g is found directly as the value of width times thickness, bt, for plates and bars and $\pi d^2/4$ for circular shapes, where d is the diameter. For structural steel shapes commonly used in construction, the Manual provides values for gross areas in Part 1. However, in lieu of using the tabulated values, A_g may be approximated as

$$A_g = \Sigma\, w_i t_i \tag{4.7}$$

where w_i and t_i are the width and thickness, respectively, of the rectangular cross-sectional element, i, of the shape. Equation 4.7 applies only to shapes that are composed of flat plate components, such as wide flanges and channels. The calculation for hollow circular shapes is similarly straightforward. The gross area of HSS shapes meeting the requirements of ASTM A500 is determined using 93% of the nominal wall thickness of the shape. Because HSS are consistently manufactured with a thickness at the low end of the tolerance limit, the values provided in the Manual are all based on this reduced thickness.

The procedure for angles requires a slight modification. The angle may be treated as an equivalent flat plate, wherein the effective width is taken as the sum of the leg dimensions less the thickness and the gross area is this effective width times the angle thickness.

An angle and its equivalent flat plate is shown in Figure 4.11.

4.5.2 Net Area

The net area is obtained by subtracting the area of any holes occurring at a particular section from the gross area of that section. Thus, holes resulting from mechanical fasteners, such as bolts, and welds, such as plug welds and slot welds, are considered. These fastening elements are normally used only to connect tension members to the adjacent parts of the structure, and the reduced areas therefore normally appear at member ends. However, if holes occur at any point along a tension member, their effect must be considered.

Plug and slot welds are made through holes in a member and are relatively uncommon in most structures in which the framing members are made from structural shapes and plates. Normal welded joints do not involve making holes in members. Thus, welds do not normally reduce the cross-sectional area of tension members so the net area is equal to the gross area.

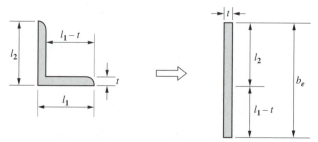

Figure 4.11 Angle and Its Equivalent Flat Plate.

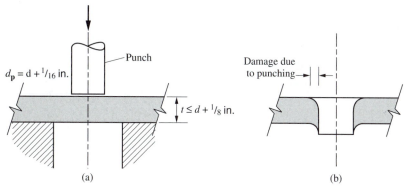

Figure 4.12 Damage Caused by Hole Punching.

In the computation of net area for a tension member with bolted end connections, determining the size of the holes is important. The criteria for standard, oversize, and slotted holes are covered in Specification Section J3.

Standard Holes

Normal steel construction requires the specification of fastener size rather than hole size. The hole is then sized according to what is required to accommodate the fastener. The manner in which the hole is fabricated is also critical.

In recognizing the needs for fabrication and erection tolerances, standard bolt holes are made $1/16$ in. larger in diameter than the bolt to be inserted in the hole. Thus, a $3/4$-in. bolt requires a hole with a $(3/4 + 1/16) = 13/16$ in. diameter. In the case of punched holes, the punching process may damage some of the material immediately adjacent to the hole. That material may not be considered fully effective in transmitting load and must also be deducted from the gross area along with the material that has actually been removed. This is schematically illustrated in Figure 4.12, for the case of punched holes. As the punch is applied to the material, the edges around the hole are deformed, as shown in Figure 4.12b. In discounting this region, the effective hole diameter is increased by another $1/16$ in. according to Specification Section D3.2.

Standard practice permits punching of holes for material thickness up to $1/8$ in. larger than the nominal bolt diameter. Otherwise, holes would be drilled, or sub-punched and reamed.

Because the decision to punch or drill a hole is a function of the steel fabricator's equipment capacity, for the design of tension members it is standard practice to deduct for holes with a diameter $1/8$ in. greater than the specified bolt size.

The following examples demonstrate gross and net area calculations for several shapes.

EXAMPLE 4.1
Gross and Net Area

GOAL: Determine the gross and net areas of a plate with a single line of holes.

GIVEN: A single line of standard holes for $3/4$-in. bolts is placed in a $6 \times 1/2$ plate, as shown in Figure 4.13a.

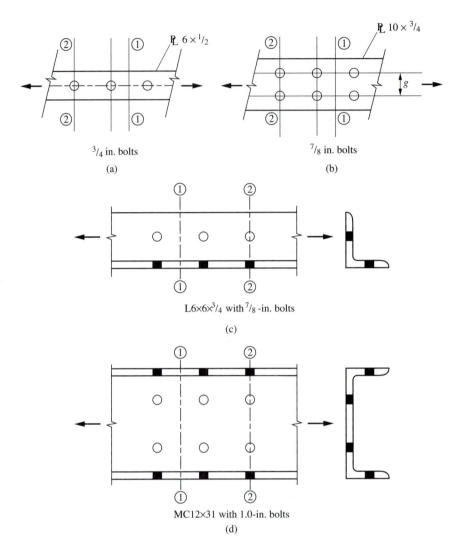

Figure 4.13 Plates and Shapes with Holes for Use with Examples 4.1 through 4.4.

SOLUTION

Step 1: Determine the gross area at Section 1-1.

$$A_g = 6(^1/_2) = 3.0 \text{ in.}^2$$

Step 2: Determine the effective hole size for a $^3/_4$-in. diameter bolt.

$$d_e = (^3/_4 + ^1/_{16} + ^1/_{16}) = ^7/_8 \text{ in.}$$

Step 3: Determine the net area at Section 2-2.

$$A_n = (b - d_e)t = (6.0 - ^7/_8)(^1/_2) = 2.56 \text{ in.}^2$$

EXAMPLE 4.2
Gross and Net Area

GOAL: Determine the gross and net areas of a plate with a double line of holes.

GIVEN: A double line of standard holes for $\frac{7}{8}$-in. bolts are placed in a $10 \times \frac{3}{4}$ plate, as shown in Figure 4.13b.

SOLUTION

Step 1:   Determine the gross area at Section 1-1.

$$A_g = 10(\tfrac{3}{4}) = 7.5 \text{ in.}^2$$

Step 2:   Determine the effective hole size for a $\frac{7}{8}$-in. diameter bolt.

$$d_e = (\tfrac{7}{8} + \tfrac{1}{16} + \tfrac{1}{16}) = 1.0 \text{ in.}$$

Step 3:   Determine the net area at Section 2-2.

$$A_n = (b - d_e)t = (10.0 - 2(1.0))(\tfrac{3}{4}) = 6.0 \text{ in.}^2$$

EXAMPLE 4.3
Gross and Net Area

GOAL: Determine the gross and net areas of an angle with a single line of holes.

GIVEN: A single line of standard holes for $\frac{7}{8}$-in. bolts is placed on each leg of a $6 \times 6 \times \frac{3}{4}$ angle as shown in Figure 4.13c.

SOLUTION

Step 1:   Determine the gross area at Section 1-1.

$$A_g = 8.46 \text{ in.}^2 \text{ from Manual Table 1-7}$$

Step 2:   Determine the effective hole size for a $\frac{7}{8}$-in. diameter bolt.

$$d_e = (\tfrac{7}{8} + \tfrac{1}{16} + \tfrac{1}{16}) = 1.0 \text{ in.}$$

Step 3:   Determine the net area at Section 2-2.

$$A_n = (8.46 - 2(1.0)(0.75)) = 6.96 \text{ in.}^2$$

EXAMPLE 4.4
Gross and Net Area

GOAL: Determine the gross and net areas of a channel with multiple lines of holes.

GIVEN: Four lines of standard holes for 1.0-in. bolts are placed in an MC12×31, as shown in Figure 4.13d. Two lines are in the web and one line is in each flange.

SOLUTION

Step 1:   Determine the gross area at Section 1-1.

$$A_g = 9.12 \text{ in.}^2 \text{ from Manual Table 1-6}$$

Step 2: Determine the effective hole size for a 1.0-in. diameter bolt.

$$d_e = (1.0 + {}^1\!/_{16} + {}^1\!/_{16}) = 1.125 \text{ in.}$$

Step 3: Determine the net area at Section 2-2.

$$A_n = (9.12 - 2(1.125)(0.37) - 2(1.125)(0.70)) = 6.71 \text{ in.}^2$$

Oversize and Slotted Holes

Section J3.2 of the Specification gives the required measurements for larger-than-standard or oversize holes, as well as for short-slotted and long-slotted holes. Figure 4.14 illustrates the criteria that apply for nominal bolt diameters of $^5\!/_8$ in. and $^3\!/_4$ in.; refer to the Specification for data for other bolt sizes. These types of holes are used to facilitate the erection of the structure and, in some cases, to permit larger rotations or deformations to take place under loading.

Short Connecting Elements

Tension members within connections are usually short connecting elements such as links, flange plates, or gusset plates. When the member is short, and the net area and gross area are close to equal, there may not be sufficient length for the entire cross section to yield uniformly. In this case, the area that is the first to yield may reach rupture at an early stage, and the rupture limit state would therefore be reached prematurely. This is an undesirable mode of failure, primarily because it is not ductile, and because it occurs suddenly, with little or no warning. Section J4.1 specifies that the net area in these splice plates and other short connecting elements may not be taken larger than 0.85 times the gross area.

4.5.3 Influence of Hole Placement

The examples shown in Section 4.5.2 represent simple cases in which the net area is found in the section that produces the largest reduction in area, usually the section with the largest number of holes. However, hole placement does not always follow simple patterns whereby every section has the same number of holes. It is sometimes advantageous to use a pattern of staggered holes, such as those shown in Figure 4.15. Figure 4.15a shows an arrangement of staggered holes for a plate and Figure 4.15b shows an example for an angle. When there are multiple holes, the center-to-center distance between adjacent holes in the direction parallel

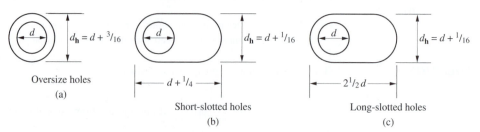

Oversize holes
(a)

Short-slotted holes
(b)

Long-slotted holes
(c)

Figure 4.14 Size Criteria for $^5\!/_8$- to $^3\!/_4$-in. Bolts in Oversize and Slotted Holes.

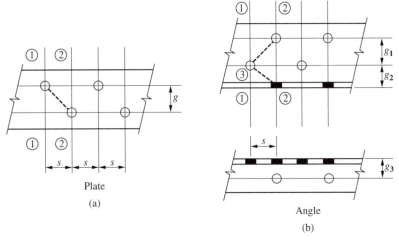

Figure 4.15 Staggered Hole Patterns in Plate and Angle.

to the primary applied force is defined as the pitch, s. When there is more than one line of holes parallel to the line of force, the center-to-center distance between adjacent holes in the direction perpendicular to the primary applied force is the gage, g.

It is not clear from Figure 4.15 what the governing net section would be for either case. For the plate, Sections 1-1 and 2-2 give identical A_n values, in which a deduction for one hole is taken for each line. Another possibility would be to follow a line that incorporates two holes, starting along line 1 and ending along line 2, as shown by the diagonal dashed line. The Specification refers to this line as a "chain" because it links together individual holes. In this case, the area of two holes would be deducted from the gross cross section. However, this approach would be no different than if both holes were along the same line. It seems reasonable in this situation that an approach that would deduct both holes would deduct too much, because these holes follow along a diagonal and not a straight line. The correct solution should be somewhere between deducting for one hole and deducting for two holes.

A simplified approach to address the interaction of staggered holes was adopted long ago by previous AISC Specifications. Although numerous studies have been conducted since this original simplification was first introduced, none have proposed a significantly more accurate approach that is equally easy to implement.

The Specification approach requires that every potential failure line be assessed with the full area of each intersected hole deducted and something added back for the increased strength provided by the diagonal path. For every diagonal on a potential failure path, the quantity $s^2/4g$ is added back into the net width to account for the overestimation of the required deduction when a full adjacent hole has been deducted. Examples 4.5 and 4.6 show the application of the staggered hole criterion.

EXAMPLE 4.5
Net Width of Plate

GOAL: Determine the net width of a plate with staggered holes.

GIVEN: The hole pattern for an 18-in.-wide plate with holes for $3/4$-in. bolts that is loaded in tension as shown in Figure 4.16. A and F represent the edges of the plate, whereas B, C, D, and E represent hole locations.

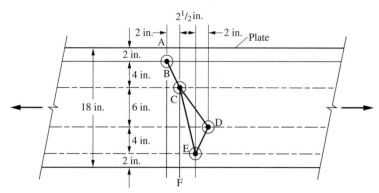

Figure 4.16 Hole Pattern for an 18-in. Plate Used in Example 4.5.

SOLUTION

Step 1: Chain A B F (a straight line through one hole).

$$\text{Deduct for 1 hole } (^3/_4 + ^1/_8) = -0.88 \text{ in.}$$

Step 2: Chain A B C F.

Deduct for 2 holes, $2(^3/_4 + ^1/_8)$	$= -1.75$
For BC add $s^2/4g = 2.0^2/(4(4.0))$	$= +0.25$
Total deduction	$= -1.50$ in.

Step 3: Chain A C E F.

Deduct for 2 holes, $2(^3/_4 + ^1/_8)$	$= -1.75$
For CE add $s^2/4g = 2.5^2/(4(10.0))$	$= +0.16$
Total deduction	$= -1.59$ in.

Step 4: Chain A B C E F.

Deduct for 3 holes, $3(^3/_4 + ^1/_8)$	$= -2.63$
For BC, add $s^2/4g = 2.0^2/(4(4.0))$	$= +0.25$
For CE, add $s^2/4g = 2.5^2/(4(10.0))$	$= +0.16$
Total deduction	$= -2.22$ in.

Step 5: Chain A B C D E F.

Deduct for 4 holes, $4(^3/_4 + ^1/_8)$	$= -3.50$
For BC, add as for previous chain	$= +0.25$
For CD, add $s^2/4g = 4.5^2/(4(6.0))$	$= +0.84$
For DE, add $s^2/4g = 2.0^2/(4(4.0))$	$= +0.25$
Total deduction	$= -2.16$ in.

Step 6: Deduct the largest quantity to obtain the least net width.

$$\boxed{\text{Net width} = b_n = 18.0 - 2.22 = 15.8 \text{ in.}}$$

EXAMPLE 4.6
Gross Area and Net Area of Angle

GOAL: Determine the gross and net areas of an angle with staggered holes.

GIVEN: A $6 \times 4 \times \frac{1}{2}$ angle with holes for $\frac{7}{8}$-in. bolts are placed as shown in Figure 4.17.

SOLUTION

Step 1:   Determine the width of the equivalent flat plate representing the width of the angle.
$$w_e = l_1 + l_2 - t = 6 + 4 - \frac{1}{2} = 9.50 \text{ in.}$$

Step 2:   Determine the gross area of the equivalent plate.

$$A_g = w_e t = 9.50(0.5) = 4.75 \text{ in.}^2 \text{ (as found in Manual Table 1-7)}$$

Step 3:   Determine the gages for each bolt line.
The gages for the holes are shown in Figure 4.17. The gage between the holes closest to the heel of the angle in the two legs must be adjusted to account for the angle thickness. Thus,
$$(g + g_1 - t) = 2.50 + 2.25 - 0.50 = 4.25 \text{ in.}$$

Step 4:   Determine the net area.
The governing net section will be Section 2-2 or Section 2-1-2. There is no need to consider Section 1-1 because b_{n1} will clearly be greater than b_{n2}.

For Section 2-2:
$$b_{n2} = \left(9.5 - 2\left(\frac{7}{8} + \frac{1}{8}\right)\right) = 7.5 \text{ in.}$$
$$A_{n2} = 7.5(0.5) = 3.75 \text{ in.}^2$$

For Section 2-1-2:
This chain has two staggers of the bolt holes, and both have the same pitch ($s = 2.50$ in). The gages are different, with one at 4.25 in. and the other at 2.50 in.
The net area for this chain becomes:

$$b_{n3} = 9.5 - 3\left(\frac{7}{8} + \frac{1}{8}\right) + \left(\frac{2.50^2}{4(2.5)}\right) + \left(\frac{2.50^2}{4(4.25)}\right) = 7.49 \text{ in.}$$
$$A_{n3} = 7.49(0.5) = 3.75 \text{ in.}^2$$

Step 5:   Select the least net area.
The lowest net area controls. In this case, both chains yield the same net area, thus

$$A_n = A_{n2} = A_{n3} = 3.75 \text{ in.}^2$$

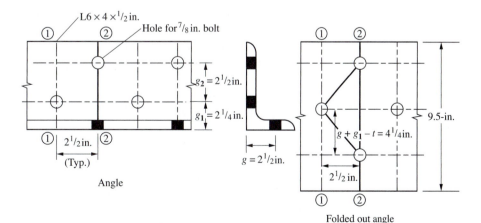

Figure 4.17 Hole Pattern for $L6 \times 4 \times \frac{1}{2}$ Used in Example 4.6.

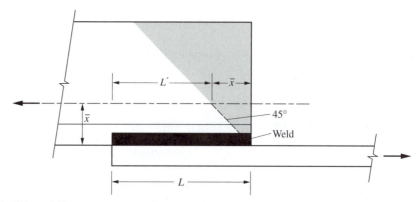

Figure 4.18 Conceptual Basis for Shear Lag Reduction Factor.

4.5.4   Effective Net Area

When all elements of a tension member are attached to connecting elements, the entire member participates fully in transferring the load to the connection. However, when not all elements are attached to connecting elements, they cannot all participate fully. Figure 4.18 shows an angle with one leg attached to a connecting element and the other, the outstanding leg, unattached. To account for the inability of this unattached leg to transfer load, the net area used in calculating the rupture strength is reduced to the effective net area, A_e.

This phenomenon occurs because the uniform stresses, occurring near the mid-length of the member at some distance from the connection, must be transferred through the more restricted area where the connection is located. The portion of the member area that is participating effectively in the transfer of force is smaller than the full net area. Thus, the net area is reduced to the effective net area.

This general behavior is called *shear lag*. Since its original introduction into the Specification, it has been approximated by the use of the shear lag reduction factor such that

$$A_e = UA_n$$

Specification Table D3.1 provides values of the shear lag factor, U, for a wide variety of elements. For all tension members, except plates and HSS, when the tension load is transmitted to some but not all of the cross-sectional elements, the effective length of the connection is reduced to $L' = L - \bar{x}$, where $\bar{x}$ is the distance from the attached face to the member centroid and L is the length of the connection, as shown in Figure 4.18. The reduction in net area is then taken in proportion to the reduction in effective length, L'/L. Thus, the reduction becomes

$$U = \frac{L'}{L} = \frac{L - \bar{x}}{L} = 1 - \frac{\bar{x}}{L} \tag{4.8}$$

Figure 4.19 shows the definition of connection length, L, for both a bolted and welded connection.

Certain shapes have the potential to significantly reduce their effective net area due to their geometry and the length of the connection. Members such as single angles, double angles, and WT shapes must be proportioned so that the shear lag factor is not less than 0.6. If these members exceed this limit, they must be designed as combined force members,

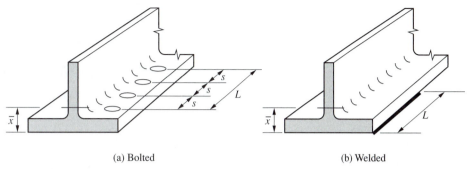

(a) Bolted   (b) Welded

Figure 4.19 Definition of Connection Length, L, for Bolted and Welded Connection.

as discussed in Chapter 8 of this book and Chapter H of the Specification. Table D3.1 of the Specification also provides a simplified approach to the shear lag factor when certain criteria are met.

For W-, M-, S-, and HP-shapes or Tees cut from these shapes, the following apply:

Flange connected with 3 or more	$b_f \geq \frac{2}{3}\,d$	$U = 0.90$
fasteners per line in the direction of loading	$b_f < \frac{2}{3}\,d$	$U = 0.85$
Web connected with 4 or more		$U = 0.70$
fasteners per line in the direction of loading		

For single angles

With 4 or more fasteners per line in the direction of loading	$U = 0.80$
With 2 or 3 fasteners per line in the direction of loading	$U = 0.60$

EXAMPLE 4.7
*Tensile Strength
of an Angle*

GOAL: Determine the design strength (LRFD) and the allowable strength (ASD) of an angle.

GIVEN: Consider an L4×4×$\frac{1}{2}$ attached through one leg to a gusset plate with $\frac{3}{4}$-in. bolts as shown in Figure 4.20. Use A36 steel.

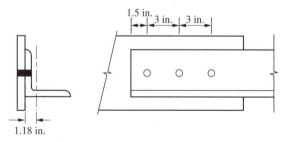

1.5 in. | 3 in. | 3 in.

1.18 in.

Figure 4.20 Single-Angle Tension Member for Example 4.7.

SOLUTION   **Step 1:**   Determine the areas needed for calculations.

$$A_g = 3.75 \text{ in.}^2 (\text{from Manual Table 1-7})$$

$$A_n = 3.75 - \left(\tfrac{3}{4} + \tfrac{1}{8}\right)\left(\tfrac{1}{2}\right) = 3.31 \text{ in.}^2$$

Shear lag factor for a 6-in. connection length and an angle with $\bar{x} = 1.18$ in. (from Manual Table 1-7)

$$U = 1 - \frac{\bar{x}}{L} = 1 - \frac{1.18}{6} = 0.80$$

$$A_e = 0.80(3.31) = 2.65 \text{ in.}^2$$

For LRFD

Step 2:   For the limit state of yielding

$$P_n = 36(3.75) = 135 \text{ kips}$$
$$\phi_t P_n = 0.9(135) = 122 \text{ kips}$$

Step 3:   For the limit state of rupture

$$P_n = 58(2.65) = 154 \text{ kips}$$
$$\phi_t P_n = 0.75(154) = 116 \text{ kips}$$

Therefore, the limit state of rupture controls and the design strength is

$$\boxed{\phi_t P_n = 116 \text{ kips}}$$

For ASD

Step 2:   For the limit state of yielding

$$P_n = 36(3.75) = 135 \text{ kips}$$
$$\frac{P_n}{\Omega_t} = \frac{135}{1.67} = 80.8 \text{ kips}$$

Step 3:   For the limit state of rupture

$$P_n = 58(2.65) = 154 \text{ kips}$$
$$\frac{P_n}{\Omega_t} = \frac{154}{2.00} = 77.0 \text{ kips}$$

Therefore, the limit state of rupture controls and the design strength is

$$\boxed{\frac{P_n}{\Omega_t} = 77.0 \text{ kips}}$$

EXAMPLE 4.8
Tensile Strength
of a Tee

GOAL: Determine the design strength (LRFD) and the allowable strength (ASD) of a WT.

GIVEN: Consider a WT6×32.5 attached to a gusset plate with welds as shown in Figure 4.21. Use A992 steel.

SOLUTION

Step 1:   Determine the needed areas

$$A_g = 9.54 \text{ in.}^2 \text{ (from Manual Table 1-8)}$$

Because the force is transferred by longitudinal welds only,

$$A_n = A_g$$

Shear lag factor for a 6-in. connection length and a Tee with $\bar{x} = 0.985$ in. (from Manual Table 1-8)

$$U = 1 - \frac{\bar{x}}{L} = 1 - \frac{0.985}{6} = 0.836$$

$$A_e = 0.836(9.54) = 7.98 \text{ in.}^2$$

For LRFD

Step 2:   For the limit state of yielding

$$P_n = 50(9.54) = 477 \text{ kips}$$
$$\phi P_n = 0.9(477) = 429 \text{ kips}$$

Step 3:   For the limit state of rupture

$$P_n = 65(7.98) = 519 \text{ kips}$$
$$\phi P_n = 0.75(519) = 389 \text{ kips}$$

Therefore, the limit state of rupture controls and the design strength is

$$\boxed{\phi P_n = 389 \text{ kips}}$$

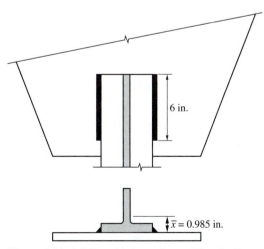

Figure 4.21 A WT welded to a Gusset Plate for Example 4.8.

For ASD

Step 2: For the limit state of yielding

$$P_n = 50(9.54) = 477 \text{ kips}$$

$$\frac{P_n}{\Omega_t} = \frac{477}{1.67} = 286 \text{ kips}$$

Step 3: For the limit state of rupture

$$P_n = 65(7.98) = 519 \text{ kips}$$

$$\frac{P_n}{\Omega_t} = \frac{519}{2.0} = 260 \text{ kips}$$

Therefore, the limit state of rupture controls and the allowable strength is

$$\frac{P_n}{\Omega_t} = 260 \text{ kips}$$

4.6   DESIGN OF TENSION MEMBERS

To design a structural steel tension member, the member size must be determined and then the appropriate limit states checked. The only additional issue to address is the slenderness of the member. For tension members, slenderness is defined as the member length divided by the least radius of gyration, L/r. The Specification has, in the past, placed a limit on the slenderness of tension members. However, there is currently no specified limitation on tension member slenderness, as indicated in Section D1. The designer should exercise caution when selecting tension members with very high slenderness ratios, that is, those near the former limit of $L/r = 300$, because these members could easily be damaged during erection and might cause other problems due to their flexibility in the transverse direction.

Because the main task in tension member design is to determine the area of the member, the two limit states of yielding and rupture can be used to determine minimum gross and net areas such that

$$A_{g\,min} = \frac{P_u}{\phi_t F_y} \text{ (LRFD)} \quad \text{or} \quad A_{g\,min} = \frac{\Omega_t P_a}{F_y} \text{ (ASD)}$$

and

$$A_{e\,min} = \frac{P_u}{\phi_t F_u} \text{ (LRFD)} \quad \text{or} \quad A_{e\,min} = \frac{\Omega_t P_a}{F_u} \text{ (ASD)}$$

Because connection details are not normally known in the early stages of member selection, it may not be possible to determine the actual deductions necessary to obtain the exact effective net area of the member being designed. One approach would be to assume a fixed percentage deduction for the effective net area. The designer would decide the magnitude of this deduction.

Part 5 of the Manual provides tables for tension member design that give the strength of tension members based on the limit states of yielding on the gross area and rupture on an effective net area equal to $0.75A_g$. If the actual effective net area differs from this assumed value, the designer can simply adjust the strength accordingly.

EXAMPLE 4.9a
Tension Member Design
by LRFD

GOAL: Select a double-angle tension member for use as a web member in a truss and determine the maximum area reduction that would be permitted for holes and shear lag.

GIVEN: The member must carry an LRFD required strength, $P_u = 405$ kips. Use equal leg angles of A36 steel.

SOLUTION

Step 1:   Determine the minimum required gross area based on the limit state of yielding

$$A_{g\ min} = 405/(0.9(36)) = 12.5 \text{ in.}^2$$

Step 2:   Based on this minimum gross area, from Manual Table 1-15, select

$$\boxed{2\text{L}6\times6\times{}^9\!/_{16} \text{ with } A_g = 12.9 \text{ in.}^2}$$

Step 3:   Determine the minimum effective net area based on the limit state of rupture

$$A_{e\ min} = 405/(0.75(58)) = 9.31 \text{ in.}^2$$

Step 4:   Thus, the combination of holes and shear lag may not reduce the area of this pair of angles by more than

$$\boxed{A_e/A_g = 9.31/12.9 = 0.722}$$

EXAMPLE 4.9b
Tension Member Design
by ASD

GOAL: Select a double-angle tension member for use as a web member in a truss and determine the maximum area reduction that would be permitted for holes and shear lag.

GIVEN: The member must carry an ASD required strength, $P_a = 270$ kips. Use equal leg angles of A36 steel.

SOLUTION

Step 1:   Determine the minimum required gross area based on the limit state of yielding

$$A_{g\ min} = 270/(36/1.67) = 12.5 \text{ in.}^2$$

Step 2:   Based on this minimum gross area, from Manual Table 1-15, select

$$\boxed{2\text{L}6\times6\times{}^9\!/_{16} \text{ with } A_g = 12.9 \text{ in.}^2}$$

Step 3:   Determine the minimum effective net area based on the limit state of rupture

$$A_{e\ min} = 270/(58/2.00) = 9.31 \text{ in.}^2$$

Step 4:   Thus, the combination of holes and shear lag may not reduce the area of this pair of angles by more than

$$\boxed{A_e/A_g = 9.31/12.9 = 0.722}$$

EXAMPLE 4.10a
Tension Member Design by LRFD

GOAL: Select a WT9 for use as a tension member and determine the maximum area reduction that would be permitted for holes and shear lag.

GIVEN: The member must carry an LRFD required strength, $P_u = 818$ kips. Use A992 steel.

SOLUTION

Step 1:   Determine the minimum required gross area based on the limit state of yielding
$$A_{g\,min} = 818/(0.9(50)) = 18.2 \text{ in.}^2$$

Step 2:   Based on the minimum gross area, from Manual Table 1-8, select

$$\boxed{\text{WT9} \times 65 \text{ with } A_g = 19.1 \text{ in.}^2}$$

Step 3:   Determine the minimum effective net area needed to resist the applied force
$$A_{e\,min} = 818/(0.75(65)) = 16.8 \text{ in.}^2$$

Step 4:   The combination of holes and shear lag may not reduce the area of this WT by more than

$$\boxed{A_e/A_g = 16.8/19.1 = 0.880}$$

EXAMPLE 4.10b
Tension Member Design by ASD

GOAL: Select a WT9 for use as a tension member and determine the maximum area reduction that would be permitted for holes and shear lag.

GIVEN: The member must carry an ASD required strength, $P_a = 545$ kips. Use A992 steel.

SOLUTION

Step 1:   Determine the minimum required gross area based on the limit state of yielding
$$A_{g\,min} = 1.67(545)/50 = 18.2 \text{ in.}^2$$

Step 2:   Based on the minimum gross area, from Manual Table 1-8, select

$$\boxed{\text{WT9} \times 65 \text{ with } A_g = 19.1 \text{ in.}^2}$$

Step 3:   Determine the minimum effective net area needed to resist the applied force
$$A_{e\,min} = 2.00(545)/65 = 16.8 \text{ in.}^2$$

Step 4:   The combination of holes and shear lag may not reduce the area of this WT by more than

$$\boxed{A_e/A_g = 16.8/19.1 = 0.880}$$

4.7   BLOCK SHEAR

When a portion of a member tears out in a combination of tension and shear as shown in Figure 4.22, the failure is known as a *block shear failure*. Even though this failure mode is primarily the result of a connection failure, it may possibly control the overall strength of a tension member. The resistance to tear-out is provided by a combination of shear on the plane parallel to the tension force and tension on the plane perpendicular to it.

Rupture will always be the controlling mode on the tension face of the failure block, due to the relatively short length of material that will be available to yield. The controlling limit states on the shear face will be either yielding or rupture, whichever has the lower strength. Unlike the situation for overall member strength, in which the yield and rupture limit states had different resistance and safety factors, block shear uses the same values for both limit states. Thus, a simple comparison of nominal strengths is appropriate to determine the controlling limit state. Section J4.3 of the Specification gives the block shear strength as

$$R_n = 0.6F_u A_{nv} + U_{bs} F_u A_{nt} \leq 0.6F_y A_{gv} + U_{bs} F_u A_{nt} \qquad (4.9)$$

where

A_{gv} = gross area in shear

A_{nv} = net area in shear

A_{nt} = net area in tension

U_{bs} = 1.0 if the tension stress is uniform and 0.5 if the tension stress is not uniform. (U_{bs} = 0.5 is addressed in Chapter 7.)

The design and allowable strengths are determined using

$$\phi = 0.75 \text{ (LRFD)} \qquad \Omega = 2.0 \text{ (ASD)}$$

For tension members, the tensile stress is assumed to be uniform. Thus, U_{bs} = 1.0 will be used.

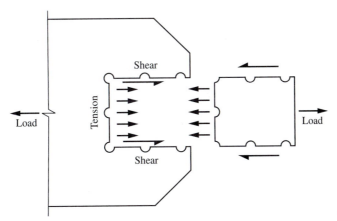

Figure 4.22 Example of a Block Shear Failure of a Plate.

EXAMPLE 4.11
*Gusset Plate Tension
Strength*

GOAL: Determine whether the gusset plate has sufficient strength in block shear.

GIVEN: The gusset plate shown in Figure 4.23 has a plate thickness of $\frac{1}{2}$ in. The required strength for LRFD is $P_u = 225$ kips and for ASD is $P_a = 150$ kips. The steel is A36 and the holes are punched for $\frac{7}{8}$-in. bolts.

SOLUTION

Step 1:   Determine the areas needed to perform the calculations.

$$A_{nt} = (6 - (\tfrac{7}{8} + \tfrac{1}{8}))(\tfrac{1}{2}) = 2.50 \text{ in.}^2$$
$$A_{gv} = 2(11)(\tfrac{1}{2}) = 11.0 \text{ in.}^2$$
$$A_{nv} = 2(11.0 - 3.5(\tfrac{7}{8} + \tfrac{1}{8}))(\tfrac{1}{2}) = 7.50 \text{ in.}^2$$

Step 2:   Determine the nominal block shear strength.

$$R_n = 0.6(58)(7.50) + 1.0(58)(2.50) = 406 \text{ kips}$$

but not greater than

$$R_n = 0.6(36)(11.0) + 1.0(58)(2.50) = 383 \text{ kips}$$

Selecting the lowest nominal strength,

$$R_n = 383 \text{ kips}$$

Step 3:   For LRFD, the design strength is

$$R_u = \phi R_n = 0.75(383) = 287 > 225 \text{ kips}$$

Because this is greater than the required strength of 225 kips, the gusset plate is adequate to resist this force based on block shear.

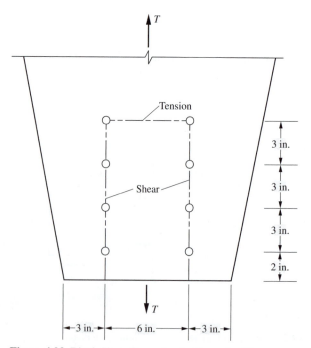

Figure 4.23 Block Shear Geometry for Example 4.9.

Step 3:   For ASD, the allowable strength is

$$R_a = \frac{R_n}{\Omega} = \frac{383}{2.00} = 192 > 150 \text{ kips}$$

Because this is greater than the required strength of 150 kips, the gusset plate is adequate to resist this force based on block shear.

EXAMPLE 4.12a
Tension Strength of Spliced Members by LRFD

GOAL: Determine the design strength of a splice between two W-shapes.

GIVEN: Two W14×43 A992 wide flanges are spliced by flange plates, as shown in Figure 4.24, with $^7/_8$-in. diameter bolts arranged as shown. The LRFD available strength of a group of six bolts is 211 kips. The plates will be selected so that they do not limit the member strength.

SOLUTION

Step 1:   Determine the design strength for the limit state of yielding.

$$A_g = 12.6 \text{ in.}^2$$
$$P_n = 50(12.6) = 630 \text{ kips}$$
$$\phi P_n = 0.9(630) = 567 \text{ kips}$$

Step 2:   Determine the net area.
Area to be deducted for each flange

$$2(^7/_8 + ^1/_8)(0.530) = 1.06 \text{ in.}^2$$

Thus, deduction for two flanges from the gross area,

$$A_n = 12.6 - 2(1.06) = 10.5 \text{ in.}^2$$

Step 3:   Determine the shear lag factor.
The W14×43 is treated as two Tee sections, each a WT7×21.5.
The $\bar{x}$ for each WT is found in Manual Table 1-8 as 1.31 in. and $L = 6.0$ in.

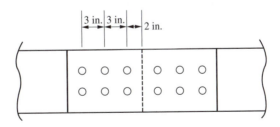

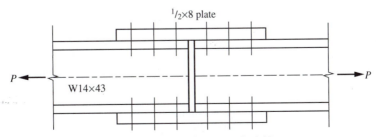

Figure 4.24 Spliced Tension Member for Example 4.12.

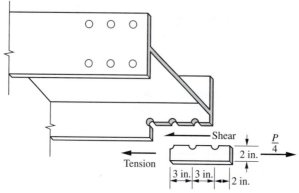

Figure 4.25 Block Shear Check for Example 4.12.

Thus,

$$U = 1 - \frac{1.31}{6.0} = 0.782$$

Specification Table D3.1 provides that for this case, with $b_f > \frac{2}{3}d$ a value of $U = 0.9$ can be used.

Step 4:   Determine the design strength for the limit state of rupture.

$$A_e = 0.9(10.5) = 9.45 \text{ in.}^2$$
$$P_n = 65(9.45) = 614 \text{ kips}$$
$$\phi P_n = 0.75(614) = 461 \text{ kips}$$

Step 5:   Determine the design block shear strength of the flanges.
The block shear limit state must be checked for tear-out of the flanges, as shown in Figure 4.25. The calculations will be carried out for one block as shown in the figure and the total obtained by adding all four flange sections.
Rupture on the tension plane

$$F_u A_{nt} = 65(2.0 - (\tfrac{1}{2})(\tfrac{7}{8}+\tfrac{1}{8}))(0.530) = 51.7 \text{ kips}$$

Yield on the shear plane

$$0.6F_y A_{gv} = 0.6(50)(8.00)(0.530) = 127 \text{ kips}$$

Rupture on the shear plane

$$0.6F_u A_{nv} = 0.6(65)\big(8.00 - 2.5(\tfrac{7}{8} + \tfrac{1}{8})\big)(0.530) = 114 \text{ kips}$$

Because shear rupture is less than shear yield, the design strength for a single block shear element is

$$R_n = (114 + 1.0(51.7)) = 165 \text{ kips}$$
$$\phi R_n = 0.75(165) = 124 \text{ kips}$$

and the total block shear strength of the W14×43 is

$$\phi R_n = 4(124) = 496 \text{ kips}$$

Step 6:   Compare the design strength for each limit state.

Bolt design strength	422 kips
Yielding of the member	567 kips

Rupture of the member 461 kips
Block shear for the member 496 kips

Step 7: The bolt design strength controls the design. Therefore, the design strength of the splice is

$$\phi R_n = 422 \text{ kips}$$

EXAMPLE 4.12b
Tension Strength of Spliced Members by ASD

GOAL: Determine the allowable strength of a splice between two W-shapes.

GIVEN: Two W14×43 A992 wide flanges are spliced by flange plates, as shown in Figure 4.24, with $7/8$-in. diameter bolts arranged as shown. The ASD available strength of a group of six bolts is 141 kips. The plates will be selected so that they do not limit the member strength.

SOLUTION

Step 1: Determine the allowable strength for the limit state of yielding.

$$A_g = 12.6 \text{ in.}^2.$$

$$P_n = 50(12.6) = 630 \text{ kips}$$

$$\frac{P_n}{\Omega} = \frac{630}{1.67} = 377 \text{ kips}$$

Step 2: Determine the net area.
Area to be deducted for each flange

$$2(7/8 + 1/8)(0.530) = 1.06 \text{ in.}^2$$

Thus, deduction for two flanges from the gross area,

$$A_n = 12.6 - 2(1.06) = 10.5 \text{ in.}^2$$

Step 3: Determine the shear lag factor.
The W14×43 is treated as two Tee sections, each a WT7×21.5.
The $\bar{x}$ for each WT is found in Manual Table 1-8 as 1.31 in. and $L = 6.0$ in.
Thus,

$$U = 1 - \frac{1.31}{6.0} = 0.782$$

Specification Table D3.1 provides that for this case, with $b_f > \frac{2}{3} d$ a value of $U = 0.9$ can be used.

Step 4: Determine the allowable strength for the limit state of rupture.

$$A_e = 0.9(10.5) = 9.45 \text{ in.}^2$$
$$P_n = 65(9.45) = 614 \text{ kips}$$
$$\frac{P_n}{\Omega} = \frac{614}{2.00} = 307 \text{ kips}$$

Step 5: Determine the design block shear strength of the flanges.

The block shear limit state must be checked for tear-out of the flanges, as shown in Figure 4.25. The calculations will be carried out for one block as shown in the figure and the total obtained by adding all four flange sections.

Rupture on the tension plane

$$F_u A_{nt} = 65(2.0 - (\frac{1}{2})(\frac{7}{8} + \frac{1}{8}))(0.530) = 51.7 \text{ kips}$$

Yield on the shear plane

$$0.6 F_y A_{gv} = 0.6(50)(8.00)(0.530) = 127 \text{ kips}$$

Rupture on the shear plane

$$0.6 F_u A_{nv} = 0.6(65)(8.00 - 2.5(\frac{7}{8} + \frac{1}{8}))(0.530) = 114 \text{ kips}$$

Because shear rupture is less than shear yield, the design strength for a single block shear element is

$$R_n = (114 + 1.0(51.7)) = 165 \text{ kips}$$

$$\frac{R_n}{\Omega} = \frac{165}{2.00} = 82.5 \text{ kips}$$

and the total block shear strength of the W14×43 is

$$\frac{R_n}{\Omega} = 4(82.5) = 330 \text{ kips}$$

Step 6: Compare the allowable strength for each limit state.

Bolt allowable strength	282 kips
Yielding of the member	377 kips
Rupture of the member	307 kips
Block shear for the member	330 kips

Step 7: The bolt allowable strength controls the design. Therefore, the allowable strength of the splice is

$$\boxed{\frac{R_n}{\Omega} = 282 \text{ kips}}$$

EXAMPLE 4.13a
Tension Strength of an Angle by LRFD

GOAL: Determine the design strength of one of a pair of angles in a tension member.

GIVEN: The truss diagonal member in Figure 4.26 consists of a pair of angles L4×3×$\frac{3}{8}$ that are loaded in tension. The bolts to be used are $\frac{3}{4}$-in. and the steel is A36. The bolt design shear strength for this connection is 47.7 kips.

SOLUTION

Step 1:   Determine the angle design strength for the limit state of yielding.

$$A_g = 2.48 \text{ in}^2.$$

$$P_n = (36)(2.48) = 89.3 \text{ kips}$$

$$\phi P_n = 0.9(89.3) = 80.4 \text{ kips}$$

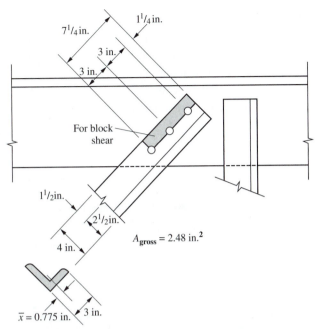

Figure 4.26 Truss Diagonal Member for Example 4.13.

Step 2: Determine the angle design strength for the limit state of rupture.

$$A_n = 2.48 - \tfrac{3}{8}(\tfrac{3}{4} + \tfrac{1}{8}) = 2.15 \text{ in}^2.$$

The shear lag coefficient is

$$U = 1 - \bar{x}/L = 1 - 0.775/6.00 = 0.871$$

$$P_n = (58)(0.871)(2.15) = 109 \text{ kips}$$

$$\phi P_n = 0.75(109) = 81.8 \text{ kips}$$

Step 3: Determine the angle design strength in block shear.
For tension rupture

$$F_u A_{nt} = 58\big(1.5 - \tfrac{1}{2}(\tfrac{3}{4} + \tfrac{1}{8})\big)(\tfrac{3}{8}) = 23.1 \text{ kips}$$

For shear yield

$$0.6F_y A_{gv} = 0.6(36)(7.25)(\tfrac{3}{8}) = 58.7 \text{ kips}$$

For shear rupture

$$0.6F_u A_{nv} = 0.6(58)\big(7.25 - 2.5(\tfrac{3}{4} + \tfrac{1}{8})\big)(\tfrac{3}{8}) = 66.1 \text{ kips}$$

Total block shear strength, using the lowest shear term, is

$$R_n = (58.7 + 1.0(23.1)) = 81.8 \text{ kips}$$

$$\phi R_n = 0.75(81.8) = 61.4 \text{ kips}$$

Step 4: Compare the design strength for each limit state.

Bolt design strength	47.7 kips
Yielding of the member	80.4 kips
Rupture of the member	81.8 kips
Block shear for the member	61.4 kips

Step 5: The bolt design strength controls the design. Therefore, the design strength of one angle is

$$\phi R_n = 47.7 \text{ kips}$$

Because the lowest design strength is that of the bolt shear, the design strength of this single angle tension member is 47.7 kips.

EXAMPLE 4.13b
Tension Strength of an Angle by ASD

GOAL: Determine the design strength of one of a pair of angles in a tension member.

GIVEN: The truss diagonal member in Figure 4.26 consists of a pair of angles L4×3×⅜ that are loaded in tension. The bolts to be used are ¾ in. and the steel is A36. The bolt allowable shear strength for this connection is 31.8 kips.

SOLUTION

Step 1: Determine the angle allowable strength for the limit state of yielding.

$$A_g = 2.48 \text{ in}^2.$$
$$P_n = (36)(2.48) = 89.3 \text{ kips}$$
$$\frac{P_n}{\Omega} = \frac{89.3}{1.67} = 53.5 \text{ kips}$$

Step 2: Determine the angle allowable strength for the limit state of rupture.

$$A_n = 2.48 - \tfrac{3}{8}(\tfrac{3}{4} + \tfrac{1}{8}) = 2.15 \text{ in}^2.$$

The shear lag coefficient is

$$U = 1 - \bar{x}/L = 1 - 0.775/6.00 = 0.871$$
$$P_n = (58)(0.871)(2.15) = 109 \text{ kips}$$
$$\frac{P_n}{\Omega} = \frac{109}{2.00} = 54.5 \text{ kips}$$

Step 3: Determine the angle allowable strength in block shear.
For tension rupture

$$F_u A_{nt} = 58\big(1.5 - \tfrac{1}{2}(\tfrac{3}{4} + \tfrac{1}{8})\big)(\tfrac{3}{8}) = 23.1 \text{ kips}$$

For shear yield

$$0.6 F_y A_{gv} = 0.6(36)(7.25)(\tfrac{3}{8}) = 58.7 \text{ kips}$$

For shear rupture

$$0.6F_u A_{nv} = 0.6(58)\left(7.25 - 2.5\left(^3/_4 + ^1/_8\right)\right)\left(^3/_8\right) = 66.1 \text{ kips}$$

Total block shear strength, using the lowest shear term, is

$$R_n = (58.7 + 1.0\,(23.1)) = 81.8 \text{ kips}$$

$$\frac{R_n}{\Omega} = \frac{81.8}{2.00} = 40.9 \text{ kips}$$

Step 4:   Compare the allowable strength for each limit state.

Bolt allowable strength	31.8 kips
Yielding of the member	53.5 kips
Rupture of the member	54.5 kips
Block shear for the member	40.9 kips

Step 5:   The bolt allowable strength controls the design. Therefore, the allowable strength of one angle is

$$\frac{R_n}{\Omega} = 31.8 \text{ kips}$$

Because the lowest allowable strength is that of the bolt shear, the allowable strength of this single angle tension member is 31.8 kips.

4.8   PIN-CONNECTED MEMBERS

When a pin connection is to be made in a tension member, a hole is cut in both the member and the parts to which it is to be attached. A pin is inserted in the hole and a mechanical means is found to keep the elements together. This type of connection is the closest to a true frictionless pin as can be made. Figure 4.27 shows the end of a pin-connected member and the dimensions needed to determine its strength. These members are not particularly common in buildings; they are used mainly for special applications, such as hangers in suspension structures or connecting links in bridge structures.

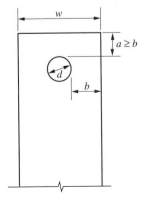

Figure 4.27 Pin-Connected Tension Member.

Specification Section D5 identifies the limit states for which pin-connected members must be designed. These are: (1) tension on the effective net area, (2) shear on the effective area, (3) bearing on the projected area of the pin, and (4) yielding on the gross section. The strength of the pin-connected tension member is taken as the lowest strength predicted by each of these limit states.

1. For tension on the effective net area, actually a rupture limit state

$$P_n = 2tb_{eff}F_u$$

$$\phi_t = 0.75\,(\text{LRFD}) \qquad \Omega_t = 2.00\,(\text{ASD})$$

where t is the thickness of the plate and b_{eff} is the effective width of the plate taken as $(2t + 0.63)$ in inches, but not more than the actual distance from the edge of the hole to the edge of the part measured perpendicular to the direction of the force.

2. For shear on the effective area, again a rupture limit state

$$P_n = 0.6\,F_u A_{sf}$$

$$\phi_t = 0.75\,(\text{LRFD}) \qquad \Omega_t = 2.00\,(\text{ASD})$$

where $A_{sf} = 2t(a + d/2)$, $a = $ the shortest distance from the edge of the pin hole to the edge of the member measured parallel to the direction of the force, and $d = $ the pin diameter.

3. For bearing on the projected area of the pin, from Section J7

$$P_n = 1.8F_y A_{pb}$$

$$\phi_t = 0.75\,(\text{LRFD}) \qquad \Omega_t = 2.00\,(\text{ASD})$$

where $A_{pb} = td$, the projected area of the pin.

4. For yielding in the gross section

$$P_n = F_y A_g$$

$$\phi_t = 0.90\,(\text{LRFD}) \qquad \Omega_t = 1.67\,(\text{ASD})$$

EXAMPLE 4.14a
Pin-Connected Member Design by LRFD

GOAL: Design a pin-connected member using LRFD.

GIVEN: A dead load of 30 kips and a live load of 70 kips. The steel has a yield stress of 50 ksi and an ultimate strength of 65 ksi. Assume a $\frac{3}{4}$-in. plate with a 4-in. pin.

SOLUTION

Step 1:   Determine the required strength.

$$P_u = 1.2(30) + 1.6(70) = 148 \text{ kips}$$

Step 2:   Determine the minimum required effective net width for the limit state of rupture.

$$b_{eff} = \frac{P_u}{\phi 2 F_u t} = \frac{148}{0.75(2)(65)(0.750)} = 2.02 \text{ in.}$$

and

$$b_{eff} = 2t + 0.63 = 2(.750) + 0.63 = 2.13 \text{ in.}$$

Therefore try a 9.0-in. plate, which will give an actual distance to the edge of the plate greater than b_{eff}. Thus, b_{eff} is used to calculate the rupture strength of the plate.

Step 3:   Determine the design strength for the limit state of tension rupture.

$$\phi P_n = 0.75(2)(2.13)(0.75)(65) = 156 \text{ kips } > 148 \text{ kips}$$

Step 4:   Determine the design strength for the limit state of shear rupture.
For a 9-in. plate and a 4-in. pin, $a = b = 2.5$ in.

$$A_{sf} = 2t(a + d/2) = 2(0.75)(2.5 + 4.0/2) = 6.75 \text{ in.}^2$$
$$\phi P_n = \phi 0.6 A_{sf} F_u = 0.75(0.6)(6.75)(65) = 197 \text{ kips} > 148 \text{ kips}$$

Step 5:   Determine the design strength for the limit state of bearing on the projected area of the pin.

$$A_{pb} = td = 0.75(4.0) = 3.0 \text{ in.}^2$$
$$\phi P_n = \phi 1.8 F_y A_{pb} = 0.75(1.8)(50)(3.0) = 203 \text{ kips} > 148 \text{ kips}$$

Step 6:   Determine the design strength for the limit state of yielding on the gross area of the member.

$$\phi P_n = \phi F_y A_g = 0.9(50)(0.75)(9.0) = 304 \text{ kips} > 148 \text{ kips}$$

Step 7:   Conclusion, the proposed

> 9-in. $\times$ $^3/_4$-in. pin-connected member with a 4-in. pin

will be sufficient to resist the applied load.

EXAMPLE 4.14b
Pin-Connected Member
Design by ASD

GOAL: Design a pin-connected member using ASD.

GIVEN: A dead load of 30 kips and a live load of 70 kips. The steel has a yield stress of 50 ksi and an ultimate strength of 65 ksi. Assume a $^3/_4$-in. plate with a 4-in. pin.

SOLUTION

Step 1:   Determine the required strength.

$$P_a = 30 + 70 = 100 \text{ kips}$$

Step 2:   Determine the minimum required effective net width for the limit state of rupture.

$$b_{eff} = \frac{P_u \Omega}{2 F_u t} = \frac{100(2.00)}{(2)(65)(0.75)} = 2.05 \text{ in.}$$

and

$$b_{eff} = 2t + 0.63 = 2(.75) + 0.63 = 2.13 \text{ in.}$$

Therefore try a 9.0 in. plate, which will give an actual distance to the edge of the plate greater than b_{eff}. Thus, b_{eff} is used to calculate the rupture strength of the plate.

Step 3:   Determine the allowable strength for the limit state of tension rupture.

$$\frac{P_n}{\Omega} = \frac{(2)(2.13)(0.75)(65)}{2.00} = 104 \text{ kips} > 100 \text{ kips}$$

Step 4: Determine the allowable strength for the limit state of shear rupture. For a 9-in. plate and a 4-in. pin, $a = b = 2.5$ in.

$$A_{sf} = 2t(a + d/2) = 2(0.75)(2.5 + 4.0/2) = 6.75 \text{ in.}^2$$

$$\frac{P_n}{\Omega} = \frac{0.6A_{sf}F_u}{\Omega} = \frac{(0.6)(6.75)(65)}{2.00} = 132 \text{ kips} > 100 \text{ kips}$$

Step 5: Determine the allowable strength for the limit state of bearing on the projected area of the pin.

$$A_{pb} = td = 0.75(4.0) = 3.0 \text{ in.}^2$$

$$\frac{P_n}{\Omega} = \frac{1.8F_y A_{pb}}{\Omega} = \frac{(1.8)(50)(3.0)}{2.00} = 135 \text{ kips} > 100 \text{ kips}$$

Step 6: Determine the allowable strength for the limit state of yielding on the gross area of the member.

$$\frac{P_n}{\Omega} = \frac{F_y A_g}{\Omega} = \frac{(50)(0.75)(9.0)}{1.67} = 202 \text{ kips} > 100 \text{ kips}$$

Step 7: Conclusion, the proposed

9-in. × ³⁄₄-in. pin-connected member with a 4-in. pin

will be sufficient to resist the applied load.

4.9 EYE-BARS AND RODS

Eye-bar tension members have not been used in new construction for many years, although the provisions for their design are still found in Section D6 of the Specification. Historically, they were commonly found as tension members in trusses and as links forming the main tension member in suspension bridges. Eye-bars are designed only for the limit state of yielding on the gross section because the dimensional requirements for the eye-bar preclude the possibility of failure at any load below that level. Figure 4.28 shows a schematic of an eyebar and Figure 4.5 shows an eyebar in a building application.

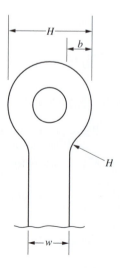

Figure 4.28 Eye-Bar Geometry.

Rods are commonly used for tension members in situations where the required tensile strength is small. These tension members would generally be considered secondary members such as sag rods, hangers, and tie rods. Rods may also be used as part of the lateral bracing system in walls and roofs.

Although it is possible to connect rods by welding to the structure, threading and bolting is the most common connection. Rods can be threaded in two ways. Standard rods have threads that reduce the cross-section area through the removal of material. The upset rod has enlarged ends with the threads reducing that area to something larger than the gross area of the rod. The strength of the rod depends on the manner in which the threads are applied.

For a standard threaded rod, the nominal strength is given in Specification Section J3.6 as $F_n = 0.75F_u$ over the area of the unthreaded body of the rod, which gives

$$P_n = 0.75F_u A_b$$

and for design,

$$\phi_t = 0.75\,(\text{LRFD}) \qquad \Omega_t = 2.00\,(\text{ASD})$$

4.10   BUILT-UP TENSION MEMBERS

Section D4 of the Specification permits tension members that are fabricated from the combination of shapes and plates. Their strength is determined in the same way as the strength for single-shape tension members. However, the designer must remember that in built-up members, bolts are usually placed along the member length to tie the various shapes together. These bolts result in holes along the member length, not just at the ends, so that rupture on the effective net section may become the controlling limit state at a location other than the member end.

Perforated cover plates or tie plates can be used to tie the separate shapes together. Limitations on the spacing of these elements are also provided in Section D4 and requirements for the placement of bolts can be found in Section J3.5.

4.11   TRUSS MEMBERS

The most common tension members found in building structures are the tension web and chord members of trusses. Trusses are normally found as roof structures and as transfer structures within a building. Depending on the particular load patterns that a truss might experience, a truss member might be called upon to always resist tension or to resist tension in some cases and compression in others. In cases in which a member is required to carry both tension and compression, it will need to be sized accordingly. Because the compression strength of a member is normally significantly less than the tension strength of that same member, as will be seen in Chapter 5, compression may actually control the design.

The typical truss member can be composed of either single shapes or a combination of shapes. When composed of a combination of shapes, the requirements discussed in Section 4.10 must be included. Otherwise, truss members are designed just as any other tension member discussed in this chapter. Examples 4.9, 4.10, and 4.13 showed the application of the tension provisions to several truss tension members.

4.12   BRACING MEMBERS

As with truss members, members used to provide lateral load resistance for a building might also be called upon to carry tension under some conditions and compression under others. They too would need to be designed to resist both loads. However, it is often more economical to provide twice as many tension members and to assume that if a tension member were called upon to resist compression, it would buckle and therefore carry no load. This would permit all bracing to be designed as tension-only members and almost certainly permit them to have a smaller cross section than if they were required to resist compression. An additional simplification that this assumption permits is the elimination of potential compression members from the analysis for member forces. This may then result in the structure being a determinate structure rather than an indeterminate one and thereby simplifying the analysis.

4.13   PROBLEMS

1. Determine the gross and net areas for an 8- × ³/₄-in. plate with a single line of standard holes for ⁷/₈-in. bolts.

2. Determine the gross and net areas for a 10- × ¹/₂-in. plate with a single line of standard holes for ³/₄-in. bolts.

3. Determine the gross and net areas for a 6 × ⁵/₈-in. plate with a single line of standard holes for 1-in. bolts.

4. Determine the gross and net areas for an L4×4×¹/₂ with two lines, one in each leg, of standard holes for ³/₄-in. bolts.

5. Determine the gross and net areas for an L5×5×⁵/₈ with two lines, one in each leg, of standard holes for ⁷/₈-in. bolts.

6. Determine the gross and net area for a WT8×20 with three lines of standard holes for ³/₄-in. bolts. Each element of the WT will be attached to the connection.

7. Determine the gross and net area for a C15×50 with five lines of standard holes for ⁷/₈-in. bolts. Each flange will contain one line of bolts and the web will contain 3 lines of bolts.

8. Determine the net width for a 10- × ¹/₂-in. plate with ³/₄-in. bolts placed in three lines as shown.

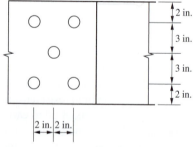

P4.8

9. Determine the net width for the L6×4×⁵/₈ shown with ⁷/₈-in. bolts.

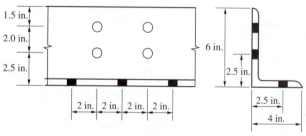

P4.9

10. Determine the gross and net areas for a double angle tension member composed of two L4×4×¹/₂ as shown below with holes for ³/₄-in. bolts staggered in each leg.

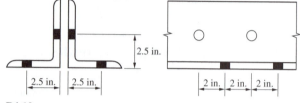

P4.10

11. For the WT8×50 attached through the flange to a 12- × ³/₄-in. plate with eight ⁷/₈-in. bolts at a spacing of 3-in. and placed in two rows as shown, determine the shear lag factor and effective net area of the WT.

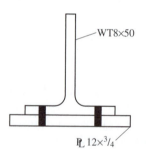

P4.11

12. A single L6×6×1 is used as a tension brace in a multistory building. One leg of the angle is attached to a gusset plate with a single line of ⁷/₈-in. bolts. Determine the shear lag factor and effective net area for three bolts with a spacing of 3-in.

13. The WT8×50 of Problem 11 is welded along the tips of the flange for a length of 12-in. on each flange. Determine the shear lag factor and effective net area for the WT.

14. Determine the available strength of a 12- × ¹/₂-in. A36 plate connected to two 12-in. plates as shown with two lines of ³/₄-in. bolts. Determine the (a) design strength by LRFD and (b) allowable strength by ASD.

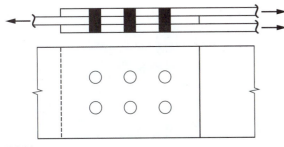

P4.14

15. Determine the available strength of an L6×4×¾ attached through the long leg to a gusset plate with ten ⅞-in. bolts at a 3-in. spacing in two lines. Use A36 steel. Determine the (a) design strength by LRFD and (b) allowable strength by ASD.

16. Determine the available strength of an 8- × ½-in. A572 Gr. 50 plate connected with three lines of ⅞-in. bolts. Determine the (a) design strength by LRFD and (b) allowable strength by ASD.

17. Determine the available strength of a WT7×15, A992 steel, with the flanges welded to a ½-in. gusset plate by a 10-in. weld along each side of the flange. Determine the (a) design strength by LRFD and (b) allowable strength by ASD.

18. Design a 10-ft-long, single-angle tension member to support a live load of 49.5 kips and a dead load of 16.5 kips (L/D = 3). The member is to be connected through one leg and only one bolt hole will occur at any cross section. Use A36 steel and limit slenderness to length/300. Design by (a) LRFD and (b) ASD.

19. Design a 10-ft-long, single-angle tension member as in Problem 18 with the same total service load, 66 kips. Use a live load of 7.3 kips and a dead load of 58.7 kips, (L/D = 0.125), (a) design by LRFD and (b) design by ASD.

20. Design a 10-ft-long, single-angle tension member as in Problem 18 with the same service load using a live load of 55.0 kips and a dead load of 11.0 kips, (L/D = 5). Design by (a) LRFD and (b) ASD.

21. Design a 27-ft long WT tension wind brace for a multistory building to resist a wind force of 380 kips. Use A992 steel and ⅞-in. bolts. Assume that no more than four bolts will occur at any particular section. The length of the connection (the number of bolts) is not known. Design by (a) LRFD and (b) ASD.

22. Design a W14 A992 tension member for a truss that will carry a dead load of 89 kips and a live load of 257 kips. The flanges will be bolted to the connecting plates with ⅞-in. bolts located so that four bolts will occur in any net section. Design by (a) LRFD and (b) ASD.

23. An L4×3×⅜ is attached to a gusset plate with three ¾-in. bolts spaced at 3 inches with an end and edge distance of 1.5-in.

as shown. Determine the available block shear strength of the A36 angle, by (a) LRFD and (b) ASD.

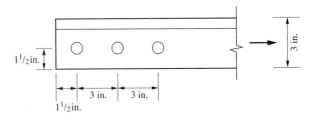

P4.23

24. Determine the available block shear strength for the 7 × ¾-in. A36 plate shown. The holes are for ¾-in. bolts. Determine by (a) LRFD and (b) ASD.

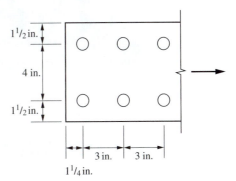

P4.24

25. Determine the available block shear strength for the WT6×20, A992 steel, attached through the flange with eight ¾-in. bolts as shown by (a) LRFD and (b) ASD.

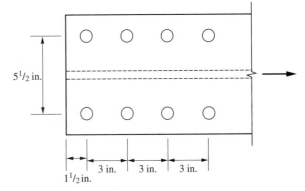

P4.25

Chapter 5

Experience Music Project, Seattle.
Photo courtesy Michael Dickter/Magnusson Klemencic Associates.

Compression Members

5.1 COMPRESSION MEMBERS IN STRUCTURES

Compression members are structural elements subjected to axial forces that tend to push the ends of the members toward each other. The most common compression member in a building structure is a *column*. Columns are vertical members that support the horizontal elements of a roof or floor system. Several columns are shown in Figure 5.1 as part of a building structure. They are the primary elements that provide the vertical space to form an occupiable volume. Other compression members are found in trusses as chord and web members and as bracing members in floors and walls. Other names often used to identify compression members are struts and posts. Throughout this chapter the terms compression member and column will be used interchangeably.

The compression members discussed in this chapter experience only axial forces. In real structures, additional load effects are often exerted on a compression member that would tend to combine bending with the axial force. These combined force members are called beam-columns and are discussed in Chapter 8. The majority of the provisions that apply to compression members are located in Chapter E of the Specification.

Table 5.1 lists the sections of the Specification and parts of the Manual discussed in this chapter.

Figure 5.1 Columns in a Multistory Building. Photo Courtesy Greg Grieco

Table 5.1 Sections of Specification and Parts of Manual Found in this Chapter

	Specification
B4	Classification of Sections for Local Buckling
E2	Slenderness Limitations and Effective Length
E3	Compressive Strength for Flexural Buckling of Members without Slender Elements
E4	Compressive Strength for Torsional and Flexural-Torsional Buckling of Members without Slender Elements
E5	Single-Angle Compression Members
E6	Built-Up Members
E7	Members with Slender Elements
	Manual
Part 1	Dimensions and Properties
Part 4	Design of Compression Members
Part 6	Design of Members Subject to Combined Loading

5.2 CROSS-SECTIONAL SHAPES FOR COMPRESSION MEMBERS

Compression members carry axial forces so the primary cross-sectional property of interest is the area. Thus, the simple relationship between force and stress,

$$f = \frac{P}{A} \tag{5.1}$$

is applicable. As long as this relationship dictates compression member strength, all cross sections with the same area will perform in the same way. In real structures, however, other factors influence the strength of the compression member and the distribution of the area becomes important.

In building structures, the typical compression member is a column and the typical column is a rolled wide flange member. Later discussions of compression member strength will show that the W-shape does not give the most efficient distribution of material for compression members. It does, however, provide a compression member that can easily be connected to other members of the system such as beams and other columns. This feature significantly influences its selection as an appropriate column cros ssection.

Figure 5.2 shows examples of rolled and built-up shapes that are used as compression members. Many of these are the same shapes used for the tension members discussed in

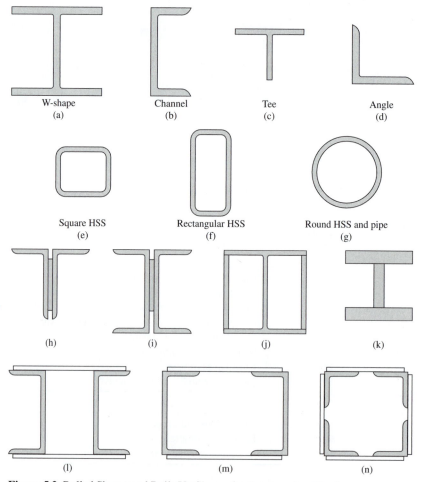

Figure 5.2 Rolled Shapes and Built-Up Shapes for Compression Members.

Chapter 4. This is reasonable because the forces being considered in these two cases are both axial, although they act in the opposite direction. However, other factors that influence the strength of compression members will dictate additional criteria for the selection of the most efficient shapes for these members.

The Tee and angle shown in Figures 5.2c and d are commonly found as chords and webs of trusses. In these applications, the geometry of the shapes helps simplify the connections between members. Angles are also used in pairs as built-up compression members with the connecting element between the two angles as shown in Figure 5.2h. The channel can be found in trusses as a single element or combined with another channel as shown in Figures 5.2b, i, l, and m. Built-up columns can also be found using back-to-back channels. The HSS shapes and pipe shown in Figures 5.2e, f, and g are commonly found as columns in buildings, particularly one-story structures where the connections to the shape can be simplified. Later you will see that the distribution of the material in these shapes is the most efficient for real columns.

5.3   COMPRESSION MEMBER STRENGTH

If no other factors were to impact the strength of a compression member, the simple axial stress relationship given in Equation 5.1 could be used to describe member strength. Thus, the maximum force that a compression member could resist at yield would be

$$P_y = F_y A_g \qquad (5.2)$$

where P_y is the yield load, sometimes called the squash load; F_y is the yield stress; and A_g is the gross area. This is the response that would be expected if a very short specimen, one whose length approximates its other two dimensions, were to be tested in compression. This type of column test specimen is called a stub column. Because most compression members will have a length that greatly exceeds its other dimensions, length effects cannot be ignored.

5.3.1   Euler Column

To address the impact of length on compression member behavior, a simple model, as shown in Figure 5.3 is used. The Swiss mathematician Leonard Euler first presented this analysis in

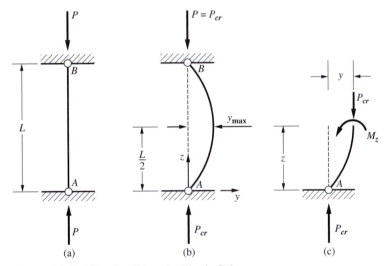

Figure 5.3 Stability Conditions for Elastic Columns.

1759. A number of assumptions are made in this column model, including: (1) the column ends are frictionless pins, (2) the column is perfectly straight, (3) the load is applied along the centroidal axis, and (4) the material behaves elastically. Based on these assumptions, this column model is usually called the *perfect column* or the *pure column*.

Figure 5.3a shows the perfect column with an applied load that will not cause any lateral displacement or yielding. In this arrangement, the load can be increased with no lateral displacement of the column. However, at a particular load, defined as the critical load or the buckling load, P_{cr}, the column will displace laterally as shown in Figure 5.3b. In this configuration, the dashed line represents the original position of the member and the solid line represents the displaced position. Note that an axis system is presented in the figure with the z-axis along the member length and the y-axis transverse to the member length. This places the x-axis perpendicular to the figure. The x- and y-axes correspond to the centroidal axes of the cross section.

A free body diagram of the lower portion of the column in its displaced position is shown in Figure 5.3c. If moments are taken about point C, equilibrium requires

$$M_z = P_{cr} y$$

From the principles of mechanics and using small displacement theory, the differential equation of the deflected member is given as

$$\frac{d^2 y}{dz^2} = -\frac{M_z}{EI_x}$$

Combining these two equations and rearranging the terms yields

$$\frac{d^2 y}{dz^2} + \frac{P_{cr}}{EI_x} y = 0$$

If the coefficient of the second term is taken as $k^2 = P_{cr}/EI_x$, the differential equation for the column becomes

$$\frac{d^2 y}{dz^2} + k^2 y = 0$$

which is a standard second-order linear ordinary differential equation. The solution to this equation is given by

$$y = A \sin kz + B \cos kz \qquad (5.3)$$

where A and B are constants of integration. To further evaluate this equation, the boundary conditions must be applied. Because at $z = 0$, $y = 0$ and at $z = L$, $y = 0$, we find that

$$B = 0$$

and

$$A \sin kL = 0$$

For Equation 5.3 to have a nontrivial solution, $(\sin kL)$ must equal zero. This requires that $kL = n\pi$ where n is any integer. Substituting for k and rearranging yields

$$P_{cr} = \frac{n^2 \pi^2 EI_x}{L^2} \qquad (5.4)$$

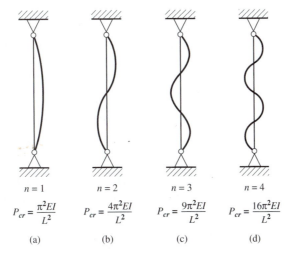

$n = 1$ $n = 2$ $n = 3$ $n = 4$

$P_{cr} = \dfrac{\pi^2 EI}{L^2}$ $P_{cr} = \dfrac{4\pi^2 EI}{L^2}$ $P_{cr} = \dfrac{9\pi^2 EI}{L^2}$ $P_{cr} = \dfrac{16\pi^2 EI}{L^2}$

(a)   (b)   (c)   (d)

Figure 5.4 Shape of Buckled Columns.

Because n can be taken as any integer, Equation 5.4 has a minimum value when $n = 1$. This is called the *Euler Buckling Load* or the *Critical Buckling Load* and is given as

$$P_{cr} = \frac{\pi^2 EI_x}{L^2} \tag{5.5}$$

If values for B and kL are substituted into Equation 5.3, the shape of the buckled column can be determined from

$$y = A \sin n\pi \tag{5.6}$$

Because any value for A will satisfy Equation 5.6, a unique magnitude of the displacement cannot be determined; however, it is clear that the shape of the buckled column is a half sine curve when $n = 1$. This is shown again in Figure 5.4a. For other values of n, different buckled shapes will result along with the higher critical buckling load. When $n > 1$, these shapes are referred to as higher modes. Several cases are shown in Figures 5.4b, c, and d. In all cases, the basic shape is the sine curve. In order for these higher modes to occur, some type of physical restraint against buckling is required at the point where the buckled shape crosses the original undeflected shape. This can be accomplished with the addition of braces, which is discussed later.

We now have two equations to predict the column strength: Equation 5.2, which does not address length; and Equation 5.5, which does. These two equations are plotted in Figure 5.5. Because the derivation of the Euler equation was based on elastic behavior and the

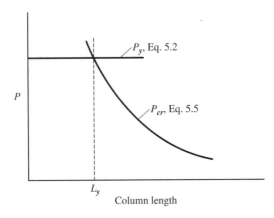

P_y, Eq. 5.2

P_{cr}, Eq. 5.5

P

L_y

Column length

Figure 5.5 Column Strength Based on Length.

column cannot carry more load than the yield load, there is an upper limit to the column strength. If the length at which this limit occurs is taken as L_y, it can be determined by setting Equation 5.2 equal to Equation 5.5 and solving for length, giving

$$L_y = \pi \sqrt{\frac{EI_x}{F_y A_g}}$$

To simplify this equation, the radius of gyration, r, will be used where

$$r = \sqrt{\frac{I}{A}}$$

Because the moment of inertia depends on the axis being considered and A is the gross area of the section, independent of axis, r will depend on the buckling axis. In the derivation just developed, the axis of buckling for the column of Figure 5.3 was taken as the x-axis, thus

$$L_y = \pi r_x \sqrt{\frac{E}{F_y}}$$

For this theoretical development, a column whose length is less than L_y would fail by yielding and could be called a short column, whereas a column with a length greater than L_y would fail by buckling and be called a long column.

It is also helpful to write Equation 5.5 in terms of stress. Dividing both sides by the area and substituting again for the radius of gyration yields

$$F_{cr} = \frac{\pi^2 E}{\left(\dfrac{L}{r}\right)^2} \tag{5.7}$$

In this equation, the radius of gyration is left unsubscripted so that it can be applied to whichever axis is determined to be the critical axis. A plot of stress versus L/r would be of the same shape as the plot of force in Figure 5.5.

5.3.2 Other Boundary Conditions

Derivation of the buckling equations presented as Equations 5.5 and 5.7 included the boundary condition of frictionless pins at both ends. For perfect columns with other boundary conditions, the moment will not be zero at the ends and will result in a nonhomogeneous differential equation. Solving the resulting differential equation and applying the appropriate boundary conditions will lead to a buckling equation of a form similar to the previous equations. To generalize the buckling equation for other end conditions, the column length, L, is replaced by the column effective length, KL, where K is the effective length factor. Thus, the general buckling equations become

$$P_{cr} = \frac{\pi^2 EI}{(KL)^2} \tag{5.8}$$

and

$$F_{cr} = \frac{\pi^2 E}{\left(\dfrac{KL}{r}\right)^2} \tag{5.9}$$

Figure 5.6 shows the original pin-ended column with several examples of columns showing the influence of different end conditions. All columns are shown with the lower support fixed against lateral translation. Three of the columns have upper ends that are also restrained from

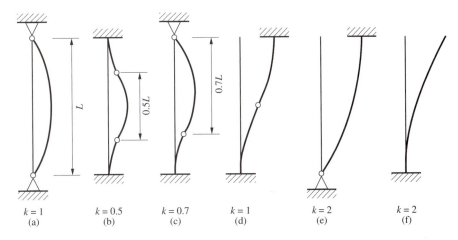

Figure 5.6 Column Buckled Shape for Different End Conditions.

lateral translation, whereas three have upper ends that are free to translate. The effective length can be visualized as the length between inflection points, where the curvature reverses. This result is similar to the original derivation when n was taken as some integer other than one. It is most easily seen in Figures 5.6b and c but can also be seen in Figure 5.6d by visualizing the extended buckled shape above the column as shown in Figure 5.7. In all cases, the buckled curve is a segment of the sine curve. The most important thing to observe is that the column with fixed ends in Figure 5.6b has an effective length of $0.5L$, whereas the column in Figure 5.6a has an effective length of L. Thus, the fixed end column will have four times the strength of the pin end column.

5.3.3   Combination of Bracing and End Conditions

The influence of intermediate bracing on the effective length was touched upon in the discussion of the higher modes of buckling. In those cases, the buckling resulted in equal length segments that reflected the mode number. Thus, a column with $n = 2$ had two equal segments, whereas a column with $n = 3$ buckled with three equal segments. If physical braces are used to provide buckling resistance to the column, the effective length will

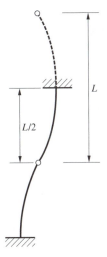

Figure 5.7 Extended Shape of Buckled Column from Figure 5.6d.

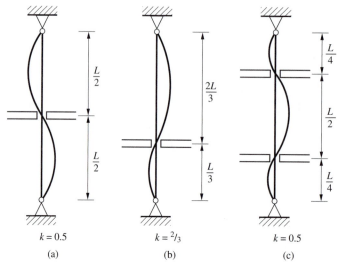

Figure 5.8 Buckled Shape for Columns with Intermediate Braces.

depend on the location of the braces. Figure 5.8 shows three columns with pin ends and intermediate supports. The column in Figure 5.8a is the same as the column in Figure 5.4b. The effective length is $0.5L$ so $K = 0.5$. The column in Figure 5.8b shows lateral braces in an unsymmetrical arrangement with one segment $L/3$ and the other $2L/3$. Although the exact location of the inflection point would be slightly into the longer segment, normal practice is to take the longest unbraced length as the effective length, thus $KL = 2L/3$ so $K = \frac{2}{3}$. The column in Figure 5.8c is braced at two locations. The longest unbraced length for this case gives an effective length $KL = 0.5L$ and a corresponding $K = 0.5$. A general rule can be stated that, when the column ends are pinned, the longest unbraced length is the effective length for buckling in that direction.

When other end conditions are present, these two influences must be combined. The columns of Figure 5.9 illustrate the influence of combinations of end supports and bracing on the column effective length. The end conditions would influence only the effective length of the end segment of the column. For the column in Figure 5.9a, the lower segment has $L = a$ and that segment would buckle with an effective length $KL = a$. The upper segment

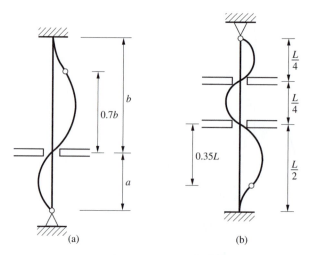

Figure 5.9 Buckled Shape for Columns with Different End Conditions and Intermediate Braces.

has $L = b$ but also has a fixed end. Thus, it would buckle with an effective length $KL = 0.7b$, obtained by combining the end conditions of Figure 5.6c with the length, b. Thus, the relationship between lengths a and b determine which end of the column dictates the overall column effective length. As an example, the column in Figure 5.9b shows that the lowest segment would set the column effective length at $0.35L$.

EXAMPLE 5.1
Theoretical Column Strength

GOAL: Determine the theoretical strength for a pin-ended column and whether it will first buckle or yield.

GIVEN: A W10×33, A992, column with a length of 20 ft.

SOLUTION

Step 1:   Determine the load that would cause buckling.
With no other information, it must be assumed that this column will buckle about its weak axis, if it buckles at all, because the effective length is 20 ft for both axes.
From Manual Table 1-1, $I_y = 36.6$ in.^{4} and $A_s = 9.71$ in.^{2}
The load that would cause it to buckle is

$$P_{cr} = \frac{\pi^2 E I_y}{L^2} = \frac{\pi^2 (29{,}000)(36.6)}{(20(12))^2} = 182 \text{ kips}$$

Step 2:   Determine the load that would cause yielding.

$$P_y = F_y A_s = 50(9.71) = 486 \text{ kips}$$

Step 3:   Conclusion.
Because $P_{cr} < P_y$, the theoretical column strength is

$$\boxed{P = 182 \text{ kips}}$$

And the column would buckle before it could reach its yield stress.

EXAMPLE 5.2
Critical Buckling Length

GOAL: Determine the overall column length that, if exceeded, would cause the column to theoretically buckle elastically before yielding.

GIVEN: A W8×31 column with fixed supports. Use steel with $F_y = 40$ ksi.

SOLUTION

Step 1:   From Manual Table 1-1, $I_y = 37.1$ in.^{4} and $A_s = 9.12$ in.^{2}

Step 2:   Determine the force that would cause the column to yield.

$$P_y = F_y A_s = 40(9.12) = 365 \text{ kips}$$

Step 3:   To determine the length that would cause this same load to be the buckling load, set this force equal to the buckling force and determine the length from

$$365 \text{ kips} = \frac{\pi^2 E I_y}{L^2} = \frac{\pi^2 (29{,}000)(37.1)}{L^2}$$

which gives

$$L = \sqrt{\frac{\pi^2 (29{,}000)(37.1)}{365}} = 171 \text{ in.}$$

So the effective length is

$$L = \frac{171}{12} = 14.3\,\text{ft}$$

Step 4:   Conclusion.

Because a fixed-end column has an effective length equal to one-half the actual length, buckling will not occur if the actual length is less than or equal to

$$\boxed{L = 2(14.3) = 28.6\,\text{ft}}$$

5.3.4   Real Column

Physical testing of specimens that effectively modeled columns found in real building structures showed that column strength was not as great as either the buckling load predicted by the Euler buckling equation or the squash load predicted by material yielding. This inability of the theory to predict actual behavior was recognized early and numerous factors were found to influence this result. Three main factors influence column strength; material inelasticity, column initial out-of-straightness, and modeling of end conditions. The influence of column end conditions has already been discussed with respect to effective length determination. Material inelasticity and initial out-of-straightness, which significantly impact real column strength, are discussed here.

Inelastic behavior of a column directly results from built-in or residual stresses in the cross section. These residual stresses are, in turn, the direct result of the manufacturing process. Steel is produced with heat, and heat is also necessary to form the steel into the shapes used in construction. Once the shape is fully formed, it is then cooled. During this cooling process residual stresses are developed. Figure 5.10 shows wide flange cross sections in various stages of cooling. Initially, as shown in Figure 5.10a, the tips of the flanges with the most surface area to give off heat begin to cool. As this material cools, it contracts, eventually reaching the ambient temperature. At this point, the fibers in this part of the section reach what is expected to be their final length.

As adjacent fibers cool, they too contract. In the process of contracting, these newly cooling fibers pull on the previously cooled fibers, placing them under some amount of compressive stress. Figure 5.10b shows a cross section with additional flange elements cooled. When the previously cooled portion of the cross section provides enough stiffness

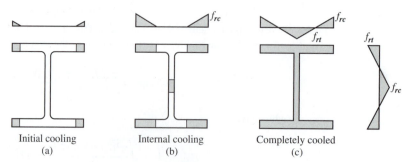

Initial cooling   Internal cooling   Completely cooled
(a)   (b)   (c)

Figure 5.10 Distribution of Residual Stresses.

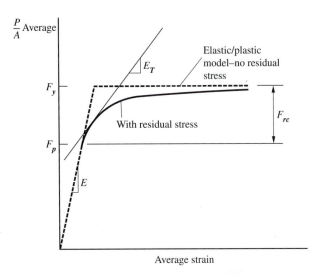

Figure 5.11 Stub Column Stress-Strain Diagrams with and without Residual Stress.

to restrain the contraction of the newly cooling material, a tensile stress is developed in the newly cooling material because it cannot contract as it otherwise would without this restraint. When completely cooled, as shown in Figure 5.10c, the tips of the flanges and the middle of the web are put into compression, and the flange web juncture is put into tension. Thus, the first fibers to cool are in compression, whereas the last to cool are in tension.

Several different representations of the residual stress distribution have been suggested. One distribution is shown in Figure 5.10c. The magnitude of the maximum residual stress does not depend on the material yield strength but is a function of material thickness. In addition, the compressive residual stress is of critical interest when considering compression members. The magnitude of this residual stress varies from 10 ksi to about 30 ksi, depending on the shape. The higher values are found in wide flanges with the thickest flange elements.

To understand the overall impact of these residual stresses on column behavior, a stub column can again be investigated. Figure 5.11 shows the stress-strain relation for a short column, one that will not buckle but exhibits the influence of residual stresses. As the column is loaded with an axial load, the member shortens and the corresponding strain and stress are developed, as if this were a perfectly elastic specimen. The response of a perfectly elastic column is shown by the dashed line in Figure 5.11. When the applied stress is added to a member with residual compressive stress, the stub column begins to shorten at a greater rate as the tips of the flange become stressed beyond the yield stress. Thus, the stress-strain curve moves off the straight line of a perfectly elastic, perfectly plastic specimen that is shown by the dashed line and follows the solid line. Continuing to add load to the column results in greater strain for a given stress and the column eventually reaches the yield stress of the perfectly elastic material. Thus, the only difference between the behavior of the actual column and the usual test specimen used to determine the stress-strain relationship is that the real column behaves inelastically as those portions of its cross section with compressive residual stresses reach the material yield stress.

If a new term, the tangent modulus, E_T, is defined as the slope of a tangent to the actual stress-strain curve at any point and shown in Figure 5.11, an improved prediction of column buckling strength can be obtained by modifying the Euler buckling equation so that

$$P_{cr} = \frac{\pi^2 E_T I_x}{(KL)^2}$$

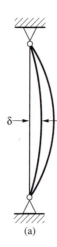 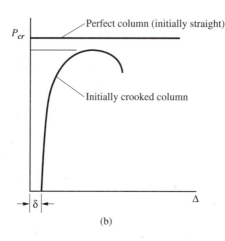

Figure 5.12 Influence of Initial Out-of-Straightness on Column Strength.

Thus, as the column is loaded beyond its elastic limit, E_T reduces and the buckling strength reduces. This partially accounts for the inability of the Euler buckling equation to accurately predict column strength.

Another factor to significantly impact column strength is the column initial out-of-straightness. Once again, the manufacturing process for steel shapes impacts the ability of the column to carry the predicted load. In this case, it is the fact that no structural steel member comes out of the production process perfectly straight. In fact, the AISC *Code of Standard Practice* permits an initial out-of-straightness of $^1/_{1000}$ of the length between points with lateral support. Although this appears to be a small variation from straightness, it impacts column strength.

Figure 5.12a shows a perfectly elastic, pin-ended column with an initial out-of-straightness, δ. A comparison of this column diagram with that used to derive the Euler column, Figure 5.3, shows that the moment along the column length will be greater for this initially crooked column in its buckled position than would have been for the initially straight column. Thus, the solution to the differential equation would be different. In addition, because the applied load works at an eccentricity from the column along its length, even before buckling, a moment is applied to the column that has not yet been accounted for. Figure 5.12b shows the load versus lateral displacement diagram for this initially crooked column compared to that of the initially straight column. This column not only exhibits greater lateral displacement, it also has a lower maximum strength.

When these two factors are combined, the Euler equation cannot properly describe column behavior on its own. Thus, the development of curves to predict column behavior has historically been a matter of curve fitting the test data with an attempt to present a simple representation of column behavior.

5.3.5 AISC Provisions

The compression members discussed thus far have either yielding or overall column buckling as the controlling limit state. Figure 5.13 plots sample column test data compared to the Euler equation and the squash load. The Structural Stability Research Council proposed three equations to predict column behavior, depending on the particular steel product used. To simplify column design, AISC selected a single curve described using two equations as their representation of column strength.

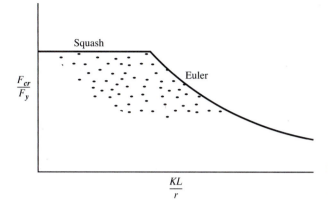

Figure 5.13 Sample Column Test Data Compared to Theoretical Column Strength.

The design basis for ASD and LRFD were presented in Sections 1.6 and 1.7, respectively. Equations 1.1 and 1.2 are repeated here in order to reinforce the relationship between the nominal strength, resistance factor, and safety factor presented throughout the Specification.

For ASD, the allowable strength is

$$R_a \leq \frac{R_n}{\Omega} \tag{1.1}$$

For LRFD, the design strength is

$$R_u \leq \phi R_n \tag{1.2}$$

As indicated earlier, the Specification provides the relationship to determine nominal strength and the corresponding resistance factor and safety factor for each limit state to be considered. The provisions for compression members with nonslender elements are given in Specification Section E3. The nominal column strength for the limit state of flexural buckling of members with nonslender elements is

$$P_n = F_{cr} A_g$$

and

$$\phi_c = 0.9 \, (\text{LRFD}) \qquad \Omega_c = 1.67 \, (\text{ASD})$$

where A_g is the gross area of the section and F_{cr} is the flexural buckling stress.

To capture column behavior when inelastic buckling dominates column strength, that is, where residual stresses become important, the Specification provides that when $KL/r \leq 4.71\sqrt{E/F_y}$ or $F_e > 0.44F_y$

$$F_{cr} = \left[0.658^{\frac{F_y}{F_e}} \right] F_y \qquad \text{(E3-2)} \tag{5.10}$$

To capture behavior when inelastic buckling is not a factor and initial crookedness is dominant, that is when $KL/r > 4.71\sqrt{E/F_y}$ or $F_e \leq 0.44F_y$

$$F_{cr} = 0.877 F_e \qquad \text{(E3-3)} \tag{5.11}$$

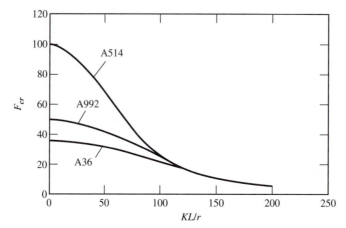

Figure 5.14 KL/r versus Critical Strength.

where F_e is the elastic buckling stress; the Euler buckling stress previously presented as Equation 5.9 and restated here is

$$F_e = \frac{\pi^2 E}{\left(\frac{KL}{r}\right)^2}$$

The flexural buckling stresses for three different steels, A36, A992, and A514, versus the slenderness ratio, KL/r, are shown in Figure 5.14. For very slender columns, the buckling stress is independent of the material yield. The division between elastic and inelastic behavior, Equations 5.10 and 5.11, corresponds to KL/r of 134, 113, and 80.2 for steels with a yield of 36, 50, and 100 ksi, respectively.

Previous editions of the Specification defined the exponent of Equation 5.10 in a slightly different form that makes the presentation a bit simpler. If a new term is defined such that

$$\lambda_c^2 = \frac{F_y}{F_e} = \left(\frac{KL}{\pi r}\right)^2 \frac{F_y}{E}$$

then the dividing point between elastic and inelastic behavior, where

$$\frac{KL}{r} = 4.71 \sqrt{\frac{E}{F_y}}$$

becomes

$$\lambda_c = \frac{KL}{\pi r} \sqrt{\frac{F_y}{E}} = \frac{4.71}{\pi} = 1.5$$

The critical flexural buckling stress becomes
for $\lambda_c \leq 1.5$

$$F_{cr} = (0.658^{\lambda_c^2}) F_y \tag{5.12}$$

and for $\lambda_c > 1.5$

$$F_{cr} = \frac{0.877}{\lambda_c^2} F_y \tag{5.13}$$

A plot of the ratio of critical flexural buckling stress to yield stress as a function of the slenderness parameter, λ_c, is given in Figure 5.15. Thus, regardless of the steel yield stress,

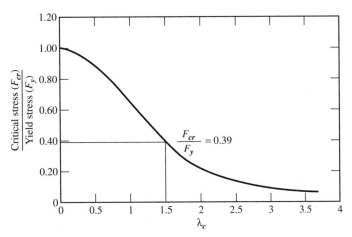

Figure 5.15 Lambda c versus Critical Stress.

the ratio of flexural buckling stress to yield stress is the same. Table 5.2 provides these numerical values in a convenient, usable form.

Previous editions of the Specification indicated that there should be an upper limit on the magnitude of the slenderness ratio at $KL/r = 200$. The intent with this limit is to have the engineer recognize that for very slender columns, the flexural buckling stress was so low as to make the column very inefficient. This limit has been removed in this edition of the Specification because there are many factors that influence column strength that would indicate that a very slender column might actually be acceptable. Section E2 simply informs the designer that column slenderness should preferably be kept to something less than 200. Table 5.3 gives the flexural buckling stress for values of KL/r from 0 to 200 for steels with three different steel yield stresses. Manual Table 4-22 provides an expanded version of this table for five different yield stresses at a slenderness ratio increment of 1.0 ft.

EXAMPLE 5.3
Column Strength by
AISC Provisions

GOAL: Determine the available column strength.

GIVEN: A W12×79 pin-ended column with a length of 10.0 ft. as shown in Figure 5.16a. Use A992 steel.

SOLUTION

Step 1: From Manual Table 1-1, $r_y = 3.05$ in. $A = 23.2$ in.^{2}

Step 2: Determine the effective slenderness ratio.
Because the length is 10.0 ft and the column has pinned ends, $KL = 10.0$ ft and

$$\frac{KL}{r} = \frac{10.0(12)}{3.05} = 39.3$$

Step 3: Determine which column strength equation to use.
Because $\frac{KL}{r} = 39.3 < 4.71\sqrt{\frac{E}{F_y}} = 4.71\sqrt{\frac{29,000}{50}} = 113$, use Equation 5.10 (E3-2).

Step 4: Determine the Euler buckling stress.

$$F_e = \frac{\pi^2(29,000)}{(39.3)^2} = 185 \text{ ksi}$$

Step 5: Determine the critical stress from Equation E3-2.

$$F_{cr} = 0.658^{(F_y/F_e)} F_y = 0.658^{(50/185)}(50) = 44.7 \text{ ksi}$$

Step 6: Determine the nominal strength.

$$P_n = 44.7(23.2) = 1040 \text{ kips}$$

Step 7: Determine the design strength for LRFD.

$$\phi P_n = 0.9(1040) = 936 \text{ kips}$$

Step 7: Determine the allowable strength for ASD.

$$\frac{P_n}{\Omega} = \frac{1040}{1.67} = 623 \text{ kips}$$

Table 5.2 Ratio of Critical Stress-to-Yield Stress

λ_c	F_{cr}/F_y	λ_c	F_{cr}/F_y	λ_c	F_{cr}/F_y
0.00	1.000	1.30	0.493	2.55	0.135
0.05	0.999	1.35	0.466	2.60	0.130
0.10	0.996	1.40	0.440	2.65	0.125
0.15	0.991	1.45	0.415	2.70	0.120
0.20	0.983	1.50	0.390	2.75	0.116
0.25	0.974	1.55	0.365	2.80	0.112
0.30	0.963	1.60	0.343	2.85	0.108
0.35	0.950	1.65	0.322	2.90	0.104
0.40	0.935	1.70	0.303	2.95	0.101
0.45	0.919	1.75	0.286	3.00	0.0974
0.50	0.901	1.80	0.271	3.05	0.0943
0.55	0.881	1.85	0.256	3.10	0.0913
0.60	0.860	1.90	0.243	3.15	0.0884
0.65	0.838	1.95	0.231	3.20	0.0856
0.70	0.815	2.00	0.219	3.25	0.0830
0.75	0.790	2.05	0.209	3.30	0.0805
0.80	0.765	2.10	0.199	3.35	0.0781
0.85	0.739	2.15	0.190	3.40	0.0759
0.90	0.712	2.20	0.181	3.45	0.0737
0.95	0.685	2.25	0.173	3.50	0.0716
1.00	0.658	2.30	0.166	3.55	0.0696
1.05	0.630	2.35	0.159	3.60	0.0677
1.10	0.603	2.40	0.152	3.65	0.0658
1.15	0.575	2.45	0.146	3.70	0.0641
1.20	0.547	2.50	0.140	3.75	0.0624
1.25	0.520				

Table 5.3   Critical Stress for Three Steels

KL/r	$F_y = 36$ ksi F_{cr} ksi	$F_y = 50$ ksi F_{cr} ksi	$F_y = 100$ ksi F_{cr} ksi	KL/r	$F_y = 36$ ksi F_{cr} ksi	$F_y = 50$ ksi F_{cr} ksi	$F_y = 100$ ksi F_{cr} ksi
0	36.0	50.0	100.0	88	23.9	28.4	32.4
2	36.0	50.0	99.9	90	23.5	27.7	31.0
4	36.0	49.9	99.8	92	23.1	26.9	29.7
6	35.9	49.9	99.5	94	22.6	26.2	28.4
8	35.9	49.8	99.1	96	22.2	25.5	27.2
10	35.8	49.6	98.5	98	21.7	24.8	26.1
12	35.7	49.5	97.9	100	21.3	24.1	25.1
14	35.6	49.3	97.2	102	20.8	23.4	24.1
16	35.5	49.1	96.3	104	20.4	22.7	23.2
18	35.4	48.8	95.4	106	19.9	22.0	22.3
20	35.2	48.6	94.3	108	19.5	21.3	21.5
22	35.1	48.3	93.2	110	19.0	20.6	20.7
24	34.9	47.9	91.9	112	18.6	20.0	20.0
26	34.7	47.6	90.6	114	18.2	19.3	19.3
28	34.5	47.2	89.2	116	17.7	18.7	18.7
30	34.3	46.8	87.7	118	17.3	18.0	18.0
32	34.1	46.4	86.1	120	16.9	17.4	17.4
34	33.9	45.9	84.4	122	16.4	16.9	16.9
36	33.6	45.5	82.7	124	16.0	16.3	16.3
38	33.4	45.0	81.0	126	15.6	15.8	15.8
40	33.1	44.5	79.1	128	15.2	15.3	15.3
42	32.8	43.9	77.3	130	14.8	14.9	14.9
44	32.5	43.4	75.3	132	14.4	14.4	14.4
46	32.2	42.8	73.4	134	14.0	14.0	14.0
48	31.9	42.2	71.4	136	13.6	13.6	13.6
50	31.6	41.6	69.4	138	13.2	13.2	13.2
52	31.2	41.0	67.3	140	12.8	12.8	12.8
54	30.9	40.4	65.3	142	12.4	12.4	12.4
56	30.5	39.8	63.2	144	12.1	12.1	12.1
58	30.2	39.1	61.1	146	11.8	11.8	11.8
60	29.8	38.4	59.1	148	11.5	11.5	11.5
62	29.4	37.7	57.0	150	11.2	11.2	11.2
64	29.0	37.1	54.9	152	10.9	10.9	10.9
66	28.6	36.4	52.9	154	10.6	10.6	10.6
68	28.2	35.7	50.9	156	10.3	10.3	10.3
70	27.8	34.9	48.8	158	10.1	10.1	10.1
72	27.4	34.2	46.9	160	9.81	9.81	9.81
74	27.0	33.5	44.9	162	9.56	9.56	9.56
76	26.6	32.8	43.0	164	9.33	9.33	9.33
78	26.1	32.0	41.1	166	9.11	9.11	9.11
80	25.7	31.3	39.2	168	8.89	8.89	8.89
82	25.3	30.6	37.3	170	8.69	8.69	8.69
84	24.8	29.8	35.6	172	8.48	8.48	8.48
86	24.4	29.1	33.9	174	8.29	8.29	8.29

(Continued)

Table 5.3 *(Continued)*

KL/r	$F_y = 36$ ksi F_{cr} ksi	$F_y = 50$ ksi F_{cr} ksi	$F_y = 100$ ksi F_{cr} ksi	KL/r	$F_y = 36$ ksi F_{cr} ksi	$F_y = 50$ ksi F_{cr} ksi	$F_y = 100$ ksi F_{cr} ksi
176	8.10	8.10	8.10	190	6.95	6.95	6.95
178	7.92	7.92	7.92	192	6.81	6.81	6.81
180	7.75	7.75	7.75	194	6.67	6.67	6.67
182	7.58	7.58	7.58	196	6.53	6.53	6.53
184	7.41	7.41	7.41	198	6.40	6.40	6.40
186	7.26	7.26	7.26	200	6.28	6.28	6.28
188	7.10	7.10	7.10				

EXAMPLE 5.4
Column Strength by
AISC Provisions

GOAL: Determine the available column strength.

GIVEN: A W10×49 column with a length of 20.0 ft, one end pinned and the other end fixed for the y-axis, and both ends pinned for the x-axis, as shown in Figure 5.16b. Use A992 steel.

SOLUTION

Step 1: From Manual Table 1-1, $r_x = 4.35$ in., $r_y = 2.54$ in., and $A = 14.4$ in.^{2}

Step 2: Determine the effective length factors from Figure 5.6.
Comparing the columns shown in Figure 5.16b with those shown in Figure 5.6, the effective length factors are $K_y = 0.7$ and $K_x = 1.0$.

Step 3: Determine the x- and y-axis slenderness ratios.

$$\frac{K_x L}{r_x} = \frac{1.0(20.0)(12)}{4.35} = 55.2$$

$$\frac{K_y L}{r_y} = \frac{0.7(20.0)(12)}{2.54} = 66.1$$

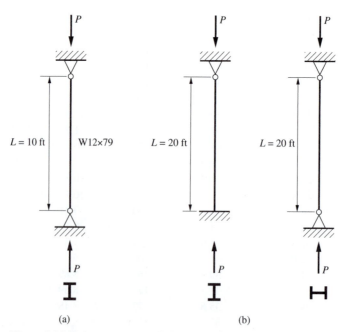

(a)　　　　　　　　　　(b)

Figure 5.16 Columns for Examples 5.3 and 5.4.

Step 4: Using the largest slenderness ratio, determine which column strength equation to use.

$$\frac{KL}{r} = 66.1 < 4.71\sqrt{\frac{E}{F_y}} = 4.71\sqrt{\frac{29,000}{50}} = 113, \text{ use Equation 5.10 (E3-2)}$$

Step 5: Determine the Euler buckling stress.

$$F_e = \frac{\pi^2(29,000)}{(66.1)^2} = 65.5 \text{ ksi}$$

Step 6: Determine the critical stress from Equation E3-2.

$$F_{cr} = 0.658^{(F_y/F_e)} F_y = 0.658^{(50/65.5)}(50) = 36.3 \text{ ksi}$$

Step 7: Determine the nominal strength.

$$P_n = 36.3(14.4) = 523 \text{ kips}$$

Step 8: Determine the design strength for LRFD.

$$\phi P_n = 0.9(523) = 471 \text{ kips}$$

Step 8: Determine the allowable strength for ASD.

$$\frac{P_n}{\Omega} = \frac{523}{1.67} = 313 \text{ kips}$$

5.4 ADDITIONAL LIMIT STATES FOR COMPRESSION

Two limit states for compression members were discussed in Section 5.3, yielding and flexural buckling. The strength equations provided in Specification Section E3 clearly show that the upper limit for column strength, $F_y A_g$, is reached only for the zero length column. Thus, the provisions are presented in the Specification as being for the limit state of flexural buckling only, even though they do consider yielding.

Singly symmetric, unsymmetric, and certain doubly symmetric members may also be limited by torsional buckling or flexural-torsional buckling. The strength provisions for these limit states are given in Section E4 of the Specification and are discussed later.

For some column profiles, another limit state may actually control overall column strength. The individual elements of a column cross section may buckle locally at a stress below the stress that would cause the overall column to buckle. If this is the case, the column is said to be a *column with slender elements*. In order to include the impact of these slender elements on column strength, the Specification provides slender element reduction factors to be incorporated into the already defined flexural, torsional, and flexural-torsional buckling provisions. The additional provisions for these types of members are presented in Sections 5.6 and 5.7.

5.5 LENGTH EFFECTS

The effective lengths that have already been discussed were all related to fairly simple columns with easily defined end conditions and bracing locations. Once a column is recognized as being a part of a real structure, determining the effective length becomes more

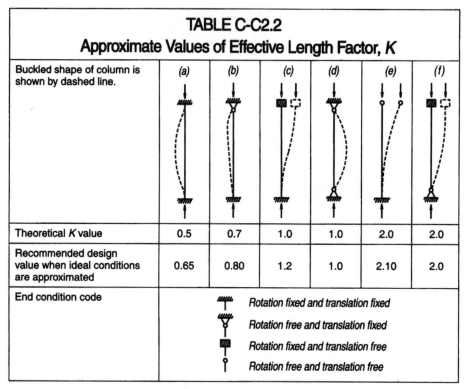

TABLE C-C2.2 Approximate Values of Effective Length Factor, *K*						
Buckled shape of column is shown by dashed line.	*(a)*	*(b)*	*(c)*	*(d)*	*(e)*	*(f)*
Theoretical *K* value	0.5	0.7	1.0	1.0	2.0	2.0
Recommended design value when ideal conditions are approximated	0.65	0.80	1.2	1.0	2.10	2.0
End condition code		Rotation fixed and translation fixed				
		Rotation free and translation fixed				
		Rotation fixed and translation free				
		Rotation free and translation free				

Figure 5.17 Values of Effective Length Factor, *K*. Copyright © American Institute of Steel Construction, Inc. Reprinted with Permission. All rights reserved.

involved. Moreover, for more complex structures, it might be simpler to determine the buckling strength of the structure through analysis. Using that analysis, the elastic buckling stress of the individual columns, F_e, can be determined. This can then be used directly in the column strength equations. However, for this book, column elastic buckling is determined through a calculation of effective length. This approach may incorporate some simplifications that would not be made in an actual buckling analysis.

A first attempt at incorporating some realistic aspects of structures is shown in Table C-C2.2 of the Commentary and here as Figure 5.17. The columns shown in this figure are the same as those shown in Figure 5.6, and the same *K*-factors are shown again and identified here as the theoretical *K*-values. What is new here is the presentation of recommended design values when ideal conditions are approximated. Most of these recommended values are based on the fact that perfectly rigid connections are difficult to obtain. Thus, for example, a fixed end column (case a) would have a theoretical $K = 0.5$ but if the end connections were to actually rotate, even just a small amount, the effective length would increase. As the end rotation increases toward what would occur for a pin-end column, *K* would approach 1.0. Thus, the recommended value of *K* is 0.65. A similar assessment of the other cases should lead to a similar understanding of the idea behind these recommended values.

When a column is part of a frame, as shown in Figure 5.18, the stiffness of the members framing into the column impact the rotation that could occur at the column ends. As with the rigid supports discussed for the columns in Figure 5.17, these end conditions permit the column end to rotate. This rotation is something between the zero rotation of a fixed support and the free rotation of a pin support. When the column under consideration is part of a frame

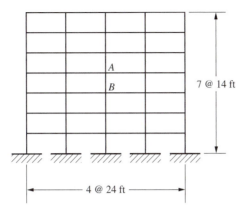

Figure 5.18 A Typical Moment Frame.

where the ends of the column are not permitted to displace laterally relative to each other, the frame is called a *braced frame*, a *sidesway prevented frame*, or a *sidesway inhibited frame*. This is shown as cases a, b, and d in Figure 5.17. For a column in a braced frame, the possible *K*-factors range from 0.5 to 1.0. In frames of this type, *K* is often taken as 1.0, a conservative approximation that simplifies design. When the column under consideration is in a frame in which the ends are permitted to move laterally, the frame is called a *moment frame*, an *unbraced frame*, a *sidesway permitted frame*, or a *sidesway uninhibited frame*. This is shown as cases c, e, and f in Figure 5.17. For the three cases shown here, the lowest value of *K* is 1.0. The other extreme case, not shown in Figure 5.17, is a pin-ended column in an unbraced frame. The effective length of this column would theoretically be infinite. Thus, the range of *K*-values for columns in moment frames ranges from 1.0 to infinity.

The determination of reliable effective length factors is a critical aspect of column design. Several approaches are presented in the literature but the most commonly used approach is through the alignment charts presented in the Commentary. The development of these charts is based on a set of assumptions that are often violated in real structures; nevertheless, the alignment charts are used extensively and often modified in an attempt to account for variations from these assumptions.

These assumptions, as given in the Commentary, are:

1. Behavior is purely elastic.

2. All members have a constant cross section.

3. All joints are rigid.

4. For columns in frames with sidesway inhibited, rotations at opposite ends of the restraining beams are equal in magnitude and opposite in direction, producing single curvature bending.

5. For columns in frames with sidesway uninhibited, rotations at opposite ends of the restraining beams are equal in magnitude and direction, producing reverse curvature bending.

6. The stiffness parameter $L\sqrt{P/EI}$ of all columns is equal.

7. Joint restraint is distributed to the column above and below the joint in proportion to EI/L for the two columns.

8. All columns buckle simultaneously.

9. No significant axial compression force exists in the girders.

EXAMPLE 5.5
Column Effective Length

GOAL: Determine the column effective length using (1) the alignment chart and (2) Equation 5.19.

GIVEN: The column *AB* in a moment frame is shown in Figure 5.21. Assume that the column has its weak axis in the plane of the frame.

SOLUTION

Part a:

Step 1:   Determine member properties from Manual Table 1-1.

$$\text{end } A: \quad W16 \times 36; \ I_{gx} = 448 \text{ in.}^4$$

$$W10 \times 88; \ I_{cx} = 534 \text{ in.}^4$$

$$\text{end } B: \quad W16 \times 77; \ I_{gx} = 1110 \text{ in.}^4$$

$$W10 \times 88; \ I_{cx} = 534 \text{ in.}^4$$

Step 2:   Determine the stiffness ratios at each end.

$$G_A = \frac{2\left(\dfrac{534}{14}\right)}{2\left(\dfrac{448}{24}\right)} = 2.04$$

$$G_B = \frac{2\left(\dfrac{534}{14}\right)}{2\left(\dfrac{1110}{24}\right)} = 0.825$$

Step 3:   Use the alignment chart shown in Figure 5.20 for a sidesway uninhibited frame. Enter G_A and G_B on the appropriate scales and construct a straight line between them, as shown in Figure 5.22. The intersection with the scale for K gives the effective length factor, in this case,

$$\boxed{K = 1.42}$$

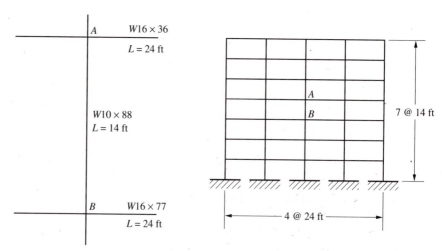

Figure 5.21 Multistory Frame for Example 5.5.

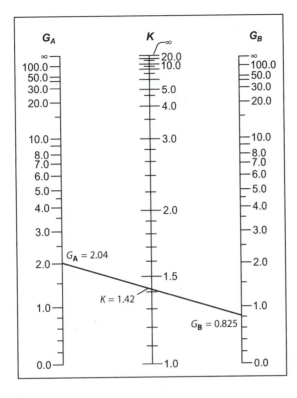

Figure 5.22 Alignment Chart for Example 5.5. Copyright © American Institute of Steel Construction, Inc. Reprinted with Permission. All rights reserved.

Part b:

Step 4:   Determine K using the stiffness ratios, G_A and G_B, determined in part (a) Step 2 and Equation 5.19.

$$K = \sqrt{\frac{1.6(2.04)(0.825) + 4(2.04 + 0.825) + 7.5}{2.04 + 0.825 + 7.5}} = 1.45$$

5.5.1   Effective Length for Inelastic Columns

The assumption of elastic behavior for all members of a frame is regularly violated. We have already seen the role that residual stresses play in determining column strength through inelastic behavior. Thus, it is useful to accommodate this inelastic behavior in the determination of K-factors. The assumption of elastic behavior is important in the calculation of G as the simplification is made to move from Equation 5.16 to Equation 5.17. Returning to Equation 5.16 and assuming that the modulus of elasticity for all columns framing into a joint have the same value and are equal to the tangent modulus, E_T, the definition of G for inelastic behavior becomes

$$G_{inelastic} = \frac{E_T(\Sigma(I/L)_c)}{E(\Sigma(I/L)_g)} \tag{5.22}$$

If G for elastic behavior is taken as $G_{elastic}$, then $G_{inelastic}$ can be formulated as

$$G_{inelastic} = \left(\frac{E_T}{E}\right) G_{elastic} \qquad (5.23)$$

Thus, the impact of including inelastic column behavior simply results in a modification of G. The ratio of tangent modulus to elastic modulus is always less than one so the actual impact of assuming elastic behavior for this application is a conservative one, as can be seen by entering the nomograph with lower G-values and determining a corresponding K-factor. Before a straightforward approach to include inelastic effects on effective length can be proposed, the relationship between the tangent modulus and the elastic modulus must be established.

The Commentary provides the following two definitions for the inelastic stiffness reduction factor, $\tau_a = E_T/E$:
If $P_n/P_y \leq 0.39$

$$\tau_a = 1.0$$

and if $P_n/P_y > 0.39$

$$\tau_a = -2.724\left(\frac{P_n}{P_y}\right)\ln\left(\frac{P_n}{P_y}\right)$$

where P_n is the column nominal strength and P_y is the yield strength. Because the column effective length is required to determine the nominal strength and τ_a is required to determine the effective length, determining the inelastic effective length becomes an iterative process. Table 5.4 provides the inelastic stiffness reduction factor based on the Commentary equation. This is similar to Manual Table 4-21 in which the stiffness reduction factor is based on the available strength of the column.

Another approach to determining the inelastic stiffness reduction factor is through the column strength equations already discussed. Elastic buckling strength is obtained through Equation 5.11. If this equation were to be used in the inelastic buckling region of the column behavior, the resulting strength prediction would be correct if the column were behaving elastically, thus using E in the inelastic region. The strength of the column in the inelastic

Table 5.4 τ_a Based on Commentary Equation

P_n/P_y	τ_a
1.00	0.000
0.95	0.133
0.90	0.258
0.85	0.376
0.80	0.486
0.75	0.588
0.70	0.680
0.65	0.763
0.60	0.835
0.55	0.896
0.50	0.944
0.45	0.979
0.40	0.998
0.39	1.000

Table 5.5 τ_a Based on Critical Stress Equations

λ_c	τ_a	λ_c	τ_a
0.00	0.00	0.80	0.558
0.05	0.00285	0.85	0.609
0.10	0.0114	0.90	0.658
0.15	0.0254	0.95	0.705
0.20	0.0449	1.00	0.750
0.25	0.0694	1.05	0.792
0.30	0.0988	1.10	0.831
0.35	0.133	1.15	0.867
0.40	0.171	1.20	0.899
0.45	0.212	1.25	0.926
0.50	0.257	1.30	0.950
0.55	0.304	1.35	0.969
0.60	0.353	1.40	0.984
0.65	0.404	1.45	0.994
0.70	0.455	1.50	1.000
0.75	0.507		

region, determined from Equation 5.10, is the strength that results because of inelastic buckling, that is, buckling using the tangent modulus, E_T. Thus, the ratio of Equation 5.10 to Equation 5.11 will yield E_T/E so that

$$\tau_a = \frac{E_T}{E} = \frac{\left(0.658\,F_y/F_e\right)F_y}{0.877\,F_e}$$

The results of this approach are presented in Table 5.5 as a function of the slenderness parameter, λ_c. The use of either Table 5.4 or 5.5 assumes that the column is loaded to its full available strength. If it is not, there is less of a reduction in both the inelastic stiffness reduction factor and the effective length.

EXAMPLE 5.6
Inelastic Column
Effective Length

GOAL: Determine the inelastic column effective length using the alignment chart.

GIVEN: Determine the inelastic effective length factor for the column in Example 5.5. Use Equation 5.19 if the column has an LRFD required strength of $P_u = 950$ kips and an ASD required strength of $P_a = 633$ kips. Use A992 steel.

SOLUTION

Step 1:   From Manual Table 1-1, for a W10x88 $A = 25.9$ in.^{2} and from Example 5.5, the elastic stiffness ratios are $G_A = 2.04$ and $G_B = 0.825$.

For LRFD

Step 2:   Determine the required stress based on the required strength.

$$\frac{P_u}{A} = \frac{950}{25.9} = 36.7 \text{ ksi}$$

Step 3:   Determine the stiffness reduction factor from Manual Table 4-21, interpolating between 36 and 37 ksi.

$$\tau_a = 0.452$$

Step 4: Determine the inelastic stiffness ratios by multiplying the elastic stiffness ratios by the stiffness reduction factor.

$$G_{iA} = 0.452(2.04) = 0.922$$
$$G_{iB} = 0.452(0.825) = 0.373$$

Step 5: Determine K from Equation 5.19.

$$K = \sqrt{\frac{1.6(0.922)(0.373) + 4(0.922 + 0.373) + 7.5}{0.922 + 0.373 + 7.5}} = 1.23$$

For ASD

Step 2: Determine the required stress based on the required strength.

$$\frac{P_a}{A} = \frac{633}{25.9} = 24.4 \text{ ksi}$$

Step 3: Determine the stiffness reduction factor from Manual Table 4-21, interpolating between 24 and 25 ksi.

$$\tau_a = 0.454$$

Step 4: Determine the inelastic stiffness ratios by multiplying the elastic stiffness ratios by the stiffness reduction factor.

$$G_{iA} = 0.454(2.04) = 0.926$$
$$G_{iB} = 0.454(0.825) = 0.375$$

Step 5: Determine K from Equation 5.19.

$$K = \sqrt{\frac{1.6(0.926)(0.375) + 4(0.926 + 0.375) + 7.5}{0.926 + 0.375 + 7.5}} = 1.23$$

5.6   SLENDER ELEMENTS IN COMPRESSION

As mentioned in Section 5.4, the columns discussed thus far are controlled by overall column buckling. For some shapes, another form of buckling may actually control column strength: local buckling of the elements that compose the column shape. Whether the shape is rolled or built-up, it can be thought of as being composed of a group of interconnected plates. Depending on how these plates are supported by each other, they could buckle at a stress below the critical buckling stress of the overall column. This is local buckling and is described through a plate critical buckling equation similar to the Euler buckling equation for columns. The critical buckling stress for an axially loaded plate is

$$F_{cr} = \frac{k\pi^2 E}{12(1 - \mu^2)\left(\frac{b}{t}\right)^2} \tag{5.24}$$

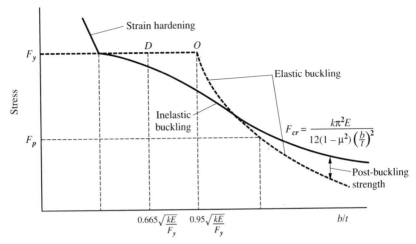

Figure 5.23 Plate Strength in Compression.

where k is a constant that depends on the plate loading, edge conditions, and length-to-width ratio; μ is Poisson's ratio; and b/t is the width perpendicular to the compression force/thickness ratio of the plate. The width/thickness ratio is called the *plate slenderness* and functions similarly to the column slenderness.

As with overall column buckling, an inelastic transition exists between elastic buckling and element yielding. This transition is due to the existence of residual stresses and imperfections in the element just as it was for overall column buckling, and results in the inelastic portion of the curve shown in Figure 5.23. In addition, for plates with low b/t ratios, strain hardening plays a critical role in their behavior and plates with large b/t ratios have significant postbuckling strength.

To insure that local buckling will not control column strength, the critical plate buckling stress for local buckling should be limited to the critical buckling stress for overall column buckling. This approach would result in a different minimum plate slenderness value for each corresponding column slenderness value, a situation that would unduly complicate column design with little value added in the process. Thus, the development of the Specification provisions starts by finding a plate slenderness that sets the plate buckling stress equal to the column yield stress. Equation 5.24 then becomes

$$\frac{b}{t} = \sqrt{\frac{k\pi^2 E}{12(1 - \mu^2)F_y}}$$

Taking $\mu = 0.3$, the standard value for steel, this plate slenderness becomes

$$\frac{b}{t} = 0.95\sqrt{\frac{kE}{F_y}}$$

which is shown as point O in Figure 5.23. This point is well above the inelastic buckling curve. In order to obtain a b/t that would bring the inelastic buckling stress closer to the yield stress, a somewhat arbitrary slenderness limit is taken as 0.7 times the limit that corresponds to the column yield stress, which gives

$$\frac{b}{t} = 0.665\sqrt{\frac{kE}{F_y}}$$

This is indicated as point D in Figure 5.23.

EXAMPLE 5.9
Column Design

GOAL: Determine the least weight section to carry the force given using the limited selection available through Figure 5.24. Design by LRFD and ASD.

GIVEN: The column is shown in Figure 5.25b. Use the loading from Example 5.8.

SOLUTION

Step 1:   Determine the effective length for each axis.
Bracing of the y-axis shown in Figure 5.25b yields $KL_y = 10.0$ ft. The unbraced x-axis has $KL_x = 30.0$ ft.

Step 2:   Determine $(KL)_{eff}$ for the x-axis.
Select a representative r_x/r_y from Figure 5.24. There are two general possibilities. Assume that the larger shapes might be needed to carry the load and try $r_x/r_y = 2.44$. Thus,

$$(KL)_{eff} = \frac{(KL)_x}{(r_x/r_y)} = \frac{30.0}{2.44} = 12.3 \text{ ft}$$

Step 3:   Determine the controlling effective length.
Because $(KL)_{eff}$ is greater than $(KL)_y = 10.0$ ft, enter the table with $KL = 12.3$ ft and interpolate between 12 ft and 13 ft.

For LRFD

Step 4:   The column must have a design strength greater than $P_u = 342$ kips. Try a W14×43, which happens to be the smallest column available with the limited selection available in Figure 5.24. This column has $r_x/r_y = 3.08$.

Step 5:   Determine the $(KL)_{eff}$ with this new r_x/r_y. Thus,

$$(KL)_{eff} = \frac{30.0}{3.08} = 9.74 \text{ ft}$$

Step 6:   Determine the new controlling effective length.
Because $(KL)_{eff}$ is less than $KL_y = 10.0$ ft, enter the table with 10.0 ft and see that the W14×43 has a design strength of 423 kips, which is greater than the required strength of 342 kips from Example 5.8.

Step 7:   Conclusion, use

> W14 × 43

Note: The W14×43 is identified in the table through a footnote as one that is slender for $F_y = 50$ ksi. This is not an issue for our design because the impact of any slender element has already been taken into account in the table.
Using the full complement of tables available in the Manual results in a smaller W12 section having the ability to carry the given load.

For ASD

Step 4:   The column must have an allowable strength greater than $P_a = 268$ kips. Try a W14×48. This column has $r_x/r_y = 3.06$.

Step 5:   Determine the $(KL)_{eff}$ with this new r_x/r_y. Thus,

$$(KL)_{eff} = \frac{30.0}{3.06} = 9.80 \text{ ft}$$

Step 6:   Determine the new controlling effective length.
Because $(KL)_{eff}$ is less than $KL_y = 10.0$ ft, enter the table with 10.0 ft and see that the W14x43 has a design strength of 281 kips, which is greater than the required strength of 268 kips from Example 5.8. Because the assumption of $r_x/r_y = 3.06$ is conservative, no additional calculation needs to be carried out.

Step 7:   Conclusion, use

$$\boxed{\text{W14×43}}$$

Note: The W14×43 is identified in the table through a footnote as one that is slender for $F_y = 50$ ksi. This is not an issue for our design because the impact of any slender element has already been taken into account in the table.
Using the full complement of tables available in the Manual results in a smaller W12 section not having the ability to carry the given load.

5.8   TORSIONAL BUCKLING AND FLEXURAL-TORSIONAL BUCKLING

Up to this point, the discussion has addressed the limit state of flexural buckling. Two additional limit states for column behavior must be addressed: torsional buckling and flexural-torsional buckling. Doubly symmetric shapes normally fail through flexural buckling, as discussed earlier in this chapter, or through torsional buckling. Singly symmetric and non-symmetric shapes can fail through flexural, torsional, or flexural-torsional buckling. Because the shapes normally used for steel members are not well suited to resist torsion, except for closed HSS shapes, it is usually most desirable to avoid any torsional limit states through proper bracing of the column.

If either of the torsional limit states must be evaluated, the Specification provisions are found in Section E4, except for single angles, which are found in Section E5. For all members except single angles, an elastic buckling stress, F_e, is determined, which is then used in Equations E3-2 and E3-3 to determine the column critical stress, F_{cr}.

The provisions for single angles take a different approach. By limiting the way that load is applied to the ends of a single angle compression member, an effective slenderness is established, which is then used in Equations E3-2 and E3-3 to determine the column critical stress, F_{cr}.

The limit states of torsional buckling and flexural torsional buckling are not normally considered in the design of W-shape columns. They generally do not govern and when they do, the critical load differs very little from the strength determined from flexural buckling. For other member types, such as WT or double angle compression elements in trusses, these limit states are quite important. Because they are so important, the Commentary provides

Table C-E4.2, giving the limiting proportions that, if satisfied, permit these members to be designed through the flexural buckling equations of Section E3. For built-up or rolled Tees, the limits require that the flange width be greater than or equal to the depth of the member. The flange thickness for built-up Tees must be equal to or greater than 1.25 times the stem thickness and for rolled Tees, equal to or greater than 1.1 times the stem thickness.

An additional factor in determining strength based on these limit states is the determination of the torsional effective length, K_z. The Commentary recommends that conservatively, $K_z = 1.0$ and provides several other possibilities if greater accuracy is desired.

5.9   SINGLE-ANGLE COMPRESSION MEMBERS

Single-angle compression members would be designed according to the provisions in Specification Section E4 except for the exclusion that permits a somewhat different approach to be taken. Studies show that the compressive strength of single angles can be reasonably predicted using the column equations of Specification Section E3 if a modified effective length is used and the member satisfies the following limiting criteria as found in Specification Section E5.

1. Members are loaded at their ends through the same leg.
2. Members are attached by either welding or a connection containing a minimum of two bolts.
3. There are no intermediate transverse loads.

Two cases are given for these provisions: (1) angles that are individual members or web members of planar trusses, and (2) angles that are web members in box or space trusses. This distinction is intended to reflect the difference in restraint provided by the elements to which the compression members are attached.

The first set of equations is for angles that:

1. Are individual members or web members of planar trusses.
2. Are equal-leg or unequal-leg connected through the longer leg.
3. Have adjacent web members attached to the same side of a gusset plate or truss chord.

In this case, buckling is assumed to occur about the x-axis where the x-axis is the geometric axis parallel to the attached leg.

If $0 \le \frac{L}{r_x} \le 80$

$$\frac{KL}{r} = 72 + 0.75\frac{L}{r_x} \qquad \text{(E5-1)} \qquad\qquad (5.26)$$

and if $\frac{L}{r_x} > 80$

$$\frac{KL}{r} = 32 + 1.25\frac{L}{r_x} \le 200 \qquad \text{(E5-2)} \qquad\qquad (5.27)$$

These effective lengths must be modified if the unequal-leg angles are attached through the shorter legs. The provisions of Specification Section E5 should be reviewed for these angles as well as similar angles in box or space trusses.

EXAMPLE 5.10
Strength of Single-Angle Compression Member

GOAL: Determine the available strength of a 10.0-ft single-angle compression member using A36 steel.

GIVEN: A 4×4×½ angle is a web member in a planar truss. It is attached by two bolts at each end through the same leg.

SOLUTION

Step 1: From Manual Table 1-7, $A = 3.75$ in.^{2} and $r_x = 1.21$.

Step 2: Determine the slenderness ratio.

$$\frac{L}{r_x} = \frac{10.0(12)}{1.21} = 99.2$$

Step 3: Determine which equation will give the effective slenderness ratio. Because

$$\frac{L}{r_x} = 99.2 > 80$$

use Equation E5-2.

Step 4: Determine the effective slenderness ratio from Equation 5.27.

$$\frac{KL}{r} = 32 + 1.25(99.2) = 156 < 200$$

Step 5: Determine which column strength equation to use.

Because $\frac{KL}{r} = 156 > 4.71\sqrt{\dfrac{29{,}000}{36}} = 134$ use Equation 5.11 (E3-3).

Step 6: Determine the Euler buckling stress.

$$F_e = \frac{\pi^2 E}{\left(\frac{KL}{r}\right)^2} = \frac{\pi^2(29{,}000)}{(156)^2} = 11.8 \text{ ksi}$$

Step 7: Determine the critical stress from Equation E3-3.

$$F_{cr} = 0.877 F_e = 0.877(11.8) = 10.3 \text{ ksi}$$

Step 8: Determine the nominal strength.

$$P_n = F_{cr} A = 10.3(3.75) = 38.6 \text{ kips}$$

For LRFD

Step 9: Determine the design strength.

$$\phi P_n = 0.9(38.6) = 34.7 \text{ kips}$$

For ASD

Step 9: Determine the allowable strength.

$$\frac{P_n}{\Omega} = \frac{38.6}{1.67} = 23.1 \text{ kips}$$

5.10   BUILT-UP MEMBERS

Members composed of more than one shape are called *built-up members.* Several of these were illustrated in Figure 5.2h through n. Built-up members are covered in Specification Section E6. Compressive strength is addressed by establishing the slenderness ratio and referring to Specification Sections E3, E4, or E7.

If a built-up section buckles so that the fasteners between the shapes are not stressed in shear but simply go "along for the ride," the only requirement is that the slenderness ratio of the shape between fasteners be no greater than 0.75 times the controlling slenderness ratio of the built-up shape. If overall buckling would put the fasteners into shear, then the controlling slenderness ratio will be somewhat greater than the slenderness ratio of the built-up shape. This modified slenderness ratio is a function of the type of connectors used and their spacing.

For intermediate connectors that are snug-tight bolted, the modified slenderness ratio is specified as

$$\left(\frac{KL}{r}\right)_m = \sqrt{\left(\frac{KL}{r}\right)_o^2 + \left(\frac{a}{r_i}\right)^2}$$

And if the intermediate connectors are welded or pretensioned bolted, the modified slenderness ratio is specified as

$$\left(\frac{KL}{r}\right)_m = \sqrt{\left(\frac{KL}{r}\right)_o^2 + 0.82\frac{\alpha^2}{(1+\alpha^2)}\left(\frac{a}{r_{ib}}\right)^2}$$

where

$\left(\frac{KL}{r}\right)_o =$ column slenderness of the built-up member acting as a unit

$a =$ distance between connectors

$r_i =$ minimum radius of gyration of the individual component

$r_{ib} =$ radius of gyration if the individual component relative to its centroidal axis parallel to the member buckling axis

$\alpha =$ separation ratio, $h/2r_{ib}$

$h =$ distance between the centroids of individual components perpendicular to the member axis of buckling

The remaining provisions in Specification Section E6 address dimensions and detailing requirements. These provisions are based on judgment and experience and are provided to insure that the built-up member behaves in a way consistent with the strength provisions already discussed. The ends of built-up compression members must be either welded or pretensioned bolted. Along the length of built-up members, the longitudinal spacing of connectors must be sufficient to provide for transfer of the required shear force in the buckled member. The spacing of connectors that satisfy the previously mentioned $^3/_4$ of the member slenderness will not necessarily satisfy this strength requirement.

The Manual provides tables of properties for double angles, double channels, and I-shapes with cap channels in Part 1 and tables of compressive strength for double angle compression members in Part 4.

5.11   PROBLEMS

1.   Determine the theoretical buckling strength, the Euler Buckling Load, for a W8×35, A992 column with an effective length of 20 ft. Will the theoretical column buckle or yield at this length?

2.   For a W12×40, A992 column, determine the effective length at which the theoretical buckling strength will equal the yield strength.

3.   A W14×68 column has an effective length for y-axis buckling equal to 24 ft. Determine the effective length for the x-axis that will provide the same theoretical buckling strength.

4.   A W14×109, A992 column has an effective length of 36 ft about both axes. Determine the available compressive strength for the column. Determine the (a) design strength by LRFD and (b) allowable strength by ASD. Is this an elastic or inelastic buckling condition?

5.   Determine the available compressive strength for a W14×120, A992 column with an effective length about both axes of 40 ft. Determine the (a) design strength by LRFD and (b) allowable strength by ASD. Is this an elastic or inelastic buckling condition?

6.   Determine the available compressive strength for a W12×45, A992 column when the effective length is 20 ft about the y-axis and 40 ft about the x-axis. Determine the (a) design strength by LRFD and (b) allowable strength by ASD. Is this an elastic or inelastic buckling condition? Describe a common condition where the effective length is different about the different axes.

7.   A W8×24, A992 column has an effective length of 12.5 ft about the y-axis and 28 ft about the x-axis. Determine the available compressive strength and indicate whether if this is due to elastic or inelastic buckling. Determine the (a) design strength by LRFD and (b) allowable strength by ASD.

8.   A W8×40 is used as a 12-ft column in a braced frame with W16×26 beams at the top and bottom as shown below. The columns above and below are also 12 ft, W8×40s. The beams provide moment restraint at each column end. Determine the effective length using the alignment chart and the available compressive strength, and the (a) design strength by LRFD and (b) allowable strength by ASD. Assume that the columns are oriented for (i) buckling about the weak axis and (ii) buckling about the strong axis. All steel is A992.

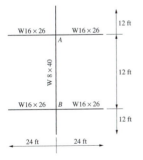

P5.8

9.   If the structure described in Problem 8 is an unbraced frame, determine the effective lengths and compressive strength as requested in Problem 8.

10.   A W12×170 column is shown with end conditions that approximate ideal conditions. Using the recommended approximate values from Commentary Table C-C2.2, determine the effective length for the y-axis and the x-axis. Which effective length will control the column strength?

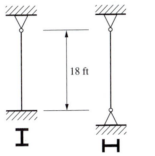

P5.10

11.   A W12×96 column is shown with end conditions that approximate ideal conditions. Using the recommended approximate values from Commentary Table C-C2.2, determine the effective length for the y-axis and the x-axis. Which effective length will control the column strength?

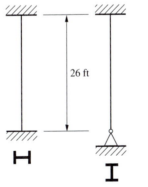

P5.11

12.   A W10×60 column with an effective length of 30 ft is called upon to carry a compressive dead load of 123 kips and a compressive live load of 170 kips. Determine whether the column will support the load by (a) LRFD and (b) ASD. Evaluate the strength for (i) $F_y = 50$ ksi and (ii) $F_y = 70$ ksi.

13.   A W14×257, A992 is used as a column in a building with an effective length of 16 ft. Determine whether the column will carry a compressive dead load of 800 kips and a compressive live load of 1100 kips by (a) LRFD and by (b) ASD.

14.   A W8×48, A992 is used in a structure to support a dead load of 60 kips and a live load of 100 kips. The column has an effective length of 20 ft. Determine whether the column will support the load by (a) LRFD and (b) ASD.

15. A W16×77, A992 is used as a column in a building to support a dead load of 130 kips and a live load of 200 kips. The column effective length is 20 ft for the *y*-axis and 30 ft for the *x*-axis. Determine whether the column will support the load by (a) LRFD and (b) ASD.

16. A W24×131, A992 is used as a column in a building to support a dead load of 245 kips and a live load of 500 kips. The column has an effective length about the *y*-axis of 18 ft and an effective length about the *x*-axis of 36 ft. Determine whether the column will support the load by (a) LRFD and (b) ASD.

17. An HSS8×8×½ A500 Gr. B is used as a column to support a dead load of 175 kips and a live load of 100 kips. The column has an effective length of 10 ft. Determine whether the column will support the load by (a) LRFD and (b) ASD.

18. For the W10×33 column, with bracing and end conditions shown below, determine the theoretical effective length for each axis and identify the axis that will limit the column strength.

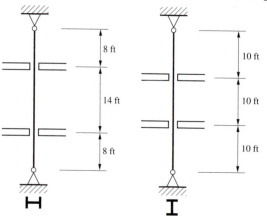

P5.18

19. A W10×45 column with end conditions and bracing is shown. Determine the least theoretical bracing and its location about the *y*-axis, in order that the *y*-axis not control the strength of the column.

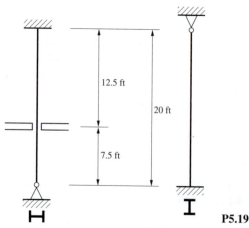

P5.19

20. A W12×50 column is an interior column with strong axis buckling in the plane of the frame in an unbraced multistory frame. The columns above and below are also W12×50. The beams framing in at the top are W16×31 and those at the bottom are W16×36. The columns are 12 ft and the beam span is 22 ft. The column carries a dead load of 75 kips and a live load of 150 kips. Determine the inelastic effective length for this condition and the corresponding compressive strength by (a) LRFD and (b) ASD. All steel is A992.

21. Select the least weight W12, A992 column to carry a live load of 130 kips and a dead load of 100 kips with an effective length about both axes of 14 ft by (a) LRFD and (b) ASD.

22. A column with pin ends for both axes must be selected to carry a compressive dead load of 95 kips and a compressive live load of 285 kips. The column is 16 ft long and is in a braced frame. Select the lightest weight W12 to support this load by (a) LRFD and (b) ASD.

23. If the column in Problem 22 had an effective length of 32 ft, select the lightest weight W12 to support this load by (a) LRFD and (b) ASD.

24. A W14 A992 column must support a dead load of 80 kips and a live load of 300 kips. The column is 22 ft long and has end conditions that approximate the ideal conditions of a fixed support at one end and a pin support at the other. Select the lightest weight W14 to support this load by (a) LRFD and (b) ASD.

25. Select the least weight W8 A992 column to support a dead load of 170 kips with an effective length of 16 ft by (a) LRFD and (b) ASD.

26. A column with an effective length of 21 ft must support a dead load of 120 kips, a live load of 175 kips, and a wind load of 84 kips. Select the lightest W14 A992 member to support the load by (a) LRFD and (b) ASD.

27. An A36 single-angle compression web member of a truss is 10 ft long and attached to gusset plates through the same leg at each end with a minimum of two bolts. The member must carry a dead load of 8 kips and a live load of 10 kips. Select the least weight equal leg angle to carry this load by (a) LRFD and (b) ASD.

28. If the compression web member of Problem 27 were loaded concentrically, determine the least weight single angle to carry the load by (a) LRFD and (b) ASD.

29. A W16×31, A992 compression member has a slender web when used in uniform compression. Determine the available strength by (a) LRFD and (b) ASD when the effective length is (i) 6 ft and (ii) 12 ft.

30. The W14×43 is the only A992 column shown in the Manual column tables that has a slender web. Determine the available strength for this column if the effective length is 5 ft and show whether the slender web impacts that strength by (a) LRFD and (b) ASD.

Chapter 6

Seattle Central Library, Seattle.
Photo courtesy Michael Dickter/Magnusson Klemencic Associates.

Bending Members

6.1 BENDING MEMBERS IN STRUCTURES

A *bending member* carries load applied normal to its longitudinal axis and transfers it to its support points through bending moments and shears. In building construction, the most common application of bending members is to provide support for floors or roofs. These beams can be either simple span or continuous span and normally transfer their load to other structural members such as columns, girders, or walls. Although the terms *beams* and *girders* are often used interchangeably, because both are bending members, the term beam normally refers to a bending member directly supporting an applied load whereas girder usually refers to a bending member that supports a beam. The distinction is not important for design because the same criteria apply to all bending members.

The most commonly used shapes for bending members are the I-shaped cross-sections and, of these, the W-shape is dominant. However, there are numerous situations where other shapes are used as bending members. L-shapes are commonly used as lintels over openings, T-shapes are found as chords of trusses that may be called upon to resist bending along with axial forces, and C-shapes may coexist with W-shapes in floor systems.

In addition to the use of the standard shapes, engineers often find it necessary to develop their own shapes by combining shapes and/or plates. Several examples of these built-up

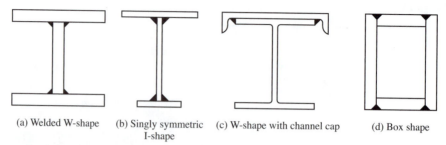

(a) Welded W-shape   (b) Singly symmetric   (c) W-shape with channel cap   (d) Box shape
                          I-shape

Figure 6.1 Built-Up Beams.

shapes are shown in Figure 6.1. Although the use of these built-up shapes is permitted by the Specification, they may not be economical because of the labor costs associated with fabrication. The complexity that results from the wide variety of possible shapes is the reason for so many separate provisions in Chapter F of the Specification.

The most common and economical bending members are those that can attain the full material yield strength without being limited by buckling of any of the cross-sectional elements. These members are referred to as compact members and are addressed first.

Table 6.1 lists the sections of the Specification and parts of the Manual discussed in this chapter.

6.2   STRENGTH OF BEAMS

As load is applied to a bending member resulting in a bending moment, stresses are developed in the cross section. For loads at or below the nominal loads, the load magnitude established in the building code, it is reasonable to expect the entire beam cross section to

Table 6.1   Sections of Specification and Parts of Manual Found in This Chapter

Specification	
B3	Design Basis
B4	Classification of Sections for Local Buckling
F1	General Provisions
F2	Doubly Symmetric Compact I-Shaped Members and Channels Bent about their Major Axis
F3	Doubly Symmetric I-Shaped Members with Compact Webs and Noncompact or Slender Flanges Bent about their Major Axis
F6	I-Shaped Members and Channels Bent about their Minor Axis
F9	Tees and Double Angles Loaded in the Plane of Symmetry
F10	Single Angles
Chapter G	Design of Members for Shear
H1	Doubly and Singly Symmetric Members Subject to Flexure and Axial Force
J10	Flanges and Webs with Concentrated Forces
Chapter L	Design for Serviceability
Appendix 1	Inelastic Analysis and Design
Manual	
Part 1	Dimensions and Properties
Part 3	Design of Flexural Members
Part 6	Design of Members Subject to Combined Loading

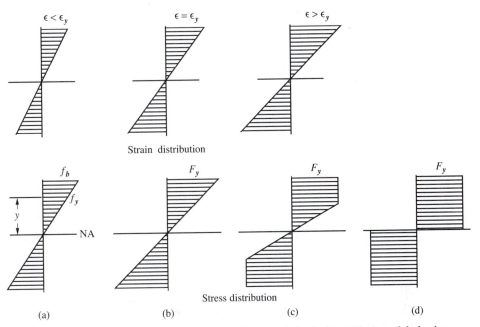

Strain distribution

Stress distribution

Figure 6.2 Cross-Sectional Bending Stresses and Strains: a) elastic; b) yield; c) partial plastic; d) plastic.

behave elastically. The stresses and strains are distributed as shown in Figure 6.2a. This elastic behavior occurs whenever the material is behaving along the initial straight line portion of the stress-strain curve of Figure 3.2.

From the basic principles of strength of materials, the relationship between the applied moment and resulting stresses is given by the familiar flexure formula:

$$f_y = \frac{My}{I} \tag{6.1}$$

where

$M =$ any applied moment that stresses the section in the elastic range

$y =$ distance from the neutral axis to the point where the stress is to be determined

$I =$ Moment of Inertia

$f_y =$ resulting bending stress at location, y

Normally the stress at the extreme fiber, that is, the fiber most distant from the neutral axis, is of interest because the largest stress occurs at this point. The distance from the neutral axis to the extreme fiber may be taken as c and the flexure formula becomes

$$f_b = \frac{Mc}{I} = \frac{M}{S} \tag{6.2}$$

where

$S =$ section modulus

$f_b =$ extreme fiber bending stress

The moment that causes the extreme fiber to reach the yield stress, F_y, is called the *yield moment*, M_y. The corresponding stress and strain diagrams are shown in Figure 6.2b. If the load is increased beyond the yield moment, the strain in the extreme fiber increases but the stress remains at F_y because these fibers are behaving as depicted by the plateau on

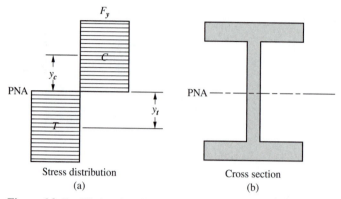

Figure 6.3 Equilibrium in a Doubly Symmetrical Wide-Flange Shape.

the stress/strain diagram, shown previously in Figures 3.2 and 3.3. The stress at some points on the cross section closer to the neutral axis also reach the yield stress whereas those even closer remain elastic as shown in Figure 6.2c.

As the moment continues to increase, the portion of the cross section experiencing the yield stress continues to increase until the entire section experiences the yield stress as shown in Figure 6.2d. Equilibrium of the cross section requires, at all times, that the total internal tension force be equal to the total internal compression force. The basic principles of strength of materials are addressed in numerous texts, such as *Mechanics of Materials*,[1]

For the doubly symmetric wide flange shape shown in Figure 6.3, equilibrium occurs when the portion of the shape above the elastic neutral axis is stressed to the yield stress in compression while the portion below the elastic neutral axis is stressed to the yield stress in tension. For a nonsymmetric shape, the area above the elastic neutral axis is not equal to the area below the elastic neutral. Thus, a new axis, which divides the tension and compression zone into equal areas, must be defined. This new axis is the *plastic neutral axis* (PNA), the axis that divides the section into two equal areas. For symmetric shapes, the elastic and plastic neutral axes coincide, as was the case for the wide flange. For nonsymmetric shapes these neutral axes are at different locations.

Because equilibrium means that the tension and compression forces are equal and opposite, they form a force couple. Although moments can be taken about any reference point for this case, it is common practice to take moments about the PNA. The moment that corresponds to this fully yielded stress distribution is called the *plastic moment*, M_p, and is given as

$$M_p = F_y(A_c y_c) + F_y(A_t y_t) \qquad (6.3)$$

where A_t and A_c are the equal tension and compression areas, respectively, and y_c and y_t are the distances from the centroid of the area to the PNA for the tension and compression areas, respectively. Equation 6.3 may be simplified to

$$M_p = F_y\left(\frac{A}{2}\right)(y_c + y_t) \qquad (6.4)$$

The two terms multiplied by the yield stress are functions of only the geometry of the cross section and are normally combined and called the *plastic section modulus*, Z. Thus, the plastic moment is given as

$$M_p = F_y Z \qquad (6.5)$$

The plastic section modulus is tabulated for all available shapes in Part 1 of the Manual.

---

[1] Pytel and Kiusalaas. *Mechanics of Materials*. Brooks/Cole, 2003.

Chapter F of the Specification contains the provisions for design of flexural members due to bending. For a given beam to attain its full plastic moment strength, it must satisfy a number of criteria as established in Section F2. If these criteria are not met, the strength is defined as something less than M_p. The criteria to be satisfied are defined by two limit states in addition to yielding: local buckling and lateral torsional buckling. Each of these limit states and their impact on beam strength are discussed in Sections 6.4 and 6.5.

EXAMPLE 6.1
Plastic Moment Strength for a Symmetric Shape

GOAL: Determine the plastic moment strength of a W-shape using the model of three rectangular plates.

GIVEN: A W24×192 is modeled, as shown in Figure 6.4. Assume $F_y = 50$ ksi.

SOLUTION

Step 1: Determine the location of the plastic neutral axis.
Because the shape is symmetric, the plastic neutral axis is located on the axis of symmetry.

Step 2: Determine the plastic section modulus as the sum of the moment of each area about the plastic neutral axis.

$$Z = \frac{A}{2}(y_c + y_t) = 2\left(A_f y_f + \frac{A_w}{2}y_w\right)$$

$$Z = 2\left[13.0(1.46)\left(\frac{22.58}{2} + \frac{1.46}{2}\right) + \frac{22.58(0.810)}{2}\left(\frac{22.58}{4}\right)\right] = 560 \text{ in.}^4$$

Step 3: Determine the plastic moment strength as the plastic section modulus times the yield stress.

$$M_p = F_y Z = 50(560) = 28,000 \text{ in.-kips}$$

or

$$\boxed{M_p = \frac{28,000}{12} = 2330 \text{ ft-kips}}$$

Step 4: Compare the calculated plastic section modulus value with that from Manual Table 1-1.

From the table, $Z_x = 559$ in.^{3}

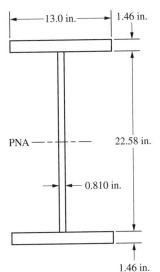

          Figure 6.4 W24×192 Model for Example 6.1.

EXAMPLE 6.2
Plastic Section Modulus for a Nonsymmetric Shape

SOLUTION

GOAL: Locate the plastic neutral axis and determine the plastic section modulus for a WT.

GIVEN: A WT12×51.5 modeled as two plates is shown in Figure 6.5. Assume that $F_y = 50\,\text{ksi}$.

Step 1: Determine the area of the T shape.

$$A_{flange} = 9.00(0.980) = 8.82 \text{ in.}^2$$
$$A_{stem} = 0.550(12.3 - 0.980) = 6.23 \text{ in.}^2$$
$$A_{total} = 8.82 + 6.23 = 15.1 \text{ in.}^2$$

Step 2: Determine one-half of the area, because one-half of the area must be above the plastic neutral axis and one-half must be below.

$$\frac{A_{total}}{2} = \frac{15.1}{2} = 7.55 \text{ in.}^2$$

Step 3: Determine whether the plastic neutral axis is in the flange or stem. Because half of the area is less than the area of the flange, the plastic neutral axis is in the flange and

$$y_p = \frac{7.55}{9.0} = 0.839 \text{ in.}$$

with the plastic neutral axis measured from the top of the flange.

Step 4: Determine the plastic section modulus as the sum of the moment of each area about the plastic neutral axis.

$$Z = 7.55\left(\frac{0.839}{2}\right) + (8.82 - 7.75)\left(\frac{0.980 - 0.839}{2}\right) + 6.23\left(0.980 - 0.839 + \frac{11.3}{2}\right)$$

$$\boxed{Z = 3.17 + 0.0754 + 36.1 = 39.3 \text{ in.}^3}$$

Step 5: Compare these values with the values in Manual Table 1.8.

$$y_p = 0.841 \text{ in.} \quad \text{and} \quad Z = 39.2 \text{ in.}^3$$

This shows the impact of the simplification in using rectangular plates and ignoring the fillets at the flange web junction.

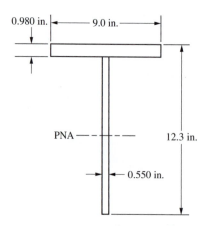

Figure 6.5 T-Beam Model for Example 6.2.

6.3   DESIGN OF COMPACT LATERALLY SUPPORTED WIDE FLANGE BEAMS

Section 6.2 showed that the nominal strength of a compact member with full lateral support is determined by the limit state of yielding. For this limit state, Specification Section F2 provides that

$$M_n = M_p = F_y Z \tag{6.6}$$

Specification Section F1 also indicates that for all flexural limit states, design strength and allowable strength are to be determined using

$$\phi = 0.90\,(\text{LRFD}) \qquad \Omega = 1.67\,(\text{ASD})$$

The design basis from Sections B3.2 and B3.4, as discussed in Chapter 1, are repeated here.

For ASD, the allowable strength is

$$R_a \leq \frac{R_n}{\Omega} \tag{1.1}$$

For LRFD, the design strength is

$$R_u \leq \phi R_n \tag{1.2}$$

EXAMPLE 6.3a **Beam Design by LRFD**	**GOAL:** Select the least-weight wide flange member for the conditions given.

GIVEN: An A992 beam, simply supported at both ends, spans 20 ft and is loaded at midspan with a dead load of 8.0 kips and a live load of 24.0 kips, as shown in Figure 6.6. Assume full lateral support and a compact section.

SOLUTION

Step 1:   Determine the required strength using the LRFD load combinations from Section 2.4.

$$P_u = 1.2P_D + 1.6P_L = 1.2(8.0) + 1.6(24.0) = 48.0\,\text{kips}$$

$$M_u = \frac{P_u L}{4} = \frac{48(20)}{4} = 240\,\text{ft-kips}$$

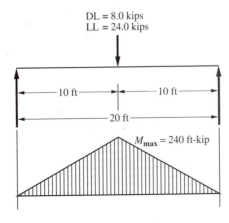

Figure 6.6 Beam Used in Example 6.3.

Step 2:   Determine the required plastic section modulus. For a compact, fully braced section

$$M_n = M_p = F_y Z$$

Thus, because Specification Section B3.3 provides that the required moment be less than the available moment, $M_u \le \phi M_n = \phi F_y Z$, and

$$Z_{req} = \frac{M_u}{\phi F_y} = \frac{240(12)}{0.90(50)} = 64.0 \text{ in.}^3$$

Step 3:   Using the required plastic section modulus, select the minimum weight W-shape from the plastic section modulus economy table, Manual Table 3-2. Start at the bottom of the Z_x column and move up until a shape in bold with at least $Z_x = 64.0 \text{ in.}^3$ is found.

> Select W18×35,   ($Z = 66.5 \text{ in.}^3$)

This is the most economical W-shape, based on section weight that provides the required plastic section modulus.

Step 4:   An alternate approach, using the same Manual Table, would be to enter the table with the required moment, $M_u = 240$ ft-kips, and proceed up the ϕM_n column of the table. The same section will be selected with this approach.

EXAMPLE 6.3b
Beam Design by ASD

GOAL: Select the least-weight wide flange member for the conditions given.

GIVEN: An A992 beam, simply supported at both ends, spans 20 ft and is loaded at midspan with a dead load of 8.0 kips and a live load of 24.0 kips, as shown in Figure 6.6. Assume full lateral support and a compact section.

SOLUTION

Step 1:   Determine the required strength using the ASD load combinations from Section 2.4.

$$P_a = P_D + P_L = (8.0) + (24.0) = 32.0 \text{ kips}$$

$$M_a = \frac{P_a L}{4} = \frac{32.0(20)}{4} = 160 \text{ ft-kips}$$

Step 2:   Determine the required plastic section modulus. For a compact, fully braced section

$$M_n = M_p = F_y Z$$

Thus, because Specification Section B3.4 provides that the required moment be less than the available moment, $M_a \le M_n / \Omega = F_y Z / \Omega$, and

$$Z_{req} = \frac{M_a}{F_y / \Omega} = \frac{160(12)}{(50/1.67)} = \frac{160(12)}{30} = 64.0 \text{ in.}^3$$

Step 3:   Using the required plastic section modulus, select the minimum weight W-shape from the plastic section modulus economy table, Manual Table 3-2. Start at the bottom of the Z_x column and move up until a shape in bold with at least $Z_x = 64.0 \text{ in.}^3$ is found.

> Select W18×35,   ($Z = 66.5 \text{ in.}^3$)

This is the most economical W-shape, based on section weight that provides the required plastic section modulus.

Step 4: An alternate approach, using the same Manual Table, would be to enter the table with the required moment, $M_a = 160$ ft-kips, and proceed up the M_n/Ω column of the table. The same section will be selected with this approach.

EXAMPLE 6.4a
Beam Design by LRFD

GOAL: Design a W-shape floor beam for the intermediate beam marked A on the floor plan shown in Figure 6.7.

GIVEN: The beam is loaded uniformly from the floor with a live load of 60 pounds per square foot (psf) and a dead load in addition to the beam self-weight of 80 psf. The beam will have full lateral support provided by the floor deck and a compact section will be selected. Use A992 steel.

SOLUTION

Step 1: Determine the required load and moment.

$$w_u = (1.2w_D + 1.6w_L)L_{trib} = (1.2(60) + 1.6(80))(10) = 2000 \text{ lb/ft}$$

$$M_u = \frac{w_u L^2}{8} = \frac{2.0(26)^2}{8} = 169 \text{ ft-kips}$$

Step 2: Determine the required plastic section modulus.
For a compact, fully braced beam, $M_n = M_p = F_y Z$. Section B3.3 of the Specification requires that

$$M_a \le \phi M_n = \phi F_y Z$$

Therefore

$$Z_{req} = \frac{M_u}{\phi F_y} = \frac{(169)(12)}{0.90(50)} = 45.1 \text{ in.}^3$$

Step 3: Using the plastic section modulus economy table, Manual Table 3-2, select the most economical W-shape based on least weight.

$$\boxed{\text{W14}\times\text{30}, \quad (Z = 47.3 \text{ in.}^3)}$$

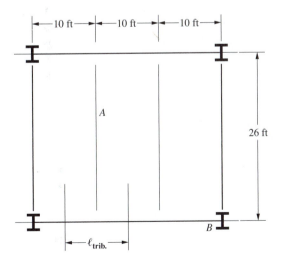

Figure 6.7 Framing Plan for Example 6.4.

Step 4: Determine the additional required strength based on the actual weight of the chosen beam. The beam weighs 30 lb/ft, which gives an additional moment of

$$M_{u(self\text{-}weight)} = 1.2\left(\frac{0.030(26)^2}{8}\right) = 1.2(2.54) = 3.04 \text{ ft-kips}$$

Step 5: Combine this moment with the moment due to superimposed load to determine the new required strength.

$$M_u = 169 + 3.04 = 172 \text{ ft-kips}$$

Step 6: Determine the new required plastic section modulus.

$$Z_{req} = \frac{M_u}{\phi F_y} = \frac{(172)(12)}{0.90(50)} = 45.9 \text{ in.}^3$$

Step 7: Make the final selection. This required plastic section modulus is less than that provided by the W14×30 already chosen. Therefore, select the

$$\boxed{\text{W14×30}}$$

Step 8: As shown in Example 6.3, an alternate approach is to use the required moment, $M_u = 172$ ft-kips, and enter the ϕM_n column to determine the same W-shape.

EXAMPLE 6.4b
Beam Design by ASD

GOAL: Design a W-shape floor beam for the intermediate beam marked A on the floor plan shown in Figure 6.7.

GIVEN: The beam is loaded uniformly from the floor with a live load of 60 pounds per square foot (psf) and a dead load in addition to the beam self-weight of 80 psf. The beam will have full lateral support provided by the floor deck and a compact section will be selected. Use A992 steel.

SOLUTION

Step 1: Determine the required load and moment.

$$w_a = (w_D + w_L)L_{trib} = (60 + 80)(10) = 1400 \text{ lb/ft}$$

$$M_a = \frac{w_a L^2}{8} = \frac{1.40(26)^2}{8} = 118 \text{ ft-kips}$$

Step 2: Determine the required plastic section modulus.
For a compact, fully braced beam, $M_n = M_p = F_y Z$. Section B3.4 of the Specification requires that

$$M_a \le \frac{M_n}{\Omega} = \frac{F_y Z}{\Omega}$$

Therefore

$$Z_{req} = \frac{M_a}{F_y/\Omega} = \frac{(118)(12)}{30} = 47.2 \text{ in.}^3$$

Step 3:   Using the plastic section modulus economy table, Manual Table 3-2, select the most economical W-shape based on least weight.

$$W14 \times 30, \quad (Z = 47.3 \text{ in.}^3)$$

Step 4:   Determine the additional required strength based on the actual weight of the chosen beam. The beam weighs 30 lb/ft, which gives an additional moment of

$$M_{a(self\text{-}weight)} = \frac{0.030(26)^2}{8} = 2.54 \text{ ft-kips}$$

Step 5:   Combine this moment with the moment due to superimposed load to determine the new required strength.

$$M_a = 118 + 2.54 = 121 \text{ ft-kips}$$

Step 6:   Determine the new required plastic section modulus.

$$Z_{req} = \frac{M_a}{F_y/\Omega} = \frac{(121)(12)}{30} = 48.4 \text{ in.}^3$$

Step 7:   Make the final selection. This required plastic section modulus is more than that provided by the W14×30. Therefore, select the

$$W16 \times 31$$

Step 8:   As shown in Example 6.3, an alternate approach is to use the required moment, $M_a = 121$ ft-kips, and enter the M_n/Ω column to determine the same W-shape.

6.4   DESIGN OF COMPACT LATERALLY UNSUPPORTED WIDE FLANGE BEAMS

6.4.1   Lateral Torsional Buckling

The compression region of a bending member cross section has a tendency to buckle similarly to how a pure compression member buckles. The major difference is that the bending tension region helps to resist that buckling. The upper half of the wide flange member in bending acts as a T in pure compression. This T is fully braced about its horizontal axis by the web so it will not buckle in that direction but it can be unbraced for some distance for buckling about its vertical axis. Thus, it will tend to try to buckle laterally. Because the tension region tends to restrain the lateral buckling, the shape actually buckles in a combined lateral and torsional mode. The beam midspan deflects in the plane down and buckles laterally, causing it to twist, as shown in Figure 6.8. The beam appears to have a tendency to fall over on its weak axis. In order to resist this tendency, the Specification requires that all bending members are restrained at their support points against rotation about their longitudinal axis. If the beam has sufficient lateral and/or torsional support along its length, the cross section can develop the yield stress before buckling. If it tends

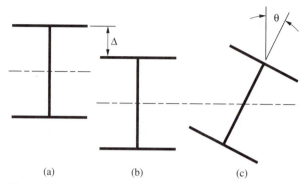

Figure 6.8 The Three Positions of a Beam Cross Section Undergoing Lateral-Torsional Buckling.

to buckle before the yield stress is reached, the nominal moment strength is less than the plastic moment.

To insure that a beam cross section can develop its full plastic moment strength without lateral torsional buckling, Specification Section F2.2, Equation F2-5, limits the slenderness to

$$\frac{L_b}{r_y} \le 1.76\sqrt{\frac{E}{F_y}} \tag{6.7}$$

where

L_b = unbraced length of the compression flange

r_y = radius of gyration for the shape about the y-axis

The practical application of this limitation is to use the unbraced length alone, rather than in combination with the radius of gyration, to form a slenderness ratio. This results in the requirement for attaining the full plastic moment strength that

$$L_b \le L_p = 1.76r_y\sqrt{\frac{E}{F_y}} \tag{6.8}$$

Thus, L_p is the maximum unbraced length that would permit the shape to reach its plastic moment strength. This value is tabulated for each shape and can be found in Manual Table 3-2 and several others.

When the unbraced length of a beam exceeds L_p, its strength is reduced due to the tendency of the member to buckle laterally at a load level below what would cause the plastic moment to be reached.

The elastic lateral torsional buckling (LTB) strength of a W-shape is given in Specification Section F2.2 as

$$M_n = F_{cr}S_x \tag{6.9}$$

where

$$F_{cr} = \frac{C_b\pi^2 E}{\left(\dfrac{L_b}{r_{ts}}\right)^2}\sqrt{1 + 0.078\frac{Jc}{S_x h_o}\left(\frac{L_b}{r_{ts}}\right)^2} \tag{6.10}$$

A beam buckles elastically if the actual stress in the member does not exceed F_y at any point. Because all hot rolled shapes have built-in residual stresses as discussed for columns

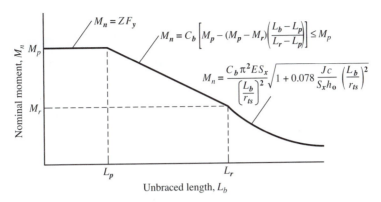

Figure 6.9 Lateral-Torsional Buckling.

in Section 5.3.4, there is a practical limit to the usefulness of this elastic LTB equation. The Specification sets the level of the residual stress at $0.3F_y$ so that only $0.7F_y$ is available to resist a bending moment elastically. This limit results in an elastic moment, $M_{rLTB} = 0.7F_y S_x$. This permits the determination of a limiting unbraced length, L_r, beyond which the member buckles elastically. The limit as provided in Specification Section F2 is

$$L_r = 1.95 r_{ts} \frac{E}{0.7F_y} \sqrt{\frac{Jc}{S_x h_o}} \sqrt{1 + \sqrt{1 + 6.76\left(\frac{0.7F_y}{E} \frac{S_x h_o}{Jc}\right)^2}} \tag{6.11}$$

Between the unbraced lengths L_p and L_r, the beam behaves inelastically. In this range, the nominal moment, M_n, is reasonably well predicted by a straight line equation. The Specification equation for the nominal moment strength, modified to use M_{rLTB}, and taking $C_b = 1$, which is discussed later, is

$$M_n = \left[M_p - (M_p - M_{rLTB})\left(\frac{L_b - L_p}{L_r - L_p}\right) \right] \tag{6.12}$$

Although the determination of F_{cr} and L_r from Equations 6.10 and 6.11 may look somewhat daunting, the Manual has extensive tables that permit their determination with little effort.

The nominal moment strength of a beam as a function of unbraced length is presented in Figure 6.9 where the curve segments are labeled according to the appropriate strength equations. Curves similar to these are available in Manual Table 3-10 for each W-shape and Table 3-11 for C- and MC-shapes. An example of these curves is given in Figure 6.10.

When M_n is to be determined through a calculation, an additional simplification can be applied to the straight line portion of the curve. From Equation 6.12, the ratio $\left(\dfrac{M_p - M_{rLTB}}{L_r - L_p}\right)$ is a constant for each beam shape. This constant is tabulated as BF in Manual Table 3-2, although it is actually given as a design value or an allowable value. Thus, for nominal strength, Equation 6.12 can be rewritten as

$$M_n = M_p - BF(L_b - L_p) \tag{6.13a}$$

and for LRFD as

$$\phi M_n = \phi M_p - BF(L_b - L_p) \tag{6.13b}$$

and for ASD as

$$\frac{M_n}{\Omega} = \frac{M_p}{\Omega} - BF(L_b - L_p) \tag{6.13c}$$

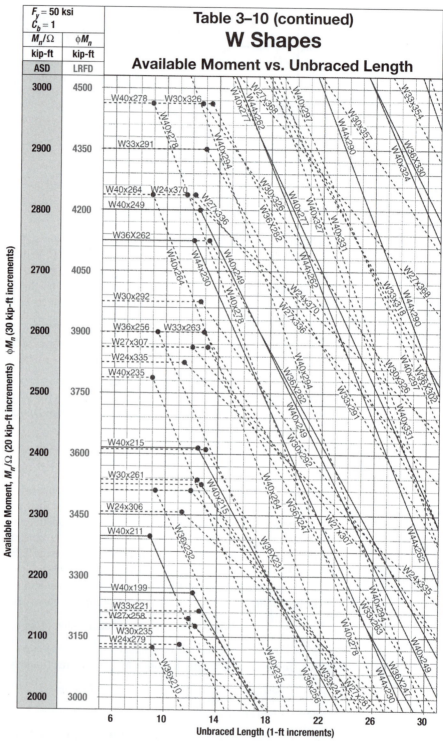

Figure 6.10 W Shapes: Available Moment versus Unbraced Length. Copyright © American Institute of Steel Construction, Inc. Reprinted with Permission. All rights reserved.

6.4.2   Moment Gradient

The nominal strength of a beam as defined in Equations 6.9, 6.12 or 6.13 assumes that the moment is uniform across the entire length of the beam as shown in Figure 6.11a. For lateral torsional buckling, this is the most severe loading case possible, because it would stress the entire length of the beam to its maximum, just as for a column. For any other loading pattern, and resulting moment diagram, the compressive force in the beam would vary with the moment diagram. Thus, the reduced stresses along the member length would result in a reduced tendency for LTB and an increase in strength. The variation in moment over a particular unbraced segment of the beam is called the *moment gradient,* which describes how the moment varies along a specific length.

For the normal case of loading that produces a moment diagram that is not constant, the nominal moment strength calculated through Equations 6.9, 6.12, or 6.13 may be increased to account for the moment gradient such that Equation 6.9 becomes

$$M_n = C_b F_{cr} S_x \qquad (6.14)$$

and Equation 6.12 becomes

$$M_n = C_b \left[M_p - (M_p - M_{rLTB}) \left(\frac{L_b - L_p}{L_r - L_p} \right) \right] \le M_p \qquad (6.15)$$

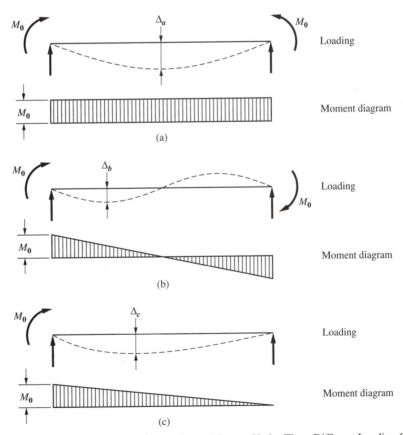

Figure 6.11 Resistance to the Maximum Moment Under Three Different Loading Conditions.

The lateral-torsional buckling modification factor, C_b, accounts for nonuniform moment diagrams over the unsupported length. It is a function of the moment gradient and provided in Specification Section F1 as

$$C_b = \frac{12.5M_{max}}{2.5M_{max} + 3M_A + 4M_B + 3M_C} R_m \le 3.0 \qquad (6.16)$$

where

M_{max} = absolute value of maximum moment in the unbraced segment

M_A = absolute value of moment at quarter point of the unbraced segment

M_B = absolute value of moment at centerline of the unbraced segment

M_C = absolute value of moment at three-quarter point of the unbraced segment

R_m = 1.0 for a doubly symmetric member

$C_b = 1.0$ for a uniform moment and can be conservatively taken as 1.0 for other cases. In doing so, however, the designer may be sacrificing significant economy. Figure 6.12 provides examples of loading conditions, bracing locations, and the corresponding C_b values.

The effect of the moment gradient factor, C_b, is to alter the nominal moment-unbraced length relationship by a constant, as shown in Figure 6.13. The shaded area shows the increase in moment capacity as a result of the use of C_b. Regardless of how small the unbraced length might be, the nominal moment strength of the member can never exceed the plastic moment strength. Thus, the upper portion of the curve in Figure 6.13 is terminated at M_p.

EXAMPLE 6.5a
Beam Strength and Design by LRFD Considering Moment Gradient

GOAL: Determine whether the W14×34 beam shown in Figure 6.14 will carry the given load. Consider the moment gradient, (a) $C_b = 1.0$, (b) C_b from Equation 6.16 and, (c) determine the least weight section to carry the load using the correct C_b.

GIVEN: Figure 6.14 shows a beam that is fixed at one support and pinned at the other. The beam has a concentrated dead load of 8 kips and a concentrated live load of 24 kips at midspan. Assume a lateral brace at both the supports and the load point.

SOLUTION

Step 1: Determine the required strength. For the load combination of 1.2D + 1.6L

$$P_u = 1.2(8.0) + 1.6(24.0) = 48.0 \text{ kips}$$

Step 2: Determine the maximum moment from an elastic analysis at the fixed end. This is given in Figure 6.14 as

$$M_u = 180 \text{ ft-kips}$$

Step 3: Determine the needed values from Manual Table 3-2 in order to use Equation 6.13b. W14×34, $Z = 54.6$ in.^{3}, $L_p = 5.40$ ft, $L_r = 15.6$ ft, $\phi M_p = 205$ ft-kips, and $BF = 7.59$ kips

Part (a) $C_b = 1.0$

Step 4: Determine the design moment strength for lateral bracing of the compression flange at the supports and the load, $L_b = 10$ ft.
Because $L_b = 10$ ft $> L_p$

$$\phi M_n = (\phi M_p - BF(L_b - L_p))$$

$$\phi M_n = (205 - 7.59(10.0 - 5.40)) = 170 \text{ ft-kips} < 180 \text{ ft-kips}$$

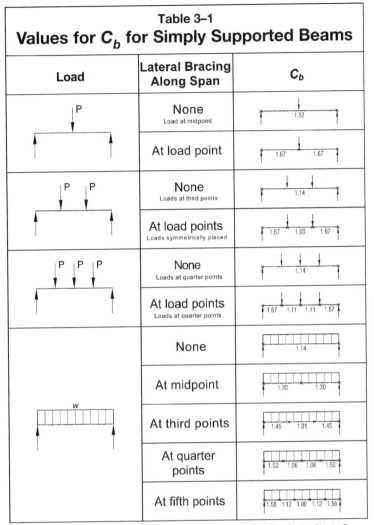

Table 3–1 Values for C_b for Simply Supported Beams		
Load	**Lateral Bracing Along Span**	**C_b**
P (load at midpoint)	None Load at midpoint	1.32
	At load point	1.67 ... 1.67
P P	None Loads at third points	1.14
	At load points Loads symmetrically placed	1.67 1.00 1.67
P P P	None Loads at quarter points	1.14
	At load points Loads at quarter points	1.67 1.11 1.11 1.67
w	None	1.14
	At midpoint	1.30 1.30
	At third points	1.45 1.01 1.45
	At quarter points	1.52 1.06 1.06 1.52
	At fifth points	1.56 1.12 1.00 1.12 1.56

Note: Lateral bracing must always be provided at points of support per AISC Specification Chapter F.

Figure 6.12 Values for C_b for Simply Supported Beams. Copyright © American Institute of Steel Construction, Inc. Reprinted with Permission. All rights reserved.

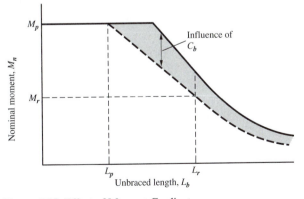

Figure 6.13 Effect of Moment Gradient.

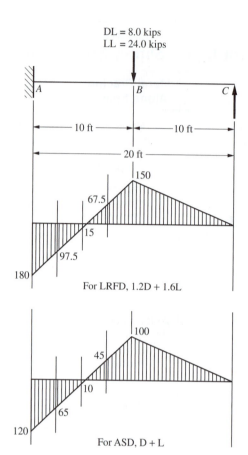

DL = 8.0 kips
LL = 24.0 kips

10 ft

10 ft

20 ft

150

67.5

15

97.5

180

For LRFD, 1.2D + 1.6L

100

45

10

65

120

For ASD, D + L

Figure 6.14 Beam Used in Example 6.5.

As an alternate approach, Manual Table 3-10 can be entered with an unbraced length of 10 ft and the design strength of the W14×34 determined to be 170 ft-kips. Therefore,

> the beam will not work if $C_b = 1.0$

Part (b) Use the calculated value of C_b.

Step 5:   Determine the correct C_b for the two unbraced segments of the beam.

For the unbraced segment BC, Figure 6.12 can be used to obtain $C_b = 1.67$. This C_b corresponds to the maximum moment of 150 ft-kips at point B on the beam. The W14×34 can resist this moment without considering C_b, as shown in Part (a) above.

For the unbraced segment AB, C_b must be calculated. Using Equation 6.16 and the moment values given in Figure 6.13,

$$C_b = \frac{12.5(180)}{2.5(180) + 3(97.5) + 4(15.0) + 3(67.5)} = 2.24$$

Step 6: Determine the design moment strength using the calculated value of C_b and the design moment strength determined from Part (a), Equation 6.13b amplified by C_b and limited to ϕM_p.

$$\phi M_n = C_b\left(\phi M_p - BF(L_b - L_p)\right) \le \phi M_p$$

$$\phi M_n = 2.24(170) = 381 \text{ ft-kips} > M_p = 205 \text{ ft-kips}$$

Therefore, the limiting strength of the beam is

$$\phi M_n = 205 \text{ ft-kips} > 180 \text{ kip-ft, and}$$

the W14×34 is adequate for bending

Part (c) Considering that $C_b = 2.24$, a smaller section can be tried.

Step 7: Assuming $\phi M_n = \phi M_p$, try a W16×31. Determine the needed values from Manual Table 3-2.

$$\phi M_p = 203 \text{ ft-kips}, \quad L_p = 4.13 \text{ ft}, \quad L_r = 11.9 \text{ ft}, \quad BF = 10.2 \text{ kips}$$

Step 8: Because $L_b = 10 \text{ ft} > L_p = 4.13 \text{ ft}$, use Equation 6.13b with C_b.

$$\phi M_n = C_b\left(\phi M_p - BF\left(L_b - L_p\right)\right)$$

$$\phi M_n = 2.24(203 - 10.2(10.0 - 4.13)) = 2.24(143) = 320 \text{ ft-kips}$$

where $\phi M_n = 320 \text{ ft-kips} > \phi M_p = 203 \text{ ft-kips}$
Thus

$$\phi M_n = 203 \text{ kip-ft} > 180 \text{ ft-kips}$$

so the W16×31 will also work

EXAMPLE 6.5b
Beam Strength and Design by ASD Considering Moment Gradient

GOAL: Determine whether the W14×34 beam shown in Figure 6.14 will carry the given load. Consider the moment gradient, (a) $C_b = 1.0$, (b) C_b from Equation 6.16 and, (c) determine the least weight section to carry the load using the correct C_b.

GIVEN: Figure 6.14 shows a beam that is fixed at one support and pinned at the other. The beam has a concentrated dead load of 8 kips and a concentrated live load of 24 kips at midspan. Assume a lateral brace at both the supports and the load point.

SOLUTION

Step 1: Determine the required strength. For the load combination of D + L

$$P_a = (8.0) + (24.0) = 32 \text{ kips}$$

Step 2: Determine the maximum moment from an elastic analysis at the fixed end. This is given in Figure 6.14 as

$$M_a = 120 \text{ ft-kips}$$

Step 3: Determine the needed values from Manual Table 3-2 in order to use Equation 6.13c.

$$W14\times34, \ Z = 54.6 \text{ in.}^3, \ L_p = 5.40 \text{ ft}, \ L_r = 15.6 \text{ ft},$$
$$M_p/\Omega = 136 \text{ ft-kips, and } BF = 5.05 \text{ kips}$$

Part (a) $C_b = 1.0$

Step 4: Determine the allowable moment strength for lateral bracing of the compression flange at the supports and the load, $L_b = 10$ ft.
Because $L_b = 10 \text{ ft} > L_p$

$$\frac{M_n}{\Omega} = \left(\frac{M_p}{\Omega} - BF(L_b - L_p)\right)$$

$$M_n/\Omega = 136 - 5.05(10.0 - 5.40) = 113 \text{ ft-kips} < 120 \text{ kip-ft}$$

As an alternate approach, Manual Table 3-10 can be entered with an unbraced length of 10 ft and the allowable strength of the W14×34 determined to be 113 ft-kips.
Therefore,

the beam will not work if $C_b = 1.0$

Part (b) Use the calculated value of C_b.

Step 5: Determine the correct C_b for the two unbraced segments of the beam.
For the unbraced segment BC, Figure 6.12 can be used to obtain $C_b = 1.67$. This C_b corresponds to the maximum moment of 100 ft-kips at point B on the beam. The W14×34 can resist this moment without considering C_b, as shown in Part (a) above.
For the unbraced segment AB, C_b must be calculated. Using Equation 6.16 and the moment values given in Figure 6.14,

$$C_b = \frac{12.5(120)}{2.5(120) + 3(65.0) + 4(10.0) + 3(45.0)} = 2.24$$

Step 6: Determine the allowable moment strength using the calculated value of C_b and the allowable moment strength determined from Part (a), Equation 6.13c amplified by C_b and limited to M_p/Ω.

$$\frac{M_n}{\Omega} = C_b\left(\frac{M_p}{\Omega} - BF(L_b - L_p)\right) \le \frac{M_p}{\Omega}$$

$$M_p/\Omega = 2.24\,(113) = 253 \text{ ft-kips} > M_p/\Omega = 136 \text{ ft-kips}$$

Therefore, the limiting strength of the beam is

$$M_p/\Omega = 136 \text{ ft-kips} > 120 \text{ kip-ft, and}$$

the W14×34 is adequate for bending

Part (c) Considering that $C_b = 2.24$, a smaller section can be tried.

Step 7:   Assuming $M_n/\Omega = M_p/\Omega$, try a W16×31. Determine the needed values from Manual Table 3-2.

$$M_p/\Omega = 135 \text{ kip-ft}, \ L_p = 4.13 \text{ ft}, \ L_r = 11.9 \text{ ft}, \ BF = 6.76 \text{ kips}$$

Step 8:   Because $L_b = 10 \text{ ft} > L_p = 4.13 \text{ ft}$, use Equation 6.13c with C_b.

$$\frac{M_n}{\Omega} = C_b\left(\frac{M_p}{\Omega} - BF(L_b - L_p)\right) \le \frac{M_p}{\Omega}$$

$$M_p/\Omega = 2.24(135 - 6.76(10.0 - 4.13)) = 2.24(95.3) = 213 \text{ ft-kips}$$

where $M_p/\Omega = 213 \text{ ft-kips} > M_p/\Omega = 135 \text{ ft-kips}$
Thus

$$M_p/\Omega = 135 \text{ kip-ft} > 120 \text{ ft-kips}$$

so the W16×31 will also work

6.5   DESIGN OF NONCOMPACT BEAMS

6.5.1   Local Buckling

Local buckling occurs when a compression element of a cross section buckles under load before it reaches the yield stress. Because this buckling occurs at a stress lower than the yield stress, the shape is not capable of reaching the plastic moment. Thus, the strength of the member is something less than M_p. Buckling of the flange and web elements, and lateral torsional buckling of the section, do not occur in isolation so it is difficult to illustrate them individually. Figure 6.15 primarily illustrates local buckling of the compression flange of a wide flange beam during loading in an experimental test. These failures occur when the flange or web are slender and can be predicted through the use of the plate buckling equation discussed in Chapter 5. The projecting flange of a wide flange member is considered an unstiffened element because the web supports only one edge whereas the other edge is unsupported and free to rotate. The wide flange web is connected at both its ends to the flanges so it is considered a "stiffened" element.

Table B4.1 of the Specification provides the limiting slenderness values, λ_p, for the flange and web in order to insure that the full plastic moment strength can be reached. When both the flange and web meet these criteria, the shapes are called *compact shapes*. If either element does not meet the criteria, the shape cannot be called compact and the nominal strength must be reduced. These shapes are discussed here.

For the flange of a W-shape to be compact, Case 1 in Table B4.1, its width-thickness ratio must satisfy the following limit:

$$\lambda_f = \frac{b}{t} \le \lambda_{pf} = 0.38\sqrt{\frac{E}{F_y}} \tag{6.17}$$

where $b/t = b_f/2t_f$. For the web to be compact, Case 9 in Table B4.1, the limiting ratio is

$$\lambda_w = \frac{h}{t_w} \le \lambda_{pw} = 3.76\sqrt{\frac{E}{F_y}} \tag{6.18}$$

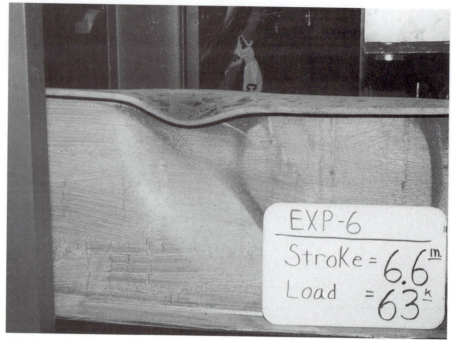

Figure 6.15 Example of Flange Local Buckling.
Photo Courtesy Donald W. White.

Using the common A992 steel with $F_y = 50$ ksi, these limits become:
for a compact flange

$$\frac{b_f}{2t_f} \leq \lambda_{pf} = 9.15$$

and for a compact web

$$\frac{h}{t_w} \leq \lambda_{pw} = 90.6$$

A comparison of these limits with the data given in Manual Table 1-1 shows that the majority of the W-shapes have compact flanges and all have compact webs.

Figure 6.16 illustrates these dimensions for several commonly used sections along with the slenderness limits as found in Specification Table B4.1. Other shapes can be found in Table B4.1 of the Specification.

6.5.2 Flange Local Buckling

The full range of nominal moment strength, M_n, of a cross section can be expressed as a function of flange slenderness, λ_f, and that relationship is shown in Figure 6.17. The three regions in the figure identify three types of behavior. The first region represents plastic behavior, in which the shape is capable of attaining its full plastic moment strength. This strength was discussed in Section 6.2. Shapes that fall into this region are called *compact*. The behavior exhibited in the middle region is inelastic and shapes that fit this category are called *noncompact*. Shapes that fall into the last region exhibit elastic buckling and are called *slender shapes*. The provisions for these last two forms of behavior are given in Specification Section F3.

Case	Description of Element	Width Thickness Ratio	λ_p (compact)	λ_r (noncompact)	Example
1	Flexure in flanges of rolled I-shaped sections and channels	b/t	$0.38\sqrt{E/F_y}$	$1.0\sqrt{E/F_y}$	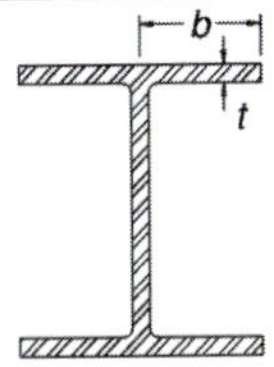
2	Flexure in flanges of doubly and singly symmetric I-shaped built-up sections	b/t	$0.38\sqrt{E/F_y}$	$0.95\sqrt{k_c E/F_L}^{[a],[b]}$	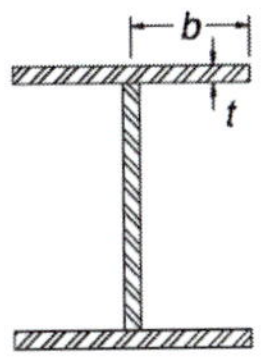
7	Flexure in flanges of tees	b/t	$0.38\sqrt{E/F_y}$	$1.0\sqrt{E/F_y}$	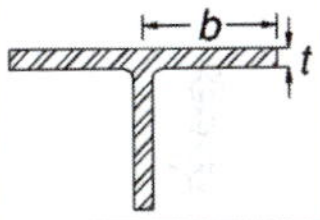
9	Flexure in webs of doubly symmetric I-shaped sections and channels	h/t_w	$3.76\sqrt{E/F_y}$	$5.70\sqrt{E/F_y}$	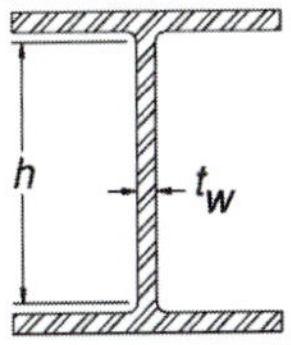
13	Flexure in webs of rectangular HSS	h/t	$2.42\sqrt{E/F_y}$	$5.70\sqrt{E/F_y}$	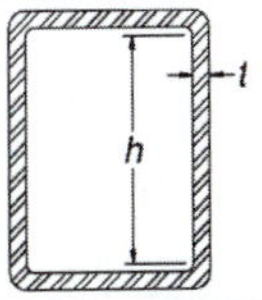

Figure 6.16 Definition of Element Slenderness from Specification Table B4.1 . Copyright © American Institute of Steel Construction, Inc. Reprinted with Permission. All rights reserved.

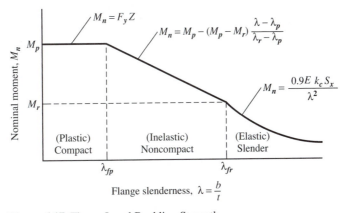

Figure 6.17 Flange Local Buckling Strength.

For I-shaped sections, the dividing line between a compact and noncompact flange was given in Equation 6.17. The division between noncompact and slender flange sections is a function of the residual stresses that exist in the hot rolled member. As was the case with lateral torsional buckling, the Specification assumes that elastic behavior continues up to the point where the elastic moment $M_{rFLB} = 0.7F_y S_x$. This corresponds to a flange slenderness, as found in Specification Table B4.1, of

$$\lambda_{rf} = 1.0\sqrt{\frac{E}{F_y}} \tag{6.19}$$

The strength at the junction of compact and noncompact behavior is

$$M_n = M_p = F_y Z$$

whereas at the junction of the noncompact and slender behavior, the moment is defined as

$$M_{rFLB} = 0.7F_y S_x$$

The strength for noncompact shapes is represented by a straight line between these points. Thus,

$$M_n = \left[M_p - \left(M_p - M_{rFLB}\right)\left(\frac{\lambda - \lambda_{pf}}{\lambda_{rf} - \lambda_{pf}}\right)\right] \tag{6.20}$$

For A992 steel with $F_y = 50$ ksi, Equation 6.19 provides an upper limit to the noncompact flange of

$$\lambda_{rf} = 1.0\sqrt{\frac{E}{F_y}} = 24.1$$

A review of Manual Table 1-1 for $b_f/2t_f$ shows that there are no W-shapes with flanges that exceed this limit. Thus, all wide flanges have either compact or noncompact flanges. A further review of the tables shows that only 10 W-shapes have noncompact flanges.

6.5.3   Web Local Buckling

A comparison of the slenderness criteria for web local buckling given in Equation 6.18 with the data available in Manual Table 1-1 for h/t_w indicates that all wide flange shapes have compact webs. Thus, the consideration of noncompact W-shapes is a consideration of only flange local buckling and there is no need to address slender elements for W-shapes. Slender webs, however, are addressed for built-up members in Chapter 7 as plate girders.

EXAMPLE 6.6
Bending Strength of Noncompact Beam

GOAL: For a W6×15, determine the (a) nominal moment strength, (b) design moment strength (LRFD), and (c) allowable moment strength (ASD).

GIVEN: A simply supported W6×15 spans 10 ft. It is braced at the ends and at the midspan ($L_b = 5$ ft). The steel is A992.

SOLUTION

Part (a) Determine the nominal strength.

Step 1:   Check the limits for flange local buckling.
For the flange, from Manual Table 1-1,

$$\frac{b_f}{2t_f} = 11.5 > \lambda_{pf} = 0.38\sqrt{\frac{E}{F_y}} = 9.15$$

Therefore, the flange is not compact. Checking for a slender flange, even though our previous review of the Manual data indicated that no W-shapes exceeded this requirement

$$\frac{b_f}{2t_f} = 11.5 < \lambda_{rf} = 1.0\sqrt{\frac{E}{F_y}} = 24.1$$

Because $\lambda_{pf} < b_f/2t_f < \lambda_{rf}$, the shape has a noncompact flange.

Step 2: Check the limit states for web local buckling.
For the web, from Manual Table 1-1

$$\frac{h}{t_w} = 21.6 < \lambda_{pw} = 3.76\sqrt{\frac{E}{F_y}} = 90.6$$

So the web is compact, as expected from our earlier evaluation.

Step 3: Because the shape is noncompact (flange), determine the nominal moment strength by Equation 6.20 with

$$M_p = F_y Z_x = 50(10.8) = 540 \text{ in.-kips}$$

$$M_{rFLB} = 0.7F_y S_x = 0.7(50)9.72 = 340 \text{ in.-kips}$$

Thus,

$$M_n = \left[540 - (540 - 340)\left(\frac{11.5 - 9.15}{24.1 - 9.15}\right)\right] = 509 \text{ in.-kips}$$

And, for flange local buckling

$$M_n = \frac{509 \text{ in.-kip}}{12 \text{ in/ft}} = 42.4 \text{ ft-kips}$$

Step 4: Check for the limit state of lateral-torsional buckling.
For this shape,

$$L_p = 1.76r_y\sqrt{\frac{E}{F_y}} = 1.76(1.45)\sqrt{\frac{29,000}{50}} = 61.5 \text{ in.}$$

Thus, $L_p = 5.13$ ft, which is greater than $L_b = 5.0$ ft, so the beam is adequately braced to resist the plastic moment. For lateral torsional buckling

$$M_n = M_p = F_y Z = 540 \text{ in.-kips}$$

or

$$M_n = 540/12 = 45.0 \text{ ft-kips}$$

Step 5: Because the moment based on flange local buckling, 42.4 ft-kips, is less than the moment based on lateral-torsional buckling, 45.0 ft-kips, local buckling controls and

$$\boxed{M_n = 42.4 \text{ ft-kips}}$$

Part (b) For LRFD,

Step 6: Determine the design moment.

$$\boxed{\phi M_n = 0.9(42.4) = 38.2 \text{ kip-ft}}$$

Part (c) For ASD,

Step 7:   Determine the allowable moment.

$$\frac{M_n}{\Omega} = \frac{42.4}{1.67} = 25.4 \text{ ft-kips}$$

6.6   DESIGN OF BEAMS FOR WEAK AXIS BENDING

Up to this point, I-shaped beams have been assumed to be bending about an axis parallel to their flanges, called the x-axis. A quick scan of the shape property tables in the Manual shows that the section modulus and plastic section modulus about the x-axis are larger than the corresponding values about the other orthogonal axis, the y-axis. Thus, bending about the x-axis is called *strong axis bending*, whereas bending about the y-axis is called *weak axis* or *minor axis bending*. Although beams are not normally oriented for bending about this weak axis, a situation may arise when it is necessary to determine the strength of a beam in this orientation.

Design of I-shaped beams for weak axis bending is relatively easy. Section F6 of the Specification applies to I-shaped members and channels bent about their minor axis. Two limit states are identified: yielding and flange local buckling. The flange and web referred to here are the same elements as when the shape is bending about its major axis. Thus, the limits on flange slenderness are the same as discussed earlier. For those few W-shapes with noncompact flanges, an equation similar to that used previously for noncompact flanges is required.

For the limit state of yielding

$$M_n = M_p = F_y Z_y \leq 1.6 F_y S_y$$

An I-shaped member bending about its weak axis has properties close to those of a rectangle. For the rectangle, the ratio of the plastic moment to the elastic yield moment, called the shape factor, equals 1.5. The addition of the web alters the elastic section modulus and plastic section modulus so that the shape factor for these weak axis bending members exceeds 1.5. To insure an appropriate level of rotational capacity at the plastic limit state, the shape factor for weak axis bending is limited to 1.6. All but four W-shapes meet this limitation.

Although I-shaped members are not often called upon to carry moment about the y-axis as pure bending members, they are called upon to participate in combined bending as discussed in Section 6.12 and combined with axial load as discussed in Chapter 8.

6.7   DESIGN OF BEAMS FOR SHEAR

Chapter G of the Specification establishes the requirements for beam shear. Although shear failures are uncommon with rolled sections, a beam can fail by shear yielding or shear buckling. Beam webs also need to be checked for shear rupture on the net area of the web when bolt holes are present. Shear rupture is addressed in the discussion of connections in Chapter 10.

The nominal shear yielding strength is based on the von Mises criterion, which states that for an unreinforced beam web that is stocky enough not to fail by buckling, the shear strength can be taken as $F_y/\sqrt{3} = 0.58F_y$. The specification rounds this stress to $0.6F_y$ and provides, in Specification Section G2, the shear strength as

$$V_n = 0.6F_y A_w C_v \tag{6.21}$$

where A_w is the area of the web, taken as the total depth times the web thickness.

The web shear coefficient, C_v, is used to account for shear web buckling. Thus, if the web is capable of reaching yield, $C_v = 1.0$. To insure that the beam web is capable of reaching yield before buckling, the Specification sets the limit on web slenderness of

$$\frac{h}{t_w} \leq 1.10\sqrt{\frac{k_v E}{F_y}}$$

where $k_v = 5$ for unstiffened webs with $h/t_w < 260$. All current ASTM A6 rolled I-shaped members have webs that meet the criteria for $k_v = 5$, and all A992 W-shapes meet the criteria for web yielding.

Thus, the nominal shear strength of a rolled W-shape can be taken as

$$V_n = 0.6F_y A_w \tag{6.22}$$

Determining the shear design strength or allowable strength is complicated by a variation in resistance and safety factors. To keep the beam shear strength provisions the same in the 2005 Specification as in earlier allowable stress specifications, the resistance and safety factors for a particular set of rolled I-shapes was liberalized. Thus, for webs of rolled I-shapes with $h/t_w \leq 2.24\sqrt{E/F_y}$

$$\phi = 1.0\,(\text{LRFD}) \qquad \Omega = 1.5\,(\text{ASD})$$

For all other shapes,

$$\phi = 0.9\,(\text{LRFD}) \qquad \Omega = 1.67\,(\text{ASD})$$

Because shear rarely controls the design of rolled beams, it may be more convenient, when not using tables from the Manual, to simply use the more conservative factors for all shear checks.

6.8   CONTINUOUS BEAMS

Beams that span over more than two supports are called *continuous beams*. Unlike simple beams, continuous beams are indeterminate and must be analyzed by applying more than the three basic equations of equilibrium. Although indeterminate analysis is not within the scope of this book, a few topics should be addressed, even if only briefly.

The Manual includes shears, moments, and deflections for several continuous beams with various uniform load patterns in Table 3-23. These results come from an elastic indeterminate analysis and can be used for the design of any beams that fit the support and loading conditions.

It has long been known that material ductility permits steel members to redistribute load. When one section of a member becomes overloaded, it can redistribute a portion of its load to a less highly loaded section. This redistribution can be accounted for through an analysis method called Plastic Analysis or a number of more modern methods capable of modeling the real behavior of the members. These methods may collectively be called *Advanced Analysis* and used in structural steel design by the provisions of Appendix 1. This appendix also provides a simplified approach to account for some of this ductility through Appendix 1.3.

Design of beams and girders that are compact and have sufficiently braced compression flanges may take advantage of this simplified redistribution approach. The compact criteria are those already discussed, whereas the unbraced length criteria are a bit more restrictive. To use the simplified redistribution, the unbraced length of the compression flange, L_b, must be less than that given in Appendix Section 1.7 as

$$L_{pd} = \left[0.12 + 0.076\left(\frac{M_1}{M_2}\right)\right]\left(\frac{E}{F_y}\right)r_y \qquad (6.23)$$

When these criteria are satisfied, the beam can be proportioned for 0.9 times the negative moments at points of support. This redistribution is permitted only for gravity-loading cases and moments determined through an elastic analysis. When this reduction in negative moment is used, the positive moment must be increased to maintain equilibrium. This can be accomplished simply by adding to the maximum positive moment, 0.1 times the average original negative moments.

EXAMPLE 6.7a
Continuous Beam Design by LRFD

GOAL: Select a compact, fully braced section for use as a continuous beam.

GIVEN: The beam must be continuous over three spans of 30 ft each. It must support a live load of 2.5 kip/ft and a dead load of 1.8 kip/ft. Use A992 steel.

SOLUTION

Step 1: Determine the required strength.

The design load is $w_u = 1.2(1.8) + 1.6(2.5) = 6.16$ kip/ft

From the beam shear, moment, and deflection diagrams in Manual Table 3-23, Case 39, the negative moment is

$$-M_{BA} = 0.100wl^2 = 0.100(6.16)(30)^2 = 554 \text{ ft-kips}$$

and the positive moment is

$$+M_{BA} = 0.0800wl^2 = 0.0800(6.16)(30)^2 = 444 \text{ ft-kips}$$

Step 2: Consider redistribution of moments according to Specification Appendix 1.

A design could be carried out for a maximum moment of 554 ft-kips but with redistribution, this moment may be reduced to

$$M_{BA} = 0.9(554) = 499 \text{ ft-kips}$$

provided that the positive moment is increased by the average negative moment reduction. Thus,

$$M_{AB} = 444 + \frac{0 + 0.1(554)}{2} = 472 \text{ ft-kips}$$

Step 3: Determine the required plastic section modulus.

Even with the increase in positive moment, the negative moment is still the maximum moment so, for a moment of 499 ft-kips,

$$Z_{req} = \frac{499(12)}{0.9(50)} = 133 \text{ in.}^3$$

Step 4: Select the least-weight W-shape from Manual Table 3-2.

Select a W24×55, with $Z = 134$ in.^{3}

EXAMPLE 6.7b *Continuous Beam* *Design by ASD*	**GOAL:** Select a compact, fully braced section for use as a continuous beam. **GIVEN:** The beam must be continuous over three spans of 30 ft each. It must support a live load of 2.5 kip/ft and a dead load of 1.8 kip/ft. Use A992 steel.
SOLUTION	**Step 1:** Determine the required strength. The design load is $w_a = (1.8) + (2.5) = 4.3$ kip/ft From the beam shear, moment, and deflection diagrams in Manual Table 3-23, Case 39, the negative moment is $$-M_{BA} = 0.100wl^2 = 0.100(4.3)(30)^2 = 387 \text{ ft-kips}$$ and the positive moment is $$+M_{BA} = 0.0800wl^2 = 0.0800(4.3)(30)^2 = 310 \text{ ft-kips}$$ **Step 2:** Consider redistribution of moments according to Specification Appendix 1. A design could be carried out for a maximum moment of 387 ft-kips but with redistribution, this moment may be reduced to $$M_{BA} = 0.9(387) = 348 \text{ ft-kips}$$ provided that the positive moment is increased by the average negative moment reduction. Thus, $$M_{AB} = 310 + \frac{0 + 0.1(387)}{2} = 329 \text{ ft-kips}$$ **Step 3:** Determine the required plastic section modulus. Even with this increase in the positive moment, the negative moment is still the maximum moment so, for a moment of 348 ft-kips $$Z_{req} = \frac{348(12)}{(50/1.67)} = 139 \text{ in.}^3$$ **Step 4:** Select the least-weight W-shape from Manual Table 3-2. > Select a W21×62, with $Z = 144$ in.^{3}

6.9   PLASTIC ANALYSIS AND DESIGN OF CONTINUOUS BEAMS

Up to this point, it has been assumed that the plastic moment strength of a bending member could be compared to the maximum elastic moment on a beam to satisfy the strength requirements of the Specification. This is accurate for determinate members in which the occurrence of the plastic moment at the single point of maximum moment results in the development of a single plastic hinge, which would lead to member failure. However, for indeterminate structures, such as continuous beams, more than one plastic hinge must form before the beam would actually collapse and this provides some additional capacity that the elastic analysis cannot capture. The formation of plastic hinges in the appropriate locations causes a collapse and the geometry of this collapse is called a *failure* or *collapse mechanism*. This is the approach referred to as *Plastic Analysis* that is permitted by Appendix 1 of the Specification for use with LRFD only.

The formation of a beam failure mechanism may best be understood by following the load history of a fixed-ended beam with a uniformly distributed load. The beam and

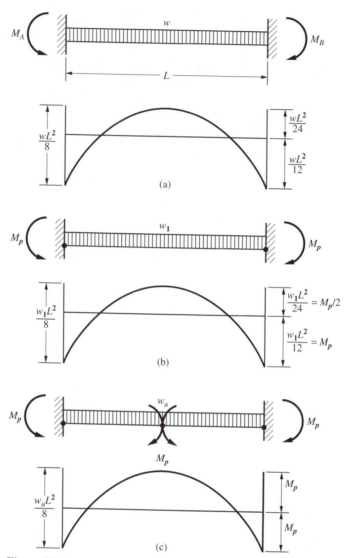

Figure 6.18 Beam and Moment Diagrams for the Development of a Plastic Mechanism.

moment diagrams that result from an elastic indeterminate analysis are given in Figure 6.18a. The largest moments occur at the fixed ends and are given by $wL^2/12$. If the load on the beam is increased, the beam behaves elastically until the moments on the ends equal the plastic moment strength of the member, as shown in Figure 6.18b. Because the application of additional load causes the member to rotate at its ends while maintaining the plastic moment, these points behave as pins. These pins are called *plastic hinges*. In this case, the load is designated as w_1. The member can continue to accept load beyond this w_1, functioning as a simple beam, until a third plastic hinge forms at the beam centerline. The formation of this third hinge makes the beam unstable, thus forming the collapse mechanism. The mechanism and corresponding moment diagram are given in Figure 6.18c.

For the collapse mechanism just described, equilibrium requires that the simple beam moment, $w_u L^2/8$, equal twice the plastic moment, thus

$$M_p = w_u L^2/16 \qquad (6.24)$$

Had this beam been designed based on an elastic analysis, it would have required a moment capacity greater than or equal to $w_u L^2 / 12$. Using a plastic analysis, a smaller plastic moment strength, equal to $w_u L^2 / 16$, must be provided for in the design. Thus, in this case of an indeterminate beam, plastic analysis has the potential to result in a smaller member being required to carry this same load.

An additional advantage to the use of plastic analysis for indeterminate beams is the simplicity of the analysis. By observation, regardless of the overall geometry of the continuous beam, each segment between supports can be evaluated independently of each other segment. This means that any beam segment, continuous at each end and loaded with a uniformly distributed load, exhibits the same collapse mechanism. Thus, the relation between the applied load and the plastic moment will be as given in Equation 6.24. Plastic analysis results for additional loading and beam configurations are given in Figure 6.19. Additional examples, as well as the development of these relations through application of energy principles, can be found in several textbooks including *Applied Plastic Design in Steel*. [2]

To insure that a given beam cross section can undergo the necessary rotation at each plastic hinge, the Specification requires that the section be compact and that the compression flange be braced such that the unbraced length in the area of the hinge is less than that already given as L_{pd} in Equation 6.23. If this limit is not satisfied, the member design must be based on an elastic analysis.

EXAMPLE 6.8
Beam Design with Plastic Analysis (LRFD only)

GOAL: Design a beam using plastic analysis and A992 steel. Plastic analysis is applicable only for LRFD load combinations.

GIVEN: A beam is simply supported at one end and fixed at the other, similar to that shown in Figure 6.14 for Example 6-5. It spans 20 ft and is loaded at its centerline with a dead load of 16.0 kips and a live load of 48.0 kips. Lateral support is provided at the ends and at each ¼ point of the span. It is assumed the final section will be compact and adequately braced.

SOLUTION

Step 1: Determine the required strength.

$$P_u = 1.2(16.0) + 1.6(48.0) = 96.0 \text{ kips}$$

Using the plastic analysis results from Figure 6.19c

$$\phi M_{preq} = \frac{Pab}{(a + 2b)} = \frac{96.0(10.0)(10.0)}{(10.0 + 2(10.0))} = 320 \text{ ft-kips}$$

Step 2: Select the required W-shape from Manual Table 3-2.

$$W21 \times 44, \quad \phi M_p = 358 \text{ ft-kips}$$

Step 3: Calculate the maximum permitted unbraced length through Equation 6.23. From Manual Table 1-1, $r_y = 1.26$ in.

$$L_{pd} = \left[0.12 + 0.076 \left(\frac{M_1}{M_2} \right) \right] \left(\frac{E}{F_y} \right) r_y$$

$$= \left[0.12 + 0.076 \left(\frac{0}{320} \right) \right] \left(\frac{29,000}{50} \right) (1.26) = 87.7 \text{ in.}$$

---

[2] Disque, R. O. *Applied Design in Steel*. New York: Van Nostrand Reinhold Company, 1971.

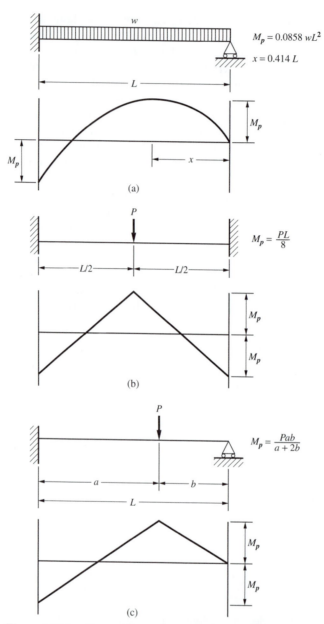

Figure 6.19 Loading and Beam Configurations Resulting from Plastic Analysis.

Step 4: Check the initial assumptions on compactness and lateral bracing.

Because the provided unbraced length equals 5 ft or 60 in., and this is less than the maximum permitted unbraced length of 87.7 in., the bracing of this W21×44 is acceptable for plastic design. A check of the compact flange and web criteria show that this shape is compact.

Thus,

use a W21×44

6.10   PROVISIONS FOR DOUBLE-ANGLE AND TEE MEMBERS

Provisions for beams formed by combining a pair of angles to form a T, and beams made from a T that has been cut from an I-shape, are found in Section F9 of the Specification. These provisions are specifically for these singly symmetric members loaded in the plane of symmetry with the stem either in tension or compression. Three limit states must be considered in the design of these T-shaped members: yielding, lateral-torsional buckling, and flange local buckling. The stem has no local buckling provisions because, when the stem is in compression, the section is limited to an elastic stress distribution.

6.10.1   Yielding

For the limit state of yielding

$$M_n = M_p = F_y Z_x$$

and M_p is limited, depending on the orientation of the section. For the stem in tension

$$M_p \leq 1.6 M_y = 1.6 F_y S_x$$

and for the stem in compression

$$M_p \leq M_y = F_y S_x$$

These limits are necessary to insure that the member is capable of rotating sufficiently to attain the plastic moment strength without the extreme fibers of the shape reaching the strain-hardening region.

6.10.2   Lateral-Torsional Buckling

Lateral-torsional buckling also must account for the orientation of the shape. The nominal strength is given by

$$M_n = M_{cr} = \frac{\pi \sqrt{E I_y G J}}{L_b} \left[B + \sqrt{1 + B^2} \right] \tag{6.25}$$

where

$$B = \pm 2.3 \left(\frac{d}{L_b} \right) \sqrt{\frac{I_y}{J}}$$

The plus sign for B applies when the stem is in tension, the more stable orientation for lateral torsional buckling. The negative sign for B is to be used if any portion of the stem is in compression along the span.

6.10.3   Flange-Local Buckling

The limit state of flange-local buckling for these shapes reflects the same behavior as for the I-shapes already considered. In fact, the limiting width/thickness ratios are the same as discussed earlier. For compact flanges, the limit state of flange local buckling does not apply. For noncompact flanges, $\lambda_p < \lambda \leq \lambda_r$

$$M_n = F_y S_{xc} \left(1.19 - 0.50 \left(\frac{b_f}{2 t_f} \right) \sqrt{\frac{F_y}{E}} \right) \tag{6.26}$$

and for slender flanges $\lambda_r < \lambda$

$$M_n = \frac{0.69 E S_{xc}}{\left(\dfrac{b_f}{2t_f}\right)^2} \tag{6.27}$$

where S_{xc} is the section modulus referred to the compression flange. If the stem is in compression, this limit state does not apply.

EXAMPLE 6.9
Bending Strength of
WT-shape

GOAL: Determine the nominal moment strength for the given WT member if the stem is in (a) tension (Figure 6.20a) and (b) compression (Figure 6.20b).

GIVEN: A WT9×17.5 is used as a beam and has lateral support provided at 5-ft intervals.

SOLUTION

Step 1:   Determine the section properties for the WT-shapes from Manual Table 1-8.

$$Z_x = 11.2 \text{ in.}^3, \quad S_x = 6.21 \text{ in.}^3, \quad d = 8.85 \text{ in.}, \quad I_y = 7.67 \text{ in.}^4, \quad J = 0.252 \text{ in.}^4$$

Part (a) Determine the nominal moment strength for the stem in tension. The WT is oriented as shown in Figure 6.20a.

Step 2:   Determine the nominal moment strength for the limit state of yielding.

$$M_p = F_y Z_x = 50(11.2) = 560 \text{ in.-kips}$$

and

$$M_y = F_y S_x = 50(6.21) = 311 \text{ in.-kips}$$

but the strength is limited by

$$M_n \leq 1.6 M_y = 1.6(311) = 498 \text{ in.-kips}$$

Thus,

$$M_n = 498 \text{ in.-kips for the limit state of yielding}$$

Step 2:   Determine the nominal moment strength for the limit state of lateral-torsional buckling. Determine B from Equation F9-5.

$$B = \pm 2.3 \left(\frac{d}{L_b}\right)\sqrt{\frac{I_y}{J}} = 2.3\left(\frac{8.85}{60.0}\right)\sqrt{\frac{7.67}{0.252}} = \pm 1.87$$

B is taken as positive for the stem in tension so that Equation 6.25 becomes

$$M_n = M_{cr} = \frac{\pi\sqrt{29,000(7.67)(11,200)(0.252)}}{60.0}\left[1.87 + \sqrt{1 + 1.87^2}\right] = 5240 \text{ in.-kips}$$

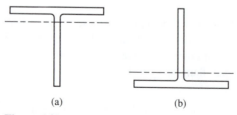

(a)   (b)

Figure 6.20 T-Beam Orientation for Example 6.9.

and

$$M_n = \frac{5240}{12} = 437 \text{ ft-kips}$$

Step 3: Consider the limit state of flange local buckling.

The limit state of flange local buckling does not apply to the WT9×17.5 because the flange is compact, so the nominal moment strength of this WT is the smaller strength given by the limit states of yielding and lateral-torsional buckling. Thus,

$$\boxed{M_n = 437 \text{ ft-kips}}$$

Part (b) Determine the nominal moment strength for the stem in compression. The WT is oriented as shown in Figure 6.20b.

Step 4: Determine the nominal moment strength for the limit state of yielding.

M_p is limited to M_y so that, from above

$$M_n = M_y = 311 \text{ in.-kips}$$

Step 5: Determine the nominal moment strength for the limit state of lateral-torsional buckling.

B is taken as negative so that

$$M_n = M_{cr} = \frac{\pi \sqrt{29,000(7.67)(11,200)(0.252)}}{60.0}\left[-1.87 + \sqrt{1 + (-1.87)^2}\right] = 329 \text{ in.-kips}$$

Step 6: Determine the controlling limit state strength for the WT with the stem in compression.

$$M_n = \frac{311}{12} = 25.9 \text{ ft-kips}$$

Note: This example shows that using a WT-shape with the stem in compression significantly penalizes the strength of the member. Even so, beams with this orientation are often easier to construct, such as lintels in masonry walls where WTs are used in this orientation.

6.11 SINGLE-ANGLE BENDING MEMBERS

When single angles are used as bending members, they can be bending about one of the geometric axis, parallel to the legs, or about the principal axes. They are often used as lintels over openings in masonry walls where they are bending about the geometric axes. Unfortunately, this most useful orientation of the single-angle bending member is also the most complex orientation for the determination of strength. Figure 6.21a shows a single angle oriented for bending about the geometric axis whereas Figure 6.21b shows the angle oriented for bending about the principal axis.

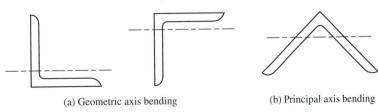

(a) Geometric axis bending   (b) Principal axis bending

Figure 6.21 Single-Angle Bending About Geometric Axis and Principal Axis.

Specification Section F10 gives the provisions for single-angle bending members. The limit states to be checked for these members are yielding, lateral torsional buckling, and leg local buckling. For the treatment here, only fully braced angles bending about a geometric axis are discussed.

6.11.1 Yielding

The ratio of the plastic section modulus to the elastic section modulus for angles can easily exceed 1.5. Thus, in order to be sure that the angle is not strained into the strain hardening region, the nominal moment for the limit state of yielding is taken as

$$M_n = 1.5M_y = 1.5F_yS$$

where S is taken as the least section modulus about the axis of bending.

6.11.2 Leg Local Buckling

Legs of angles in compression have the same tendency to buckle as other compression elements. Specification Table B4-1 defines the slenderness as b/t, in Case 6, as

$$\lambda_p = 0.54\sqrt{\frac{E}{F_y}}$$

and

$$\lambda_r = 0.91\sqrt{\frac{E}{F_y}}$$

The strength of noncompact and slender angles is shown in Figure 6.22. In the region of noncompact behavior, the straight-line transition is given in the Specification as Equation F10-7

$$M_n = F_yS_c\left(2.43 - 1.72\left(\frac{b}{t}\right)\sqrt{\frac{F_y}{E}}\right)$$

and the elastic buckling strength is given as

$$M_n = \frac{0.71ES_c}{\left(\dfrac{b}{t}\right)^2}$$

where S_c is the elastic section modulus to the toe in compression, relative to the axis of bending.

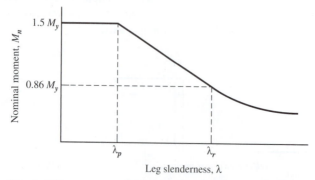

Figure 6.22 Strength of a Single Angle in a Function of Leg Slenderness.

6.11.3   Lateral-Torsional Buckling

The limit state of lateral torsional buckling depends on whether the toe of the angle is in tension or compression. It is also a function of the axes of bending. The Specification gives the strength provisions for four different bending orientations. You are encouraged to study these provisions and to work toward orienting the sections and providing lateral restraint so that the angle can carry its greatest bending moment.

6.12   MEMBERS IN BIAXIAL BENDING

Bending members are often called upon to resist forces that result in bending about two orthogonal axes. Examples of this member type are crane girders and roof purlins in industrial buildings. Regardless of the actual orientation of an applied moment, it is possible to brake the moment into components about the two principal axes, as shown in Figure 6.23. Once this is accomplished, the ability of the section to resist the combined moments can be determined through the interaction equation.

The Specification, Chapter H, addresses the interaction of forces. For the combination of moments, a simple linear interaction equation is used, as shown in Figure 6.24. This is taken from the equation provided in Specification Section H1 for combined axial load and moment. When the axial load is zero, Equation H1-1b reduces to

$$\frac{M_{rx}}{M_{cx}} + \frac{M_{ry}}{M_{cy}} \leq 1.0$$

where the moment terms relate to the x- and y-axes, the numerator is the required strength, and the denominator is the available strength, determined as though the member was bending about only one axis at a time. Thus, if the required x-axis moment is 79% of the x-axis strength, only 21% of the y-axis strength is available to resist moment. More attention is given to the use of interaction equations when axial load is combined with the bending moment in Chapter 8.

6.13   SERVICEABILITY CRITERIA FOR BEAMS

There are several serviceability considerations that the designer must address. A general set of provisions are found in Specification Chapter L. Although failure to satisfy these criteria may not impact the strength of the member or overall structure, it may lead to the

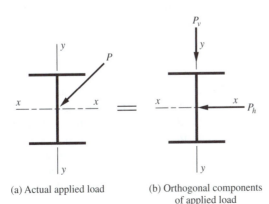

(a) Actual applied load   (b) Orthogonal components
                                      of applied load

Figure 6.23 Biaxial Bending of I-Shaped Beam.

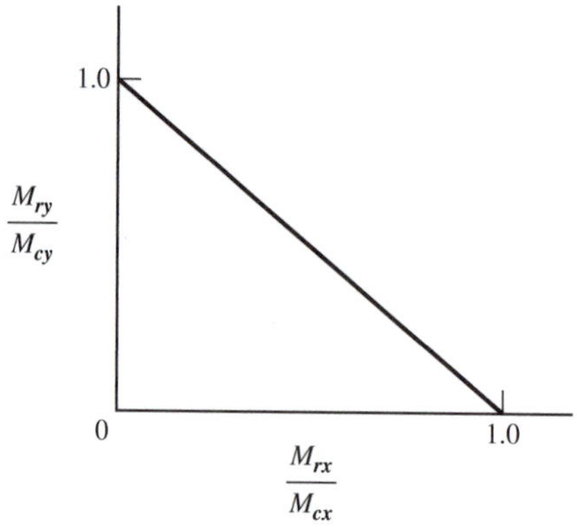

Figure 6.24 Simple Linear Interaction Diagram for Biaxial Bending.

first signs of difficulty for successful completion of a project. The specific criteria should be discussed in detail with the designer's client so the quality of the final product is consistent with the expectations of the owner. Experience may indicate that a certain amount of floor vibration may be annoying at first but occupants become used to it with time. The client may be unwilling to deal with this period of dissatisfaction and insist that the system be designed so that there are no vibration complaints. This must be known at the beginning of a project, not after the occupants move in and find the floor response objectionable. The engineer must be sure to identify these considerations for the owner so that the decisions made are appropriate to meet the expected outcome.

Beams generally have three serviceability issues to be addressed:

6.13.1   Deflection

Deflection is the normal response of a beam to its imposed load. It is impossible to erect a beam with zero deflection under load but the designer will be able to limit that deflection with proper attention to this limit state. Deflections must be addressed for a variety of loading cases. Deflection under dead load is critical because it impacts the construction process, including the amount of concrete fill needed to form a flat and level floor. Live load deflection is critical because it impacts the finishes of elements attached to the floor, such as ceilings and walls, and may be visible to the occupants. Experience has demonstrated that live load deflection is not a problem if it is limited to $1/360$ of the span. Dead load deflection limitations are a function of the particular structural element and loading. *Design Guide 3—Serviceability Design Considerations for Steel Buildings* from the American Institute of Steel Construction covers deflection and other serviceability design criteria.

6.13.2   Vibration

Although vibration of floor systems is not a safety consideration, it can be a very annoying response and very difficult to correct after the building is erected. The most common problem is with wide-open spaces with very little damping, such as the jewelry department in a department store. To reduce the risk of annoyance, a general rule is to space the beams or joists sufficiently far apart so that the slab thickness is large enough to provide the needed stiffness and damping. *Design Guide 11—Floor Vibrations Due to Human Activity* from the American Institute of Steel Construction covers the design of steel-framed floor systems for human comfort.

6.13.3 Drift

Under lateral loading, a building will sway sideways. This lateral displacement is called *drift*. As with deflection and vibration, drift is usually not a safety consideration but it can be annoying and have a negative impact on nonstructural elements, causing cracks in finishes. Beams and girders are important in reducing the drift and their final size might actually be determined by drift considerations. However, the impact of drift considerations on beams cannot be determined for the beams alone without also looking at the other parts of the lateral load resisting system. This serviceability limit state is treated in Chapter 8. Drift is also discussed in *Design Guide 3*.

Because beam deflection is a serviceability consideration, calculations are carried out using the specific loads under which the serviceability consideration is are to be checked. This can be live load, deal load, or some combination of loads, but normally does not include any load factors. Thus, regardless of whether a design is completed using LRFD or ASD, serviceability considerations are checked for the same loads. Numerous elastic analysis techniques are available to determine the maximum deflection of a given beam and loading. Some common loading conditions with their corresponding maximum deflection are shown in Figure 6.25. These and many others are given in Manual Table 3-23.

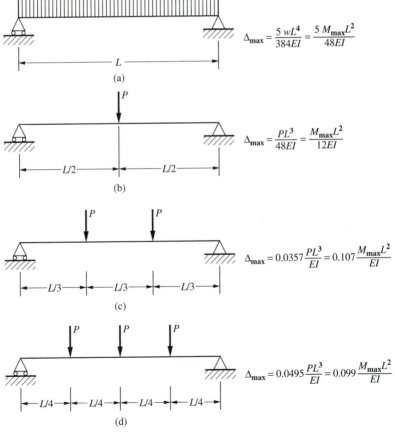

Figure 6.25 Some Common Loading Conditions with their Corresponding Maximum Deflections.

EXAMPLE 6.10
Live Load Deflection

GOAL: Check the live load deflection of a previously designed beam.

GIVEN: Use the information from Example 6.3 where a W18×35 was selected. Limit the live load deflection for an acceptable design to $^1/_{360}$ of the span.

SOLUTION

Step 1:   Collect the required information from Example 6.3.
For the W18×35, $I = 510$ in.^{4}. The live load is 24 kips applied at the center of a 20-ft span.

Step 2:   Determine the live load deflection. Using the deflection equation found in Figure 6.25 for case (b).

$$\Delta = \frac{PL^3}{48EI} = \frac{24(20.0)^3(12)^3}{48(29,000)(510)} = 0.467 \text{ in.}$$

Step 3:   Compare the calculated deflection to the given limit.
The deflection limit is

$$\Delta_{max} = \frac{20.0(12)}{360} = 0.667 \text{ in.}$$

Because

$$\boxed{\Delta = 0.467 \text{ in.} < \Delta_{max} = 0.667 \text{ in.}}$$
the deflection satisfies the set criteria

EXAMPLE 6.11
Beam Design through Deflection Limit

GOAL: Select a W-shape to satisfy a live load deflection limit.

GIVEN: Use the data from Example 6.10 except that the deflection limit is set to a more severe level of $^1/_{1000}$ of the span. If the selected member does not meet the established criteria, select a W-shape that satisfies the limitation.

SOLUTION

Step 1:   Check the new deflection limit.

$$\Delta_{max} = \frac{20.0(12)}{1000} = 0.240 \text{ in.}$$

From Example 6.10 we know already that the given beam deflects too much.

Step 2:   Determine the minimum acceptable moment of inertia, I_{min}, to insure that the deflection does not exceed the given limit.
Rearranging the maximum deflection equation to solve for I_{min}

$$I_{min} = \frac{PL^3}{48E\,\Delta_{max}} = \frac{24(20.0)^3(12)^3}{48(29,000)(0.240)} = 993 \text{ in.}^4$$

Step 3:   Select a beam with $I \geq 993$ in.^{4} and, from Example 6.3, one that satisfies the strength limit, $Z \geq 64.0$ in.^{3}.
From the moment of inertia tables, Manual Table 3-3, select a

$$\boxed{\text{W21×55, with } I = 1140 \text{ in.}^4 \text{ and } Z = 126 \text{ in.}^3}$$

This is the lightest-weight W-shape that will satisfy the required moment of inertia.

6.14   CONCENTRATED FORCES ON BEAMS

Before a beam can be called upon to carry a given load, that load must be transferred to the beam through some type of connection. In a similar manner, the beam reactions must be carried to its supporting structure through some type of connection. Although the majority of beams are loaded through connections to their webs, some may be loaded by applying a concentrated force to the top flange and some will have their reactions resisted by bearing on a supporting element. In these cases, a check must be made to establish that the beam web has sufficient strength to resist the applied forces.

Four limit states determine the load carrying strength of the web to resist these concentrated forces: web local yielding, web crippling, web sidesway buckling, and flange local bending. These limit states are all described in Section J10 of the Specification. Although it is possible to select a beam with a sufficiently thick web so that these limit states do not control, it is normally more economical to add bearing stiffeners under the concentrated loads to provide the necessary strength. The design of these stiffeners is covered in Section 7.4 of this book under the discussion of plate girders, because they are much more commonly found in that application.

6.15   PROBLEMS

1. Determine the plastic section modulus for a W44×335 modeled as three rectangles forming the flanges and the web. Compare the calculated value to that given in the Manual.

2. Determine the plastic section modulus for a W36×800 modeled as three rectangles forming the flanges and the web. Compare the calculated value to that given in the Manual.

3. Determine the plastic section modulus for a W33×118 modeled as three rectangles forming the flanges and the web. Compare the calculated value to that given in the Manual.

4. Determine the plastic section modulus for a W21×44 modeled as three rectangles forming the flanges and the web. Compare the calculated value to that given in the Manual.

5. Determine the plastic section modulus for a W18×50 modeled as three rectangles forming the flanges and the web. Compare the calculated value to that given in the Manual.

6. Determine the elastic neutral axis, elastic section modulus, plastic neutral axis, and plastic section modulus for a WT15×66 modeled as two rectangles forming the flange and the stem. Compare the calculated values to those given in the Manual.

7. Determine the plastic section modulus for an HSS8×4×$\frac{1}{2}$ modeled as four rectangles forming the flanges and webs. Remember to use the design wall thickness for the plate thickness and ignore the corner radius. Compare the calculated value to that given in the Manual.

8. Determine the elastic neutral axis, elastic section modulus, plastic neutral axis, and plastic section modulus for a C15×50 modeled as three rectangles forming the flanges and the web. Compare the calculated values to those given in the Manual.

9. Determine the elastic neutral axis, elastic section modulus, plastic neutral axis, and plastic section modulus, all about

the geometric axis, for a L4×4×$\frac{1}{2}$ modeled as two rectangles. Compare the calculated values to those given in the Manual.

10. A simply supported beam spans 20 ft and carries a uniformly distributed dead load of 0.8 kip/ft including the beam self-weight and a live load of 2.3 kip/ft. Determine the minimum required plastic section modulus and select the lightest-weight W-shape to carry the moment. Assume full lateral support and A992 steel. Design by (a) LRFD and (b) ASD.

11. Considering only bending, determine the lightest-weight W-shape to carry a uniform dead load of 1.2 kip/ft including the beam self-weight and a live load of 3.2 kip/ft on a simple span of 24 ft. Assume full lateral support and A992 steel. Design by (a) LRFD and (b) ASD.

12. A beam is required to carry a uniform dead load of 0.85 kip/ft including its self-weight and a concentrated live load of 12 kips at the center of a 30-ft span. Considering only bending, determine the least-weight W-shape to carry the load. Assume full lateral support and A992 steel. Design by (a) LRFD and (b) ASD.

13. Considering both shear and bending, determine the lightest-weight W-shape to carry the following loads: a uniform dead load of 0.6 kip/ft plus self-weight, a concentrated dead load of 2.1 kips, and a concentrated live load of 6.4 kips, located at the center of a 16-ft span. Assume full lateral support and A992 steel. Design by (a) LRFD and (b) ASD.

14. Considering both shear and bending, determine the lightest W-shape to carry a uniform dead load of 4.0 kip/ft plus the self-weight and a uniform live load of 2.3 kip/ft on a simple span of 10.0 ft. Assume full lateral support and A992 steel. Design by (a) LRFD and (b) ASD.

15. A 24-ft simple span laterally supported beam is required to carry a total uniformly distributed service load of 8.0 k/ft. Determine the lightest, A992, W-shape to carry this load if it is broken down as follows: Use LRFD.

    a. Live load = 1.0 k/ft; dead load = 7.0 k/ft

    b. Live load = 3.0 k/ft; dead load = 5.0 k/ft

    c. Live load = 5.0 k/ft; dead load = 3.0 k/ft

    d. Live load = 7.0 k/ft; dead load = 1.0 k/ft.

16. Repeat the designs specified in Problem 15 using ASD.

17. A 30-ft simply supported beam is loaded at the third points of the span with concentrated dead loads of 12.0 kips and live loads of 18.0 kips. Lateral supports are provided at the load points and the supports. The self-weight of the beam can be ignored. Determine the least weight W-shape to carry the load. Use A992 steel and assume $C_b = 1.0$. Design by (a) LRFD and (b) ASD.

18. An 18-ft simple span beam is loaded with a uniform dead load of 1.4 kip/ft, including the beam self-weight and a uniform live load of 2.3 kip/ft. The lateral supports are located at 6.0-ft intervals. Determine the least weight W-shape to carry the load. Use A992 steel. Design by (a) LRFD and (b) ASD.

19. An A992 W18×60 is used on a 36-ft simple span to carry a uniformly distributed load. Determine the locations of lateral supports in order to provide just enough strength to carry (a) a design moment of 435 ft-kips and (b) an allowable moment of 290 ft-kips.

20. A girder that carries a uniformly distributed dead load of 1.7 k/ft plus its self-weight and three 15-kip concentrated live loads at the quarter points of the 36-ft span is to be sized. Using A992 steel, determine the lightest W-shape to carry the load with lateral supports provided at the supports and load points. Limit deflection to $^1/_{360}$ of span. Design by (a) LRFD and (b) ASD.

21. A 32-ft simple span beam carries a uniform dead load of 2.3 k/ft plus its self-weight and a uniform live load of 3.1 k/ft. The beam is laterally supported at the supports only. Determine the minimum weight W-shape to carry the load using A992 steel. Limit live load deflection to $^1/_{360}$ of span and check shear strength. Design by (a) LRFD and (b) ASD.

22. A 36-ft simple span beam carries a uniformly distributed dead load of 3.4 kip/ft plus its self-weight and a uniformly distributed live load of 2.4 kip/ft. Determine the least-weight W-shape to carry the load while limiting the live load deflection to $^1/_{360}$ of the span. Use A992 steel and assume full lateral support. Design by (a) LRFD and (b) ASD.

23. A simple span beam with a uniformly distributed dead load of 1.1 k/ft, including the self-weight and concentrated dead loads of 3.4 kips and live loads of 6.0 kips at the third points of a 24-ft span, is to be designed with lateral supports at the third points and live load deflection limited to $^1/_{360}$ of the span. Be sure to check shear. Determine the least-weight W-shape to carry the loads. Use A992 steel. Design by (a) LRFD and (b) ASD.

24. A fixed-ended beam on a 28-ft span is required to carry a total ultimate uniformly distributed load of 32.0 kips. Using plastic analysis and A992 steel, determine the design moment and select the lightest W-shape. Assume (a) full lateral support and (b) lateral support at the ends and center line.

25. A beam is fixed at one support and simply supported at the other. A concentrated ultimate load of 32.0 kips is applied at the center of the 40-ft span. Using plastic analysis and A992 steel, determine the lightest W-shape to carry the load when the nominal depth of the beam is limited to 18 in. Assume (a) full lateral support and (b) lateral supports at the ends and the load.

26. A fixed-ended beam on a 40-ft. span is required to carry a total ultimate uniformly distributed load of 72.5 kips. Using plastic analysis and A992 steel, determine the lightest-weight W-shape to carry the load. Assume full lateral support.

27. A 3-span continuous beam is to be selected to carry a uniformly distributed dead load of 4.7 kip/ft, including its self-weight and a uniformly distributed live load of 10.5 kip/ft. Be sure to check the shear strength of this beam. Use A992 steel and assume full lateral support. Design by (a) LRFD and (b) ASD.

28. Determine the available bending strength of a WT8×25, A992 steel, if the stem is in compression. Determine by (a) LRFD and (b) ASD.

Chapter 7

United States Courthouse, Seattle.
Photo courtesy Michael Dickter/Magnusson Klemencic
Associates.

Plate Girders

7.1 BACKGROUND

A plate girder is a bending member composed of individual steel plates. Although they are normally the member of choice for situations where the available rolled shapes are not large enough to carry the intended load, there is no requirement that they will always be at the large end of the spectrum of member sizes. Beams fabricated from individual steel plates to meet a specific requirement are generally identified in the field as *plate girders*.

Plate girders are used in building structures for special situations such as very long spans or very large loads. Perhaps their most common application is as a transfer girder, a bending member that supports a structure above and permits the column spacing to be changed below. They are also very common in industrial structures for use as crane girders and as support for large pieces of equipment. In commercial buildings, they are often used to span large open areas to meet particular architectural requirements and, because of their normally greater depth and resulting stiffness, they tend to deflect less than other potential long span solutions. An example of a building application of the plate girder is shown in Figure 7.1.

The cross section of a typical plate girder is shown in Figure 7.2. Although it is possible to combine steel plates into numerous geometries, the plate girders addressed here are those formed from three plates, one for the web and two for the flanges. Because the web and flanges of the plate girder are fabricated from individual plates, they can be designed with

181

Figure 7.1 Application of a Plate Girder.

web and flanges from the same grade of steel (a homogeneous plate girder) or from different grades of steel (a hybrid plate girder). For hybrid girders, the flanges are usually fabricated with a higher grade of steel than that used in the web. This takes advantage of the higher stresses that can be developed in the flanges, which are located a greater distance from the neutral axis than the web, resulting in a higher moment strength contribution. Hybrid girders are relatively common in bridge construction, though they are rarely used in buildings. In the past, hybrid girders have been included in the AISC Specification but they are not specifically addressed in this edition. Thus, they are not discussed in this book.

Another type of plate girder is the singly symmetric girder, one with flanges that are not of the same size, as seen in Figure 7.3. Although singly symmetric plate girders are addressed in the AISC Specification, they are not particularly common in buildings and are not specifically addressed here. However, the principles for all of these plate girders are the same and the careful application of the Specification provisions will lead to an economical and safe design for each of them.

Built-up plate girders with compact webs are designed according to the same provisions as rolled I-shaped members presented in Specification Sections F2 and F3 and discussed in Chapter 6 in this book. The discussion of plate girders in this chapter addresses these built-up I-shapes with noncompact or slender webs.

Table 7.1 lists the sections of the Specification and parts of the Manual discussed in this chapter.

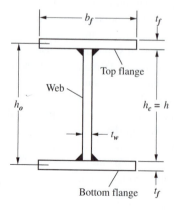

Figure 7.2 Typical Plate Girder Definitions.

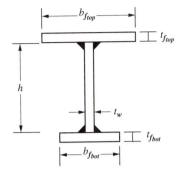

Figure 7.3 Singly Symmetric Plate Girder.

7.2   HOMOGENEOUS PLATE GIRDERS IN BENDING

The behavior of plate girders can best be understood by considering flexure and shear separately. In flexure, a plate girder is considered in this book as either noncompact or slender according to the proportions of the web. Flanges can be compact, noncompact, or slender and the moment strength for these limit states is the same as that discussed in Chapter 6. Thus, it is possible, for example, for a noncompact web plate girder to have a slender flange, potentially controlling the capacity of the member. The design rules for each type of girder are considered separately.

With our discussion limited to doubly symmetric plate girders, the limit states that must be considered are compression flange yielding, compression flange local buckling, web local buckling, and lateral torsional buckling. These are the same limit states considered for the rolled I-shapes in Chapter 6 with the addition of web local buckling. The additional limit state of tension flange yielding can be ignored because the compression flange always controls over the tension flange in these doubly symmetric members. Plate girders with noncompact webs are addressed in Specification Section F4 and those with slender webs in Section F5. The nominal strength of plate girders, for all limit states, can be described through the use of Figure 7.4. As with the rolled I-shaped members discussed in Chapter 6, the behavior is plastic, inelastic, or elastic. Figure 7.4 shows that the plastic behavior corresponds to an area of the figure described as compact. Inelastic behavior corresponds to the area identified as noncompact and elastic behavior corresponds to the area identified as slender.

Applying the same figure to the lateral torsional buckling limit state, the fully braced region corresponds to plastic behavior, whereas the partially braced region corresponds to either inelastic or elastic buckling.

The flexural design strength (LRFD) and allowable strength (ASD) are determined just as they were for the flexural members discussed in Chapter 6. Thus

$$\phi = 0.9 \, (\text{LRFD}) \qquad \Omega = 1.67 \, (\text{ASD})$$

Table 7.1   Sections of Specification and Parts of Manual Found in This Chapter

	Specification
B4	Classification of Sections for Local Buckling
F4	Other I-Shaped Members with Compact or Noncompact Webs, Bent about their Major Axis
F5	Doubly Symmetric and Singly Symmetric I-Shaped Members with Slender Webs Bent about their Major Axis
G	Design of Members for Shear
J4.4	Strength of Elements in Compression
J10	Flanges and Webs with Concentrated Forces

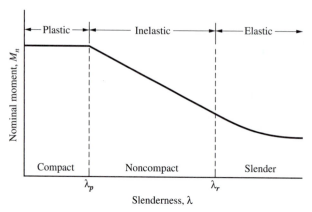

Figure 7.4 Plate Girder Nominal Flexural Strength.

and for ASD, the allowable strength is

$$R_a \le \frac{R_n}{\Omega} \tag{1.1}$$

For LRFD, the design strength is

$$R_u \le \phi R_n \tag{1.2}$$

7.2.1   Noncompact Web Plate Girders

The influence of web slenderness on the strength of plate girders is not treated as a separate limit state to be assessed through its own set of requirements. Rather, web slenderness comes into play as it influences the flange yielding or flange local buckling strength and the lateral-torsional buckling strength as provided in Specification Section F4.

The slenderness parameter of the web is defined as $\lambda_w = h_c/t_w$. For the doubly symmetric plate girder, Case 9 in Specification Table B4.1, this can be simplified to $\lambda_w = h/t_w$, where h is the distance between the flanges. For a plate girder to be noncompact, the following requirements are set:

$$\lambda_p < \lambda_w \le \lambda_r$$

where, according to Table B4.1 of the Specification

$$\lambda_p = 3.76\sqrt{\frac{E}{F_y}}$$

and

$$\lambda_r = 5.70\sqrt{\frac{E}{F_y}}$$

The influence of the noncompact web is characterized through the web plastification factor, R_{pc}. This factor is used to assess the ability of the section to reach its full plastic capacity and modifies the flange yielding or local buckling and lateral torsional buckling limit states. The web plastification factor is given in the Specification as Equation F4.9 and is shown here as

$$R_{pc} = \left[\frac{M_p}{M_{yc}} - \left(\frac{M_p}{M_{yc}} - 1 \right) \left(\frac{\lambda - \lambda_{pw}}{\lambda_{rw} - \lambda_{pw}} \right) \right] \le \frac{M_p}{M_{yc}} \tag{7.1}$$

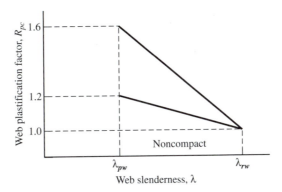

Figure 7.5 Web Plastification Factor.

where

$$M_p = F_y Z_x \leq 1.6 F_y S_x$$

and

$$M_{yc} = F_y S_x$$

Equation 7.1 is shown in Figure 7.5 for two values of M_p/M_{yc}, one with a maximum value of 1.6 to account for the limit on M_p, and one at 1.2. The ratio M_p/M_{yc} is the shape factor that was discussed in Chapter 6. As was the case in that discussion, it must be limited to 1.6 in order to insure that the necessary rotation can take place before strain hardening occurs as the section undergoes plastic deformation. The minimum R_{pc} is seen to be 1.0, regardless of M_p/M_{yc}. Thus, a conservative approach would be to take $R_{pc} = 1.0$. Because a plate girder with a web that is only slightly noncompact would have significant additional strength reflected through the use of R_{pc} and the calculation of R_{pc} is not particularly difficult, there is no advantage to this simplification. Thus, Equation 7.1 will be used throughout this chapter, as appropriate.

The strength of a plate girder as a function of flange slenderness is shown in Figure 7.6. For a compact flange girder, the impact of the noncompact web is to modify the capacity by the factor R_{pc}, as given in Equation 7.2.

$$M_n = R_{pc} F_y S_x \tag{7.2}$$

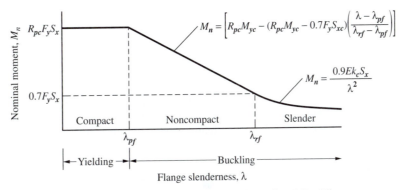

Figure 7.6 Nominal Flexural Strength Based on Flange Local Buckling.

At the lower limit for a noncompact web, λ_{pw}

$$R_{pc} = \frac{M_p}{M_{yc}}$$

and

$$M_n = \frac{M_p}{M_{yc}} F_y S_x = M_p$$

At the upper limit of a noncompact web, λ_{rw}, $R_{pc} = 1.0$ and the nominal strength becomes

$$M_n = 1.0 F_y S_x = M_{yc} \tag{7.3}$$

At the juncture between the noncompact and slender flange, λ_{rf}, the nominal moment strength is given as

$$M_n = 0.7 F_y S_x \tag{7.4}$$

where the $0.7F_y$ accounts for the residual stresses in the member. This is the same residual stress assumed to have occurred in the hot rolled shapes, even though this is a welded shape.

The moment strength for a noncompact flange plate girder is found through the linear interpolation between the end points as shown in Figure 7.6 and is given by Equation 7.5.

$$M_n = \left[R_{pc} M_{yc} - \left(R_{pc} M_{yc} - 0.7 F_y S_{xc} \right) \left(\frac{\lambda - \lambda_{pf}}{\lambda_{rf} - \lambda_{pf}} \right) \right] \tag{7.5}$$

The influence of the noncompact web is diminished as the flange becomes more and more noncompact. Once the flange becomes slender, R_{pc} is no longer needed and flange local buckling controls the strength of the girder. Behavior of the slender flange girder is an elastic buckling phenomena depicted in Figure 7.6 and given as

$$M_n = \frac{0.9 E k_c S_x}{\lambda^2} \tag{7.6}$$

where the plate buckling factor is

$$k_c = \frac{4}{\sqrt{h/t_w}}$$

This plate buckling factor must be taken in the range from 0.35 to 0.76.

Lateral torsional buckling behavior for noncompact web plate girders is, in principle, the same as for rolled beams. However, the equations in the Specification are slightly altered and the web plastification factor must be included. As was the case when considering the noncompact flange, the influence of the noncompact web is diminished as the lateral torsional buckling response becomes more dominant. Figure 7.7 shows the strength of a noncompact web plate girder when considering lateral torsional buckling.

For a plate girder with lateral supports at a spacing no greater than L_p, Equation 7.2, which includes the influence of R_{pc}, again defines the girder strength. This girder would be considered to have full lateral support. The definition of this limiting unbraced length is slightly different than it was for a compact web member. This difference is slight but has been used in the Specification because it gives more accurate results when used for singly symmetric girders.

$$L_p = 1.1 r_t \sqrt{\frac{E}{F_y}}$$

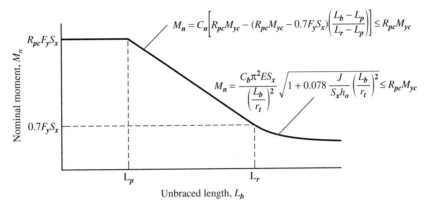

Figure 7.7 Nominal Flexural Strength Based on Unbraced Length.

For I-shapes with a rectangular compression flange, the effective radius of gyration for lateral-torsional buckling is

$$r_t = \frac{b_{fc}}{\sqrt{12\left(\dfrac{h_o}{d} + \dfrac{1}{6}a_w\dfrac{h^2}{h_o d}\right)}}$$

This can conservatively be taken as the radius of gyration of the compression flange plus one-third the compression portion of the web.

The strength of a section undergoing elastic lateral-torsional buckling can be obtained through plate buckling theory. The Specification gives

$$M_n = \frac{C_b \pi^2 E S_x}{\left(\dfrac{L_b}{r_t}\right)^2}\sqrt{1 + 0.078\frac{J}{S_x h_o}\left(\frac{L_b}{r_t}\right)^2} \le R_{pc}M_{yc} \tag{7.7}$$

Because residual stresses occur in plate girders just as they do for a rolled shape, elastic buckling cannot occur if the residual stress pushes the actual stress on the shape beyond the yield stress. With the residual stress taken as $0.3F_y$, the available elastic stress is again taken as $0.7F_y$. Thus, Equation 7.4 again gives the limiting strength, this time for elastic lateral-torsional buckling. Using this strength with Equation 7.7, the unbraced length that defines the limit of elastic lateral-torsional buckling is obtained as

$$L_r = 1.95r_t\frac{E}{0.7F_y}\sqrt{\frac{J}{S_x h_o}}\sqrt{1 + \sqrt{1 + 6.76\left(\frac{0.7F_y}{E}\frac{S_x h_o}{J}\right)^2}} \tag{7.8}$$

Equation 7.8 differs slightly from the corresponding equation in the Specification because it has been modified here to reflect only the doubly symmetric girders considered in this chapter.

The lateral-torsional buckling strength when the member has an unbraced length between L_p and L_r is given by the same straight-line equation as used previously. This time, however, it accounts for the noncompact web by including R_{pc} at the upper limit. Thus

$$M_n = C_b\left[R_{pc}M_{yc} - (R_{pc}M_{yc} - 0.7F_yS_x)\left(\frac{L_b - L_p}{L_r - L_p}\right)\right] \le R_{pc}M_{yc} \tag{7.9}$$

7.2.2   Slender Web Plate Girders

Slender web plate girders are covered in Specification Section F5. They are those built-up members with web slenderness, $\lambda_w = \dfrac{h}{t_w}$, exceeding the limit

$$\lambda_{rw} = 5.70\sqrt{\frac{E}{F_y}}$$

as given in Table B4.1.

As was the case for noncompact web members, web slenderness is not a limit state to evaluate on its own. Its impact on member strength is characterized through the bending strength reduction factor, R_{pg}. The bending strength reduction factor is given by Equation 7.10 and shown in Figure 7.8.

$$R_{pg} = 1 - \frac{a_w}{1200 + 300a_w}\left(\frac{h}{t_w} - 5.7\sqrt{\frac{E}{F_y}}\right) \leq 1.0 \tag{7.10}$$

where

$$a_w = \frac{A_w}{A_f}$$

A_w = area of the web

A_f = area of flange

The Specification places limits on the proportions of members that can be designed according to its provisions. The web-to-flange area ratio, a_w, is limited to 10 to prevent the designer from using these provisions for members that are essentially webs with small stiffeners. In addition, the web slenderness is limited so that $h/t_w \leq 260$. This insures that the girder is not so slender that the stated provisions do not properly reflect its behavior.

Equation 7.10 shows h_c from the Specification equation replaced by h in order to take advantage of the doubly symmetric limitations of this discussion. The bending strength reduction factor reduces the strength of the girder uniformly in all ranges of flange compactness and lateral bracing. Thus, it could simply be held as a final reduction, as shown in the Specification, or used within the primary equations as shown below. Because it is uniformly applied to all of the limit states, the influence of a slender web on flexural strength is easily visualized.

The strength of a plate girder as a function of flange slenderness is shown in Figure 7.9. The slenderness parameters for the flange are defined in Table B4.1 and are the same as those used in Chapter 6 and Section 7.2.1. For a compact flange plate girder, $\lambda \leq \lambda_p$

$$M_n = R_{pg}F_yS_x \tag{7.11}$$

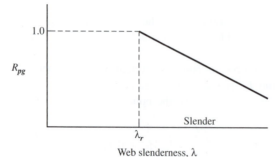

Figure 7.8 Bending Strength Reduction Factor.

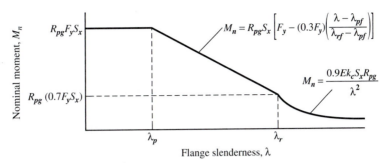

Figure 7.9 Nominal Flexural Strength Based on Flange Slenderness.

Because R_{pg} will not exceed 1.0, the bending strength of the slender web girder is limited to the yield moment.

At the juncture between the noncompact and slender flange, λ_{rf}, the strength is limited to elastic behavior, after accounting for residual stresses. Thus

$$M_n = R_{pg}(0.7F_yS_x) \tag{7.12}$$

The moment capacity for the slender web-noncompact flange plate girder is given by the linear transition

$$M_n = R_{pg}S_x\left[F_y - (0.3F_y)\left(\frac{\lambda - \lambda_{pf}}{\lambda_{rf} - \lambda_{pf}}\right)\right] \tag{7.13}$$

For the slender web-slender flange member, the strength is the same as it was for the noncompact web-slender flange member and given in Equation 7-6, except for the use of R_{pg}. Thus

$$M_n = \frac{0.9Ek_cS_xR_{pg}}{\lambda^2} \tag{7.14}$$

The plate buckling factor, k_c, is as previously defined.

Lateral torsional buckling for the slender web girder appears to be quite similar to the noncompact web girder and is shown in Figure 7.10. However, some differences must be noted. For a member to be considered as having full lateral support, its unbraced length is

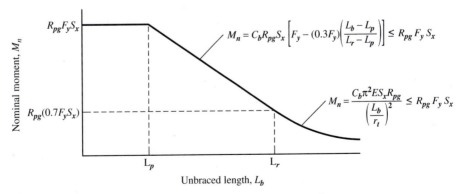

Figure 7.10 Nominal Flexural Strength Based on Unbraced Length.

limited to a spacing not greater than L_p, where

$$L_p = 1.1 r_t \sqrt{\frac{E}{F_y}}$$

This is the same limit used for noncompact web girders but is different than that used for compact web members.

The elastic lateral-torsional buckling strength of the slender web girder is given in the Specification as

$$M_n = \frac{C_b \pi^2 E S_x R_{pg}}{\left(\frac{L_b}{r_t}\right)^2} \leq R_{pg} F_y S_x \tag{7.15}$$

When this strength is set equal to the corresponding strength limit given by Equation 7.12, the limiting unbraced length, L_r, becomes

$$L_r = \pi r_t \sqrt{\frac{E}{0.7 F_y}}$$

This limit for elastic lateral-torsional buckling is not the same as was used for the noncompact web girder. Thus, L_r is different for the three types of plate girder: compact web, noncompact web, and slender web.

For the inelastic lateral torsional buckling region, the strength is given by a linear equation similar to those used previously, with the addition of the R_{pg} multiplier; thus

$$M_n = C_b R_{pg} S_x \left[F_y - (0.3 F_y) \left(\frac{L_b - L_p}{L_r - L_p} \right) \right] \leq R_{pg} F_y S_x \tag{7.16}$$

EXAMPLE 7.1
Plate Girder Flexural
Strength

GOAL: Determine the available flexural strength for two different plate girder designs, a web thickness of (a) $^3/_8$ in. (noncompact) and (b) $^1/_4$ in. (slender).

GIVEN: The cross section of a homogeneous, A36, plate girder is shown in Figure 7.11. The span is 120 ft and the unbraced length of the compression flange is 20 ft. Assume $C_b = 1.0$.

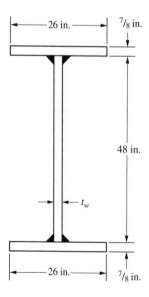

Figure 7.11 Plate Girder for Example 7.1.

SOLUTION

Step 1: Determine the section properties for both plate girders.

(a) $^3/_8$ in. web
$t_w = 0.375$ in.
$A = 63.5$ in.^{2}
$I_x = 30,600$ in.^{4}
$I_y = 2560$ in.^{4}
$S_x = 1230$ in.^{3}
$Z_x = 1330$ in.^{3}
$r_t = 7.13$ in.
$r_y = 6.35$ in.

(b) $^1/_4$ in. web
$t_w = 0.25$ in.
$A = 57.5$ in.^{2}
$I_x = 29,500$ in.^{4}
$I_y = 2560$ in.^{4}
$S_x = 1190$ in.^{3}
$Z_x = 1260$ in.^{3}
$r_t = 7.27$ in.
$r_y = 6.67$ in.

Part (a) For the plate girder with a $^3/_8$-in. web plate

Step 2: Check the web slenderness in order to determine which sections of the Specification must be followed.

$$\lambda_w = \frac{h_c}{t_w} = \frac{48}{0.375} = 128 > \lambda_p = 3.76\sqrt{\frac{E}{F_y}} = 107$$

$$< \lambda_r = 5.7\sqrt{\frac{E}{F_y}} = 162$$

Thus, this is a noncompact web girder and the provisions of Section F4 must be followed. The web plastification factor must be determined.

Step 3: Determine the shape factor and calculate R_{pc}.

$$\frac{M_p}{M_{yc}} = \frac{Z}{S} = \frac{1330}{1230} = 1.08 < 1.6$$

Therefore, use 1.08 in the calculation of R_{pc}, Equation 7.1.

$$R_{pc} = \left[1.08 - (1.08 - 1.0)\left(\frac{128-107}{162-107}\right)\right] = 1.05 \le 1.08$$

Step 4: Determine the nominal bending strength for compression flange yielding.

$$M_n = R_{pc}F_yS_x$$
$$= 1.05(36)(1230) = 46,500 \text{ in.-kips}$$
$$= \frac{46,500}{12} = 3880 \text{ ft-kips}$$

Step 5: Check the unbraced length for lateral torsional buckling, with $L_b = 20$ ft or 240 in.

$$L_p = 1.1(7.13)\sqrt{\frac{29,000}{36}} = 223 \text{ in.}$$
$$= \frac{223}{12} = 18.6 \text{ ft}$$

and

$$L_r = 1.95(7.13)\left(\frac{29,000}{0.7(36)}\right)\sqrt{\frac{12.5}{1230(48.9)}}\sqrt{1+\sqrt{1+6.76\left(\frac{0.7(36)(1230)(48.9)}{29,000(12.5)}\right)^2}}$$
$$= 796 \text{ in.} = \frac{796}{12} = 66.3 \text{ ft}$$

where

$$J = \sum \frac{1}{3}bt^3 = 2\left(\frac{26(0.875)^3}{3}\right) + \frac{48(0.375)^3}{3} = 12.5 \text{ in.}^4$$

Step 6:   Determine the nominal strength based on lateral-torsional buckling.

Because the unbraced length is between L_p and L_r, the straight line equation, Equation 7.9, is used. Thus

$$M_n = \left[3880 - \left(3880 - \frac{0.7(36)(1230)}{12} \right) \left(\frac{20.0 - 18.6}{66.3 - 18.6} \right) \right] = 3840 \text{ ft-kips}$$

Step 7:   Check for the limit state of compression flange local buckling.

$$\lambda_f = \frac{b_f}{2t_f} = \frac{26}{2(0.875)} = 14.9$$

and the limits are

$$\lambda_p = 0.38 \sqrt{\frac{29,000}{36}} = 10.8$$

and, with $k_c = \dfrac{4}{\sqrt{48/0.375}} = 0.354$

$$\lambda_r = 0.95 \sqrt{\frac{0.354(29,000)}{36}} = 16.0$$

Step 8:   Determine the nominal moment strength for the limit state of flange local buckling.

Because $\lambda_p < \lambda_f < \lambda_r$, the shape has noncompact flanges so that, from Equation 7.5

$$M_n = \left[3880 - \left(3880 - \frac{0.7(36)(1230)}{12} \right) \left(\frac{14.9 - 10.8}{16.0 - 10.8} \right) \right] = 2860 \text{ ft-kips}$$

Step 9:   Determine the lowest available moment for the limit states checked.

For compression flange local buckling

Part (a) for LRFD

$$M_u = 0.9(2860) = 2570 \text{ ft-kips}$$

Part (a) for ASD

$$M_a = \frac{2860}{1.67} = 1710 \text{ ft-kips}$$

Part (b) For the plate girder with a $\frac{1}{4}$-in. web plate

Step 10:   Check the web slenderness in order to determine which sections of the Specification must be followed.

$$\lambda_w = \frac{h_c}{t_w} = \frac{48}{0.25} = 192 > \lambda_r = 162$$

Therefore, this is a slender web plate girder and the provisions of Section F5 must be followed.

Step 11:   Determine the bending strength reduction factor, Equation 7.10.

$$a_w = \frac{12.0}{22.8} = 0.526$$

$$R_{pg} = 1 - \frac{0.526}{1200 + 300(0.526)}\left(\frac{48.0}{0.250} - 5.7\sqrt{\frac{29,000}{36}}\right) = 0.988$$

Step 12:   Determine the nominal moment strength for the limit state of yielding.

$$M_n = \frac{R_{pg}F_yS_x}{12} = \frac{0.988(36)(1190)}{12} = 3530 \text{ ft-kips}$$

Step 13:   Check the unbraced length for the limit state of lateral torsional buckling.

$$L_b = 20 \text{ ft}$$

$$L_p = 1.1(7.27)\sqrt{\frac{29,000}{36}} = 227 \text{ in.}$$

$$= \frac{227}{12} = 19.0 \text{ ft}$$

and

$$L_r = \pi(7.27)\sqrt{\frac{29,000}{0.7(36)}} = 775 \text{ in.}$$

$$= \frac{775}{12} = 64.6 \text{ ft}$$

Step 14:   Determine the nominal moment strength for the limit state of lateral torsional buckling. Because the unbraced length is between L_p and L_r, the nominal moment for lateral torsional buckling, Equation 7.16, is

$$M_n = 0.988(1190)\left[36 - 0.3(36)\left(\frac{20.0 - 19.0}{64.6 - 19.0}\right)\right]\left(\frac{1}{12}\right) = 3500 \text{ ft-kips}$$

Step 15:   Check compression flange local buckling.
For compression flange local buckling, the flange slenderness is the same as it was for Part (a) of this problem, $\lambda_f = 14.9$, and the compact flange limit is also the same, $\lambda_p = 10.8$. However, the limiting flange slenderness for the noncompact flange is different because it is influenced by the web thickness through k_c. For the $\frac{1}{4}$-in. web plate

$$k_c = \frac{4}{\sqrt{48/0.250}} = 0.289 < 0.35$$

therefore

$$k_c = 0.350$$

and

$$\lambda_r = 0.95\sqrt{\frac{0.350(29,000)}{36}} = 16.0$$

Step 16:   Determine the nominal moment strength for flange local buckling, Equation 7.13.

$$M_n = \left(\frac{0.988(1190)}{12}\right)\left(36 - 0.3(36)\left(\frac{14.9 - 10.8}{16.0 - 10.8}\right)\right) = 2690 \text{ ft-kips}$$

Step 17:   Determine the lowest nominal moment for the limit states checked.
For compression flange local buckling

Part (b) for LRFD

$$M_u = 0.9(2690) = 2420 \text{ ft-kips}$$

Part (b) for ASD

$$M_a = \frac{2690}{1.67} = 1610 \text{ ft-kips}$$

EXAMPLE 7.2
Plate Girder
Flexural Strength

GOAL: Determine the available moment strength for an A572 Gr. 50 I-shaped built-up member.

GIVEN: The girder is shown in Figure 7.12. It has lateral supports for the compression flange at 8-ft intervals. $r_y = 2.81$ in.

SOLUTION

Step 1:   Check the web slenderness in order to determine which sections of the Specification must be followed.

$$\lambda_w = \frac{h_c}{t_w} = \frac{20}{0.375} = 53.3 < \lambda_p = 3.76\sqrt{\frac{E}{F_y}} = 90.6$$

Therefore, the girder is a compact web girder and should be designed according to Section F3. These are the provisions that were discussed in Chapter 6.

Step 2:   Check the unbraced length limits for lateral-torsional buckling.

$$L_b = 8.0 \text{ ft}$$

$$L_p = 1.76r_y\sqrt{\frac{E}{F_y}} = 1.76(2.81)\sqrt{\frac{29,000}{50}} = 119 \text{ in.}$$

$$= \frac{119}{12} = 9.92 \text{ ft} > L_b = 8 \text{ ft}$$

Therefore, lateral torsional buckling is not a factor.

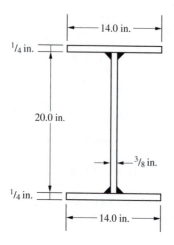

Figure 7.12 Plate Girder for Example 7.2.

Step 3:   Check the slenderness limits for flange local buckling.

$$\lambda_f = \frac{b}{t} = \frac{7.0}{0.250} = 28.0$$

The limiting slenderness is

$$\lambda_p = 0.38\sqrt{\frac{29{,}000}{50}} = 9.15$$

and with

$$k_c = \frac{4}{\sqrt{20.0/0.375}} = 0.548$$

$$\lambda_r = 0.95\sqrt{\frac{0.548(29{,}000)}{50}} = 16.9$$

Step 4:   Determine the nominal strength for the limit state of flange local buckling. The flange is slender. Thus, for flange local buckling

$$M_n = \frac{0.9Ek_c S_x}{\lambda_f^2}$$

$$M_n = \frac{0.9(29{,}000)(0.548)(94.4)}{(28.0)^2(12)} = 144 \text{ ft-kip}$$

Step 5:   Determine the lowest nominal moment for the limit states checked. Flange local buckling is the controlling limit state. Thus

For LRFD

$$M_u = 0.9(144) = 130 \text{ ft-kips}$$

For ASD

$$M_a = \frac{144}{1.67} = 86.2 \text{ ft-kips}$$

7.3   HOMOGENEOUS PLATE GIRDERS IN SHEAR

Shear is an important factor in the behavior and design of plate girders because the webs have the potential to be relatively thin. Two design procedures are available for shear design of plate girders. One accounts for the postbuckling strength available through tension field action, whereas the other uses only the buckling strength of the web without relying on any available postbuckling strength. Transverse stiffeners can be used to increase web shear strength but are not required unless tension field action is to be counted on.

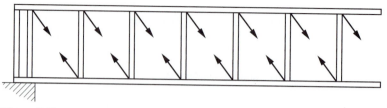

Figure 7.13 Plate Girder Showing Tension Field Action.

The limit states for web shear are web yielding and web buckling. If tension field action is not considered, these limit states are evaluated and the strength of the web determined. Under certain circumstances it is possible to take advantage of the postbuckling strength of the girder web to determine a higher strength limit. Research has demonstrated that a plate girder with transverse stiffeners and a thin web can act as a Pratt truss once the web buckles, thus providing additional postbuckling strength. This truss behavior is illustrated in Figure 7.13, where the buckled panel of the girder simulates the tension diagonal of the truss and the stiffener represents the vertical web member. The designer must decide whether to use this tension field action or to design a conventional, nontension field girder. It will be seen that web yielding controls the maximum strength of the girder web. If the size of the girder web permits web yielding, there will be no advantage to considering stiffeners, with or without tension field action.

Shear design strength (LRFD) and allowable strength (ASD) are determined with $\phi = 0.9$ and $\Omega = 1.67$, as was the case for flexure.

7.3.1 Nontension Field Action

The nominal shear strength for a nontension field plate girder is a function of the slenderness of the web. This slenderness is defined as $\lambda_{wv} = h/t_w$ and the limits used to describe the behavior are

$$\lambda_{wvp} = 1.1 \sqrt{\frac{k_v E}{F_y}}$$

and

$$\lambda_{wvr} = 1.37 \sqrt{\frac{k_v E}{F_y}}$$

The web plate-buckling coefficient, k_v, for unstiffened webs of I-shaped members that meet the proportioning criteria of the Specification, that is, $\lambda_{wv} < 260$, is taken as $k_v = 5.0$. For stiffened webs

$$k_v = 5 + \frac{5}{(a/h)^2}$$

but is taken as 5.0 when $a/h > 3.0$ or $a/h > [260/(h/t_w)]^2$

The nominal shear strength of a nontension field girder is given by

$$V_n = 0.6 F_y A_w C_v \qquad (7.17)$$

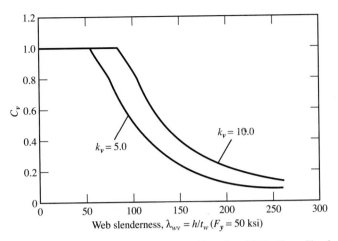

Figure 7.14 Web Shear Coefficient as a Functional Web Shear Slenderness.

where

A_w = the overall depth times the web thickness and the web shear coefficient, C_v, is a function of web shear slenderness.

For $\lambda_{wv} \le \lambda_{wvp} = 1.10\sqrt{k_v E/F_y}$

$$C_v = 1.0$$

for $\lambda_{wvp} = 1.10\sqrt{k_v E/F_y} < \lambda_{wv} \le \lambda_{wvr} = 1.37\sqrt{k_v E/F_y}$

$$C_v = \frac{1.10\sqrt{k_v E/F_y}}{\lambda_{wv}}$$

and for $\lambda_{wv} > \lambda_{wvr} = 1.37\sqrt{k_v E/F_y}$

$$C_v = \frac{1.51 E k_v}{(\lambda_{wv})^2 F_y}$$

The web shear coefficient is shown in Figure 7.14 for two cases of the web plate buckling coefficient, $k_v = 5.0$ and 10.0. For a web with shear slenderness less than λ_{wvp}, $C_v = 1.0$ and the web reaches its full plastic strength. For a web with shear slenderness greater than λ_{wvr}, the web buckles elastically and for a web with shear slenderness between these, the web buckles inelastically. Comparing the two curves in Figure 7.14 shows the impact of adding stiffeners to a nontension field girder; a girder with no stiffeners, $k_v = 5.0$; and one with stiffeners spaced so that the panels are square, $a/h = 1.0$ and $k_v = 10.0$.

7.3.2 Tension Field Action

Although the Specification does not require that tension field action be considered, a designer may take advantage of tension field action when stiffeners are present. The impact of tension field action is to increase the web shear strength, with the nominal shear strength determined as a combination of web buckling strength and web postbuckling strength. Both of these strength components are functions of stiffener spacing.

To include tension field action in the strength calculation, the plate girder must meet four limitations of Specification Section G3. Figure 7.15 illustrates these four limitations as described on the next page.

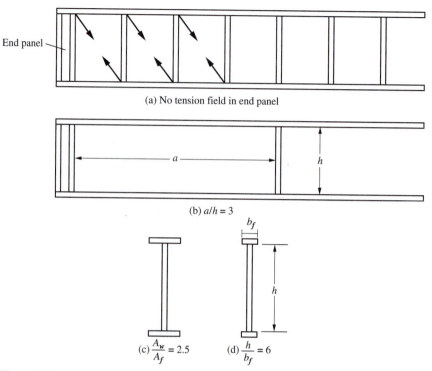

Figure 7.15 Limitations on Plate Girder to Permit Tension Field Action.

No Tension Field Action in End Panels

Figure 7.15a shows a plate girder with a potentially buckled web. The diagonal tension that is developed in the web brings two orthogonal components of force to the flange-stiffener intersection. The vertical stiffener resists the vertical component and the flange resists the horizontal component, just as for the Pratt truss. The end panel has no next panel to help resist the horizontal component; thus, this last panel must resist the shear force through beam shear, not tension field action. Every stiffened plate girder has end panels that must be designed as nontension field panels. This usually results in narrower panels at the ends of tension field girders.

Proportions of Panels

The Specification provides two limits on the proportions of stiffened panels. Tension field action may not be considered if

$$\frac{a}{h} > 3$$

or

$$\frac{a}{h} > \left(\frac{260}{h/t_w}\right)^2$$

Figure 7.15b shows a portion of a stiffened plate girder with stiffeners placed at the limit of $a/h = 3$. The panel is quite elongated and its effectiveness to resist vertical forces is significantly reduced when compared to a panel with a smaller aspect ratio such as that shown in Figure 7.15a.

Proportion of Web Area to Flange Area

For doubly symmetric plate girders, the ratio of web area to flange area cannot exceed 2.5. If this limit is exceeded, the flanges are not sufficient to resist the developed diagonal tension forces. Figure 7.15c shows a plate girder with a ratio of areas at this limit.

Proportion of Flange Width to Web Height

The Specification limits the proportions of a tension field plate girder so that it retains its ability to resist lateral buckling due to the compression forces developed in the flange. The web height, h, cannot exceed six times the flange width, b_f. Figure 7.15d shows a plate girder at this limit. It is easily seen that this is a rather slender member.

The maximum shear strength of a girder web is for the limit state of yielding. Thus, tension field action is effective only if $h/t_w > \lambda_{wvp}$ as defined for nontension field action girders. Otherwise, the strength of the girder is already at its yield strength. Thus, for $\dfrac{h}{t_w} > \lambda_{wvp} = 1.1\sqrt{\dfrac{k_v E}{F_y}}$ the nominal shear strength is the shear buckling strength plus the postbuckling strength as given by

$$V_n = 0.6F_y A_w \left(C_v + \frac{1 - C_v}{1.15\sqrt{1 + (a/h)^2}} \right) \tag{7.18}$$

Figure 7.16 shows the web shear strength for tension field and nontension field plate girders, in terms of $V_n/(0.6F_y A_w)$, as a function of web shear slenderness for a variety of panel sizes. Equation 7.18 can be rewritten to show that the strength due to tension field action is simply the combination of the prebuckling strength and the postbuckling strength as

$$V_n = 0.6F_y A_w C_v + 0.6F_y A_w \left[\frac{1 - C_v}{1.15\sqrt{1 + (a/h)^2}} \right]$$

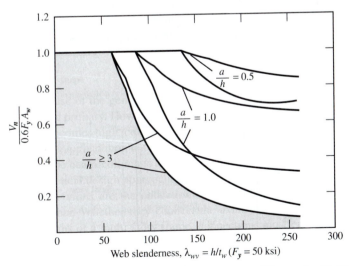

Figure 7.16 Web Shear Strength for Tension Field and Nontension Field Plate Girder.

The prebuckling strength can be seen in Figure 7.16 as the strength of the nontension field girder. The addition of the postbuckling strength shifts the curves for each particular *a/h* shown in the figure.

The end panel in a tension field plate girder must be especially rigid in order for the remainder of the web to properly function as a Pratt truss. Thus, the stiffener spacing for the panel next to the support must be less than that within the span and shear in the end panel must conform to the rules for a nontension field girder.

7.4 STIFFENERS FOR PLATE GIRDERS

When stiffeners are required for a plate girder, they can be either intermediate stiffeners or bearing stiffeners. Intermediate stiffeners purpose is to increase girder shear strength, either by controlling the buckling strength of the girder web or by permitting the postbuckling strength to be reached. These stiffeners are distributed along the girder length and result in panel sizes with aspect ratios, *a/h*, that impact girder shear strength. Bearing stiffeners usually occur at the locations of concentrated loads or reactions. They permit the transfer of concentrated forces that could not already be transferred through direct bearing on the girder web.

7.4.1 Intermediate Stiffeners

The Specification requirements for intermediate stiffeners are prescriptive in nature. There are no forces for which these stiffeners must be sized; they are simply sized to meet the specific limitations provided in Sections G2.2 and G3.3. As already discussed, stiffeners are not required if the nontension field girder web strength is determined using $k_v = 5$. This is shown as the area under the lowest curve in Figure 7.16. The increase in strength indicated by the other curves in Figure 7.16, any area to the right of the shaded area, is the result of increasing k_v to values greater than 5 and this in turn is the result of having panel aspect ratios of 3 or less. Because stiffeners are required to produce a panel with this aspect ratio, intermediate stiffeners are required for these girders.

The only other size requirement for intermediate stiffeners in nontension field girders is a limit on their moment of inertia. Specification Section G2.2 states that transverse stiffeners used to develop the available web shear strength shall have a moment of inertia about an axis in the web center for stiffener pairs or about the face in contact with the web plate for single stiffeners shown in Figure 7.17, I_{st}, such that

$$I_{st} \geq at_w^3 j$$

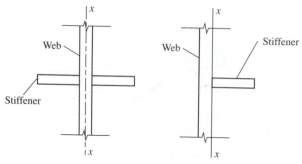

Figure 7.17 Web Stiffener Minimum Moment of Inertia.

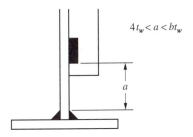

Figure 7.18 Detailing Requirement for Intermediate Stiffeners.

where

$$j = \frac{2.5}{(a/h)^2} - 2 \geq 0.5$$

In addition, the Specification provides detailing requirements for intermediate stiffeners. They can be stopped short of the tension flange and, when used in pairs, do not need to be attached to the compression flange. The weld by which they are attached to the web shall be terminated between four and six times the web thickness from the near toe to the web-to-flange weld, as shown in Figure 7.18, but there is no specific requirement for sizing that weld. Normally it would be sized based on the plate thickness. When single stiffeners are used, they must be attached to the compression flange, if it consists of a rectangular plate, to resist any uplift tendency due to torsion in the flange. Because intermediate stiffeners provide a convenient mechanism to transfer bracing forces to the girder, these stiffeners also must be connected to the compression flange and must be capable of transmitting 1% of the total flange force.

Intermediate stiffeners for tension field girders must meet the requirements already discussed and the slenderness and area requirements of Specification Section G3.3 where

$$\left(\frac{b}{t}\right)_{st} \leq 0.56\sqrt{\frac{E}{F_{yst}}}$$

and

$$A_{st} \geq \frac{F_y}{F_{yst}}\left[0.15D_s h t_w (1 - C_v)\frac{V_r}{V_c} - 18t_w^2\right] \geq 0$$

where

$(b/t)_{st}$ = width thickness ratio of the stiffener

F_{yst} = yield stress of the stiffener

D_s = 1.0 for stiffeners in pairs

= 1.8 for single-angle stiffeners

= 2.4 for single-plate stiffeners

V_r = required shear strength at the location of the stiffener

V_c = available shear strength as defined in Section G3.2

This stiffener area is based on the force that must be carried by the stiffener. It accounts for any difference in web and stiffener material strength, the ratio of the required-to-available shear strength, and a reduction for the contribution of the web in resisting the vertical force.

7.4.2 Bearing Stiffeners

Bearing stiffeners are required when the strength of the girder web is not sufficient to resist the concentrated forces exerted on it. Although bearing stiffeners can be required for rolled I-shaped members, they are much more likely to be required for plate girders, particularly at the girder supports. Specification Section J10 addresses the appropriate limit states. Normally the forces to be resisted are compressive in nature. For those cases, the limit states of web local yielding, web crippling, and web sidesway buckling must be checked. When the applied load is tensile, web local yielding and flange local bending must be considered. If the strength of the web is insufficient to resist the applied force, bearing stiffeners can be used.

The relationship between available strength and nominal strength varies for each limit state associated with web strength. Thus, either design strengths or allowable strengths must be compared, not nominal strengths, to determine the minimum web strength. The appropriate resistance factors and safety factors are given with the following discussion of the limit states.

Web Local Yielding

When a single concentrated force is applied to a girder, as shown in Figure 7.19, the force is assumed to be delivered to the girder over a length of bearing, N, and is then distributed through the flange and into the web. The narrowest portion of the web is the critical section. This occurs below the web to flange weld, dimensioned as k in Figure 7.19a. The distribution takes place along a line with a slope of 1:2.5. So when the critical section is reached, the force has been distributed over a length of N plus $2.5k$ in each direction. If the concentrated force is applied so that the force distributes along the web in both directions, this distribution increases the bearing length by $5k$ as shown in Figure 7.19b. If the bearing is close to the end of the member, distribution takes place only in one direction, toward the midspan. The Specification defines "close to the member end" as being within the member depth from the end. Thus, the available length of the web is $(N + 2.5k)$, as shown in Figure 7.19c.

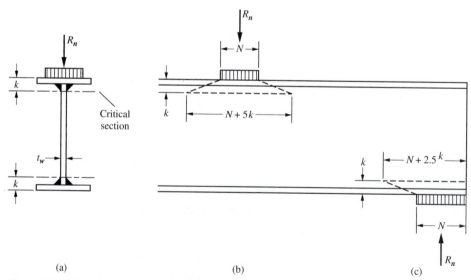

(a)   (b)   (c)

Figure 7.19 Single Concentrated Force Applied to Beam.

The nominal strength of the girder web when the concentrated force to be resisted is applied at a distance from the member end that is greater than the depth of the member, d, is

$$R_n = (5k + N)F_{yw}t_w$$

When the concentrated force to be resisted is applied at a distance from the member end that is less than or equal to the depth of the member, d, the nominal strength is

$$R_n = (2.5k + N)F_{yw}t_w$$

where

F_{yw} = yield stress of the web

N = length of bearing

k = distance from the outer face of the flange to the web toe of the fillet weld

t_w = web thickness

For web local yielding

$$\phi = 1.0 \,(\text{LRFD}) \qquad \Omega = 1.50 \,(\text{ASD})$$

Web Crippling

The criteria for the limit state of web crippling also depend on the location of the force with respect to the end of the girder.

When the concentrated compressive force is applied at a distance from the member end that is greater than $d/2$

$$R_n = 0.80t_w^2 \left[1 + 3\left(\frac{N}{d}\right)\left(\frac{t_w}{t_f}\right)^{1.5} \right] \sqrt{\frac{EF_{yw}t_f}{t_w}}$$

When the force is applied at a distance less than $d/2$ and $N/d \leq 0.2$

$$R_n = 0.40t_w^2 \left[1 + 3\left(\frac{N}{d}\right)\left(\frac{t_w}{t_f}\right)^{1.5} \right] \sqrt{\frac{EF_{yw}t_f}{t_w}}$$

and when $N/d > 0.2$

$$R_n = 0.40t_w^2 \left[1 + \left(\frac{4N}{d} - 0.2\right)\left(\frac{t_w}{t_f}\right)^{1.5} \right] \sqrt{\frac{EF_{yw}t_f}{t_w}}$$

For web crippling

$$\phi = 0.75 \,(\text{LRFD}) \qquad \Omega = 2.0 \,(\text{ASD})$$

Web Sidesway Buckling

The web of a plate girder is generally a slender element, as has already been discussed. If the tension and compression flanges of the girder are not prevented from displacing laterally with respect to each other at the point of load, web sidesway buckling must be assessed. Two provisions are provided for sidesway buckling: (1) if the compression flange is restrained against rotation, and (2) if it is not.

When the compression flange is restrained against rotation and the ratio of web slenderness to lateral buckling slenderness, $(h/t_w)/(l/b_f) \leq 2.3$, the nominal strength is given as

$$R_n = \frac{C_r t_w^3 t_f}{h^2}\left[1 + 0.4\left(\frac{h/t_w}{l/b_f}\right)^3\right]$$

If, $(h/t_w)/(l/b_f) > 2.3$, the limit state does not apply.

When the compression flange is not restrained against rotation and the ratio of web slenderness to lateral buckling slenderness, $(h/t_w)/(l/b_f) \leq 1.7$, the nominal strength is given as

$$R_n = \frac{C_r t_w^3 t_f}{h^2}\left[0.4\left(\frac{h/t_w}{l/b_f}\right)^3\right]$$

and if $(h/t_w)/(l/b_f) > 1.7$, the limit state does not apply.

In the above equations

l = largest laterally unbraced length along either flange at the point of load

C_r = 960,000 ksi when $M_u < M_y$ or $1.5M_a < M_y$ and 480,000 ksi when $M_u \geq M_y$ or $1.5M_a \geq M_y$

For web sidesway buckling

$$\phi = 0.85\,(\text{LRFD}) \qquad \Omega = 1.76\,(\text{ASD})$$

Flange Local Bending

This limit state applies when a single tensile concentrated force is applied to the flange and the length of loading across the member flange is greater than $0.15b_f$, as shown in Figure 7.20. The nominal strength is

$$R_n = 6.25t_f^2 F_{yf}$$

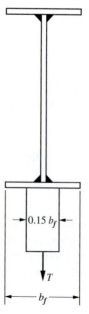

Figure 7.20 Flange Local Bending for an Applied Tension Load.

If the force is applied at a distance less than $10t_f$ from the member end, the strength must be reduced by 50%.

For flange local bending,

$$\phi = 0.9\,(\text{LRFD}) \qquad \Omega = 1.67\,(\text{ASD})$$

7.4.3   Bearing Stiffener Design

Once the appropriate limit states are checked, a decision is made as to the need for bearing stiffeners. Although it is possible to select a web plate that would not require stiffeners, this is not usually the most economical approach, even though the addition of stiffeners is a high labor and thus high cost element. When bearing stiffeners are to be sized, Section J10.8 of the Specification requires that they be sized according to the provisions for tension members or compression members as appropriate.

Stiffeners designed to resist tensile forces must be designed according to the requirements of Chapter D, for the difference between the required strength and the minimum available limit state strength. The stiffener must be welded to the flange and web and these welds must be sized to resist the force being transferred to the stiffeners.

Stiffeners required to resist compressive forces must be designed according to the provisions of Chapter E, except for stiffeners with $KL/r \leq 25$, which may be designed with $F_{cr} = F_y$, according to Section J4.4, for the difference between the required strength and the minimum available limit state strength. These stiffeners must also be welded to the flange and web and these welds must be sized to resist the force being transferred to the stiffeners.

EXAMPLE 7.3
Plate Girder Flexural and Shear Strength

GOAL: Determine the available moment and shear strength using tension field action.

GIVEN: A built-up member is shown in Figure 7.21. Assume that the beam is laterally braced continuously. Use A572 Gr. 50 for the member plates and A36 for the stiffeners. $S_x = 464$ in³.

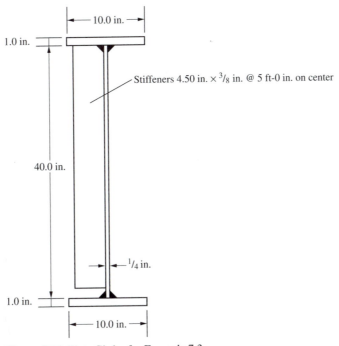

Figure 7.21 Plate Girder for Example 7.3.

SOLUTION

Step 1: Check the web slenderness to determine which section of the Specification must be used.

$$\lambda_w = \frac{h}{t_w} = \frac{40}{0.25} = 160 > \lambda_r = 5.7\sqrt{\frac{29,000}{50}} = 137$$

Therefore, this is a slender web plate girder and the provisions of Section F5 must be applied.

Step 2: Determine the bending strength reduction factor.
The ratio of web area to flange area is given as

$$a_w = \frac{40.0(0.250)}{10.0(1.00)} = 1.00$$

and

$$R_{pg} = 1 - \frac{1.00}{1200 + 300(1.00)}\left(\frac{40.0}{0.250} - 5.7\sqrt{\frac{29,000}{50}}\right) = 0.985$$

Step 3: Determine the nominal strength for the limit state of yielding.

$$M_n = \frac{0.985(50)(464)}{12} = 1900 \text{ ft-kips}$$

Step 4: Determine the nominal moment strength for lateral-torsional buckling.
The flange is fully braced, so this limit state does not apply.

Step 5: Determine the nominal moment strength for the limit state of flange local buckling.
The flange slenderness is

$$\lambda_f = \frac{b}{t} = \frac{5.0}{1.0} = 5.0 < 0.38\sqrt{\frac{29,000}{50}} = 9.15$$

Therefore, the flange is compact and there is no reduction in strength.

Step 6: Determine the lowest nominal moment strength for the limit states calculated.

$$M_n = 1900 \text{ kip-ft}$$

Step 7: Determine the available moment strength.

For LRFD

$$M_u = 0.9(1900) = 1710 \text{ ft-kips}$$

For ASD

$$M_a = \frac{1900}{1.67} = 1140 \text{ ft-kips}$$

Step 8: Determine the shear strength with tension field action. First check the intermediate stiffeners against the prescriptive requirements.
Proportions of panel

$$\frac{a}{h} = \frac{60.0}{40.0} = 1.50 < 3$$

$$< \left(\frac{260}{h/t_w}\right)^2 = \left(\frac{260}{40.0/0.250}\right)^2 = 2.64$$

Proportions of web to flange

$$\frac{A_w}{A_f} = \frac{42.0(0.250)}{10.0(1.00)} = 1.05 < 2.5$$

Proportions of flange width to web height

$$\frac{h}{b_f} = \frac{40.0}{10.0} = 4.00 < 6.0$$

Because the criteria have been satisfied, it is permissible to use tension field action in all but the last panel.

Step 9:   Determine whether tension field action increases shear strength by checking the web shear slenderness.

$$k_v = 5 + \frac{5}{(a/h)^2} = 5 + \frac{5}{(60.0/40.0)^2} = 7.22$$

thus

$$\lambda_{wvp} = 1.1\sqrt{\frac{k_v E}{F_y}} = 1.1\sqrt{\frac{7.22(29,000)}{50}} = 71.2$$

and

$$\lambda_{wvr} = 1.37\sqrt{\frac{k_v E}{F_y}} = 1.37\sqrt{\frac{7.22(29,000)}{50}} = 88.7$$

Because $\lambda_{wv} = \dfrac{h}{t_w} = \dfrac{40.0}{0.250} = 160$ is greater than λ_{wvp} and λ_{wvr}, tension-field action provides an increase in web shear strength.

Step 10:   Determine the shear strength coefficient.

$$C_v = \frac{1.51(29,000)(7.22)}{(160)^2(50)} = 0.247$$

The nominal shear strength is

$$V_n = 0.6(50)(42.0)(0.250)\left(0.247 + \left[\frac{1 - 0.247}{1.15\sqrt{1 + (1.50)^2}}\right]\right) = 192 \text{ kips}$$

Step 11:   Determine the available shear strength.

For LRFD

$$V_u = 0.9(192) = 173 \text{ kips}$$

For ASD

$$V_a = \frac{192}{1.67} = 115 \text{ kips}$$

Step 12:   Check the intermediate stiffener size for meeting the criteria.
For a single plate stiffener

$$I_{st} = \frac{bh^3}{3} = \frac{0.375(4.50)^3}{3} = 11.4 \text{ in.}^4$$

the length of the member that results from the moment gradient along the member. In this case, the member ends must remain in their original position relative to each other, thus, no sway is considered. The moment created by the load, P, acting at an eccentricity, δ, from the deformed member is superimposed on the moment gradient resulting from the applied end moments. The magnitude of this additional, second-order moment depends on the properties of the column itself. Thus, this is called the *member effect*.

When the beam-column is part of a structure that is permitted to sway, the displacements of the overall structure also influence the moments in the member. For a beam-column that is permitted to sway an amount Δ, as shown in Figure 8.2b, the additional moment is given by $P\Delta$. Because the lateral displacement of a given member is a function of the properties of all of the members in a given story, this moment is called the *structure effect*.

Both of these second-order effects must be accounted for in the design of beam-columns. Procedures for incorporating these effects will be addressed once an overall approach to beam-column design is established.

8.3   INTERACTION PRINCIPLES

The interaction of axial load and bending within the elastic response range of a beam-column can be investigated through the straightforward techniques of *superposition*. This is the approach normally considered in elementary strength of materials in which the normal stress due to an axial force is added to the normal stress due to a bending moment.

Although the superposition of individual stress effects is both simple and correct for elastic stresses, there are significant limitations when applying this approach to the limit states of real structures. These include:

1. Superposition of stress is correct only for behavior within the elastic range and only for similar stress types.
2. Superposition of strain can be extended only into the inelastic range when deformations are small.
3. Superposition cannot account for member deformations or stability effects such as local buckling.
4. Superposition cannot account for structural deflections and system stability.

With these limitations in mind, it is desirable to develop interaction equations that will reflect the true limit state's behavior of beam-columns. Any limit state interaction equation must reflect the following characteristics:

Axial Load
1. Maximum column strength
2. Individual column slenderness

Bending Moment
1. Lateral support conditions
2. Sidesway conditions
3. Member second-order effects
4. Structure second-order effects
5. Moment gradient along the member

The resulting equations must also provide a close correlation with test results and theoretical analyses for beam-columns, including the two limiting cases of pure bending and pure compression.

Application of the resulting interaction equations can be regarded as a process of determining available axial strength in the presence of a given bending moment or determining the available moment strength in the presence of a given axial load. An applied bending moment consumes a portion of the column strength, leaving a reduced axial load strength. When the two actions are added together, the resulting total load must not exceed the total column strength. Conversely, the axial load can be regarded as consuming a fraction of the moment strength. This fraction, plus the applied moments, must not exceed the maximum beam strength.

8.4   INTERACTION EQUATIONS

A simple form of the three-dimensional interaction equation is

$$\frac{P_r}{P_c} + \frac{M_{rx}}{M_{cx}} + \frac{M_{ry}}{M_{cy}} \leq 1.0 \tag{8.1}$$

where the terms with the subscript r represent the required strength and those with the subscript c represent the available strength.

This interaction equation is plotted in Figure 8.3. The figure shows that this results in a straight line representation of the interaction between any two of the load components. The horizontal plane of Figure 8.3 represents biaxial bending interaction, whereas the vertical planes represent the interaction of axial compression plus either strong or weak axis bending. It should also be apparent that the three-dimensional aspect is represented by a plane with intercepts given by the straight lines on the three coordinate planes.

The Specification interaction equations in Chapter H of the Specification result from fitting interaction equations of the form of Equation 8.1 to a set of data developed from an analysis of forces and moments for various plastic stress distributions on a stub column. Figure 8.4a shows the actual analysis results for a W14×82 stub column. Figure 8.4b shows

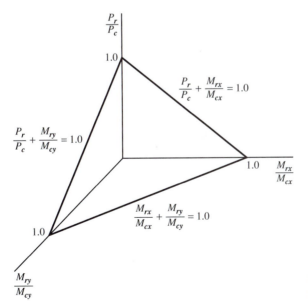

Figure 8.3 Simplified Interaction Surface.

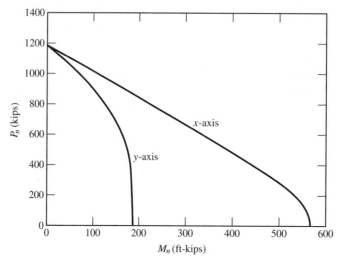

Figure 8.4a Interaction Diagram for Stub W14×82 Column.

the same data plotted as functions of the normalized axial and flexural strength. In both cases, the influence of length on the axial or flexural strength is not included. Using curves of this type, developed for a wide variety of steel beam-column shapes, two equations were developed that are conservative and accurate for x-axis bending. When applied to y-axis bending, they are significantly more conservative. Simplicity of design and the low level use of weak axis bending justify this extra level of conservatism.

An additional modification to these equations is required to account for the length effects. Rather than normalizing the curves on the yield load and the plastic moment as was done in Figure 8.4b, the equations were developed around the nominal strength of the column and the nominal strength of the beam. The resulting equations, given as Equations H1-1a and H1-1b in the Specification, are given here as Equations 8.2 and 8.3 and are plotted in Figure 8.5. The equations shown here consider bending about both principal axes, whereas the plot in Figure 8.5 is for single axis bending.

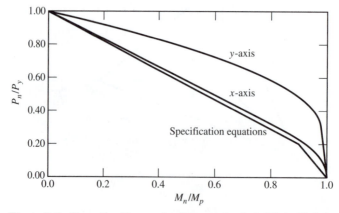

Figure 8.4b Normalized Interaction Diagram for Stub W14×82 Column.

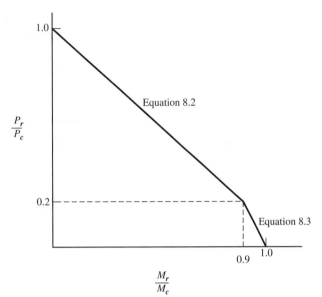

Figure 8.5 Interaction Equations 8.2 and 8.3.

For $\dfrac{P_r}{P_c} \geq 0.2$

$$\frac{P_r}{P_c} + \frac{8}{9}\left(\frac{M_{rx}}{M_{cx}} + \frac{M_{ry}}{M_{cy}}\right) \leq 1.0 \qquad \text{(H1-1a)} \qquad (8.2)$$

For $\dfrac{P_r}{P_c} < 0.2$

$$\frac{P_r}{2P_c} + \left(\frac{M_{rx}}{M_{cx}} + \frac{M_{ry}}{M_{cy}}\right) \leq 1.0 \qquad \text{(H1-1b)} \qquad (8.3)$$

where

P_r = required compressive strength, kips
P_c = available compressive strength, kips
M_r = required flexural strength, ft-kips
M_c = available flexural strength, ft-kips
x = subscript relating symbol to strong axis bending
y = subscript relating symbol to weak axis bending

It is important to note that

1. The available column strength, P_c, is based on the axis of the column with the largest slenderness ratio. This is not necessarily the axis about which bending takes place.
2. The available bending strength, M_c, is based on the bending strength of the beam without axial load, including the influence of all the beam limit states.
3. The required compressive strength, P_r, is the second-order force on the member.
4. The required flexural strength, M_r, is the second-order bending moment on the member.

Second-order forces and moments can be determined through a second-order analysis or by a modification of the results of a first-order analysis using amplification factors. These amplification factors will be discussed as they relate to braced frames (Section 8.5) and moment frames (Section 8.6).

Additional provisions are available for cases where the axial strength limit state is out-of-plane buckling and the flexural limit state is in-plane bending. Equations H1-1a and 1-1b are conservative for this situation but an additional approach is available. Specification Section H1.3 provides that (1) for in-plane instability, Equations H1-1a and H1-1b should be used where the compressive strength is determined for buckling in the plane of the frame, and (2) for out-of-plane buckling

$$\frac{P_r}{P_{co}} + \left(\frac{M_r}{M_{cx}}\right)^2 \leq 1.0$$

where

P_{co} = available compressive strength out of the plane of bending

M_{cx} = available flexural strength for strong axis bending

If bending is about only the weak axis, the moment term is neglected. If there is significant biaxial bending, meaning both axes exhibit a required-moment-to-available-moment ratio greater than or equal to 0.05, then this option is not available. Although this optional approach can provide a more economical solution in some cases, it is not used in the examples or problems of this book.

8.5   BRACED FRAMES

A frame is considered to be braced if a positive system, that is, an actual system such as a shear wall (masonry, concrete, steel, or other material) or diagonal steel member, as illustrated in Figure 8.6, serves to resist the lateral loads, to stabilize the frame under gravity loads, and resist lateral displacements. In these cases, columns are considered braced against

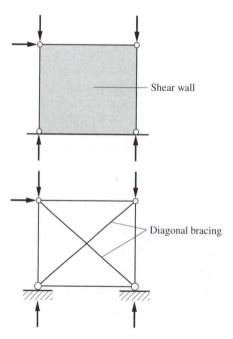

Figure 8.6 Braced Frame.

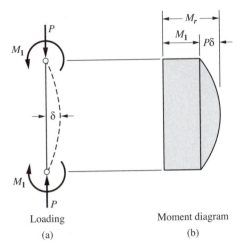

| Loading | Moment diagram | **Figure 8.7** An Axially Loaded Column with |
| (a) | (b) | Equal and Opposite End Moments. |

lateral translation and the in-plane K-factor is taken as 1.0 or less. This is the type of column that was discussed in Chapter 6. Later in this chapter the requirements for bracing to insure that a structure can be considered a braced frame, as found in Appendix 6, are discussed.

If the column in a braced frame is rigidly connected to a girder, bending moments result from the application of the gravity loads to the girder. These moments can be determined through a first-order elastic analysis. The additional second-order moments resulting from the displacement along the column length can be determined through the application of an amplification factor.

The full derivation of the amplification factor has been presented by various authors.[1,2] Although this derivation is quite complex, a somewhat simplified derivation is presented here to help establish the background. An axially loaded column with equal and opposite end moments is shown in Figure 8.7a. The resulting moment diagram is shown in Figure 8.7b where the moments from both the end moments and the secondary effects are given. The maximum moment occurring at the midheight of the column, M_r, is shown to be

$$M_r = M_1 + P\delta$$

The amplification factor is defined as

$$AF = \frac{M_r}{M_1} = \frac{M_1 + P\delta}{M_1}$$

Rearranging terms yields

$$AF = \frac{1}{1 - \dfrac{P\delta}{M_1 + P\delta}}$$

Two simplifying assumptions will be made. The first is based on the assumption that δ is sufficiently small that

$$\frac{\delta}{M_1 + P\delta} = \frac{\delta}{M_1}$$

[1]Galambos, T. V., *Structural Members and Frames*. Englewood Cliffs, NJ: Prentice Hall, Inc., 1968.

[2]Johnson, B. G., Ed., *Guide to Stability Design Criteria for Metal Structures*, 3rd ed., SSRC, New York: Wiley, 1976.

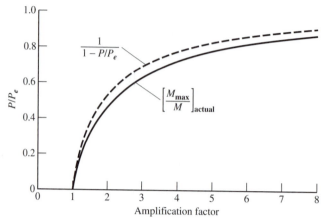

Figure 8.8 Amplified Moment: Exact and Approximate.

and the second, using the beam deflection, $\delta = M_1 L^2 / 8EI$, assumes that

$$\frac{M_1}{\delta} = \frac{8EI}{L^2} \approx \frac{\pi^2 EI}{L^2} = P_e$$

Because these simplifying assumptions are in error in opposite directions, they tend to be offsetting. This results in a fairly accurate prediction of the amplification. Thus,

$$AF = \frac{1}{1 - P/P_e} \qquad (8.4)$$

A comparison between the actual amplification and that given by Equation 8.4 is shown in Figure 8.8.

　　The discussion so far has assumed that the moments at each end of the column are equal and opposite and that the resulting moment diagram is uniform. This is the most severe loading case for a beam-column. If the moment is not uniformly distributed, the displacement along the member is less than previously considered and the resulting amplified moment is less than indicated. It has been customary in design practice to use the case of uniform moment as a base and to provide for other moment gradients by converting them to an equivalent uniform moment through the use of an additional factor, C_m.

　　Numerous studies have shown that a reasonably accurate correction results for beam-columns braced against translation and not subject to transverse loading between their supports, if the moment is reduced through its multiplication by C_m, where

$$C_m = 0.6 - 0.4(M_1/M_2) \qquad (8.5)$$

M_1/M_2 is the ratio of the smaller to larger moments at the ends of the member unbraced length in the plane of bending. M_1/M_2 is positive when the member is bent in reverse curvature and negative when bent in single curvature.

　　For beam-columns in braced frames where the member is subjected to transverse loading between supports, C_m may be taken from Commentary Table C-C2.1 or conservatively taken as 1.0.

　　The combination of the amplification factor, AF, and the equivalent moment factor, C_m, accounts for the total member secondary effects. This combined factor is given as B_1 in the Specification and is shown here as Equation 8.6.

$$B_1 = \frac{C_m}{1 - \dfrac{\alpha P_r}{P_{e1}}} \geq 1.0 \qquad (8.6)$$

where

α = 1.6 for ASD and 1.0 for LRFD to account for the nonlinear behavior of the structure at its ultimate strength

P_r = required strength

P_{e1} = Euler buckling load for the column with an effective length factor, $K = 1.0$

Thus, the value of M_r in Equations 8.2 and 8.3 is taken as

$$M_r = B_1 M$$

where M is the maximum moment on the beam-column. It is possible for C_m to be less than 1.0 and for Equation 8.6 to give an amplification factor less than 1.0. This indicates that the combination of the $P\delta$ effects and the nonuniform moment gradient result in a moment less than the maximum moment on the beam-column from first-order effects. In this case, the amplification factor $B_1 = 1.0$.

EXAMPLE 8.1a
Column Design for
Combined Axial and
Bending by LRFD

GOAL: Design column A1 in Figure 8.9 for the given loads using the LRFD provisions and the second-order amplification factor provided in the Specification.

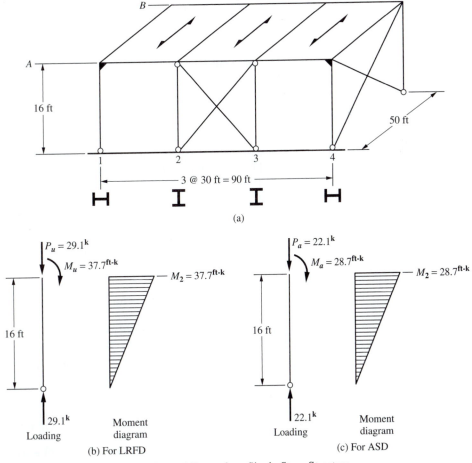

Figure 8.9 Three-Dimensional Braced Frame for a Single-Story Structure.

GIVEN: The three-dimensional braced frame for a single-story structure is given in Figure 8.9. Rigid connections are provided at the roof level for columns A1, B1, A4, and B4. All other column connections are pinned. Dead Load = 50 psf, Snow Load = 20 psf, Roof Live Load = 10 psf, and Wind load = 20 psf horizontal. Use A992 steel. Assume that the X-bracing is so much stiffer than the rigid frames that it resists all lateral load.

SOLUTION

Step 1:   Determine the appropriate load combinations. From ASCE 7, Section 2.3, the following two combinations are considered.

$$1.2D + 1.6(L_r \text{ or } S \text{ or } R) + (0.5L \text{ or } 0.8W) \qquad \text{(ASCE 7-3)}$$

$$1.2D + 1.6W + 0.5L + 0.5(L_r \text{ or } S \text{ or } R) \qquad \text{(ASCE 7-4)}$$

Step 2:   Determine the factored roof gravity loads for each load combination.

$$1.2(50) + 1.6(20) = 92 \text{ psf} \qquad \text{(ASCE 7-3)}$$

$$1.2(50) + 0.5(20) = 70 \text{ psf} \qquad \text{(ASCE 7-4)}$$

Because column A1 does not participate in the lateral load resistance, use a uniformly distributed roof load of 92 psf.

Step 3:   Carry out a preliminary first-order analysis. Because the structure is indeterminate, a number of approaches can be taken. If an arbitrary 6-to-1 ratio of moment of inertia for beams to columns is assumed, a moment distribution analysis yields the moment and force given in Figure 8.9b. Thus, the column will be designed to carry

$$P_u = 29.1 \text{ kips and } M_u = 37.7 \text{ ft-kips}$$

Step 4:   Select a trial size for column A1 and determine its compressive strength and bending strength.
   Try W10×33. (Section 8.8 addresses trial section selection.)

$$A = 9.71 \text{ in.}^2, \quad r_x/r_y = 2.16, \quad r_x = 4.19 \text{ in.}, \quad r_y = 1.94 \text{ in.}, \quad I_x = 171 \text{ in.}^4$$

The column is oriented so that bending is about the x-axis of the column. It is braced against sidesway by the diagonal braces in panel A2–A3 and is pinned at the bottom and rigidly connected at the top in the plane of bending. Because this column is part of a braced frame, $K = 1.0$ should be used. Although the Specification permits the use of a lower K-factor if justified by analysis, this is not recommended because it would likely require significantly more stiffness in the braced panel.
   From Manual Table 4-1

$$\phi P_n = 213 \text{ kips for } KL = 16.0 \text{ ft}$$

From Manual Table 3-2

$$\phi M_n = 113 \text{ ft-kips for } L_b = 16.0 \text{ ft}$$

Step 5:   Check the W10×33 for combined axial load and bending in-plane.
   For an unbraced length of 16 ft, the Euler load is

$$P_{e1} = \frac{\pi^2 EI}{(KL)^2} = \frac{\pi^2 (29{,}000)(171)}{(16.0(12))^2} = 1330 \text{ kips}$$

The column is bent in single curvature between bracing points, the end points, and the moment at the base is zero; therefore $M_1/M_2 = 0.0$. Thus

$$C_m = 0.6 - 0.4(0.0) = 0.6$$

Therefore, the amplification factor becomes

$$B_1 = \frac{0.6}{1 - \dfrac{29.1}{1330}} = 0.61 < 1.0$$

The specification requires that B_1 not be less than 1.0. Therefore, taking $B_1 = 1.0$

$$M_{rx} = B_1(M_x) = 1.0(37.7) = 37.7 \text{ ft-kips}$$

To determine which equation to use, calculate

$$\frac{P_u}{\phi P_n} = \frac{29.1}{213} = 0.137 < 0.2$$

Therefore, use Equation 8.3 (H1-1b)

$$0.5(0.137) + \frac{37.7}{113} = 0.402 < 1.0$$

thus, the W10×33 will easily carry the given loads

The solution to Equation H1-1b indicates that there is a fairly wide extra margin of safety. It would be appropriate to consider a smaller column for a more economical design.

EXAMPLE 8.1b
Column Design for
Combined Axial and
Bending by ASD

GOAL: Design column A1 in Figure 8.9 for the given loads using the ASD provisions and the second-order amplification factor provided in the Specification.

GIVEN: The three-dimensional braced frame for a single-story structure is given in Figure 8.9. Rigid connections are provided at the roof level for columns A1, B1, A4, and B4. All other column connections are pinned. Dead Load = 50 psf, Snow Load = 20 psf, Roof Live Load = 10 psf, and Wind load = 20 psf horizontal. Use A992 steel. Assume that the X-bracing is so much stiffer than the rigid frames that it resists all lateral load.

SOLUTION

Step 1:   Determine the appropriate load combinations. From ASCE 7, Section 2.4, the following two combinations are considered.

$$D + (L_r \text{ or } S \text{ or } R) \qquad\qquad \text{(ASCE 7-3)}$$

$$D + 0.75W + 0.75(L_r \text{ or } S \text{ or } R) \qquad\qquad \text{(ASCE 7-4)}$$

Step 2:   Determine the factored roof gravity loads for each load combination.

$$(50) + (20) = 70 \text{ psf} \qquad\qquad \text{(ASCE 7-3)}$$

$$(50) + 0.75(20) = 65 \text{ psf} \qquad\qquad \text{(ASCE 7-4)}$$

Because column A1 does not participate in the lateral load resistance, use a uniformly distributed roof load of 70 psf.

Step 3:   Carry out a preliminary first-order analysis. Because the structure is indeterminate, a number of approaches can be taken. If an arbitrary 6-to-1 ratio of moment of inertia for beams to columns is assumed, a moment distribution analysis yields the moment and force given in Figure 8.9c. Thus, the column will be designed to carry

$$P_u = 22.1 \text{ kips and } M_u = 28.7 \text{ ft-kips}$$

Step 4:   Select a trial size for column A1 and determine its compressive strength and bending strength.

Try W10×33. (Section 8.8 addresses trial section selection.)

$$A = 9.71 \text{ in.}^2, \quad r_x/r_y = 2.16, \quad r_x = 4.19 \text{ in.}, \quad r_y = 1.94 \text{ in.}, \quad I_x = 171 \text{ in.}^4$$

The column is oriented so that bending is about the x-axis of the column. It is braced against sidesway by the diagonal braces in panel A2–A3 and is pinned at the bottom and rigidly connected at the top in the plane of bending. Because this column is part of a braced frame, $K = 1.0$ should be used. Although the Specification permits the use of a lower K-factor if justified by analysis, this is not recommended because it would likely require significantly more stiffness in the braced panel.

From Manual Table 4-1

$$P_n/\Omega = 142 \text{ kips for } KL = 16.0 \text{ ft}$$

From Manual Table 3-2

$$M_n/\Omega = 74.9 \text{ ft-kips for } L_b = 16.0 \text{ ft}$$

Step 5:   Check the W10×33 for combined axial load and bending in-plane.

For an unbraced length of 16.0 ft, the Euler load is

$$P_{e1} = \frac{\pi^2 EI}{(KL)^2} = \frac{\pi^2 (29,000)(171)}{(16.0(12))^2} = 1330 \text{ kips}$$

The column is bent in single curvature between bracing points, the end points, and the moment at the base is zero; therefore $M_1/M_2 = 0.0$. Thus

$$C_m = 0.6 - 0.4(0.0) = 0.6$$

Therefore, the amplification factor with $\alpha = 1.6$ becomes

$$B_1 = \frac{0.6}{1 - \dfrac{1.6(22.1)}{1330}} = 0.616 < 1.0$$

The specification requires that B_1 not be less than 1.0. Therefore, taking $B_1 = 1.0$

$$M_{rx} = B_1(M_x) = 1.0(28.7) = 28.7 \text{ ft-kips}$$

To determine which equation to use, calculate

$$\frac{P_a}{P_n/\Omega} = \frac{22.1}{142} = 0.156 < 0.2$$

Therefore, use Equation 8.3 (H1-1b)

$$0.5(0.156) + \frac{28.7}{74.9} = 0.461 < 1.0$$

> thus, the W10×33 will easily carry the given loads

The solution to Equation H1-1b indicates that there is a fairly wide extra margin of safety. It would be appropriate to consider a smaller column for a more economical design.

8.6 MOMENT FRAMES

A moment frame depends on the stiffness of the beams and columns that compose the frame for stability under gravity loads and under combined gravity and lateral loads. Unlike braced frames, there is no external structure to lean against for stability. Columns in moment frames are subjected to both axial load and moment and experience lateral translation.

The same interaction equations, Equations 8.2 and 8.3, are used to design beam-columns in moment frames as were previously used for braced frames. However, in addition to the member second-order effects discussed in Section 8.5, there is the additional second-order effect that results from the sway or lateral displacement of the frame.

Figure 8.10 shows a cantilever or flag pole column under the action of an axial load and a lateral load. Figure 8.10a is the column as viewed for a first-order elastic analysis where equilibrium requires a moment at the bottom, $M_{lt} = HL$. The deflection that results at the top of the column, Δ_1, is the elastic deflection of a cantilever beam, thus

$$\Delta_1 = \frac{HL^3}{3EI} \tag{8.7}$$

A second-order analysis yields the forces and displacements as shown in Figure 8.10b. The displacement, Δ_2, is the total displacement including second-order effects and the moment including second-order effects is

$$B_2 M_{lt} = HL + P\Delta_2 \tag{8.8}$$

An equivalent lateral load can be determined that results in the same moment at the bottom of the column as in the second-order analysis. This load is $H + P\Delta_2/L$ and is shown in Figure 8.10c.

It may be assumed, with only slight error, that the displacement at the top of the column for the cases in Figures 8.10b and 8.10c are the same. Thus, using the equivalent lateral load

$$\Delta_2 = \frac{(H + P\Delta_2/L)L^3}{3EI} = \frac{HL^3}{3EI}\left(1 + \frac{P\Delta_2}{HL}\right) = \Delta_1\left(1 + \frac{P\Delta_2}{HL}\right) \tag{8.9}$$

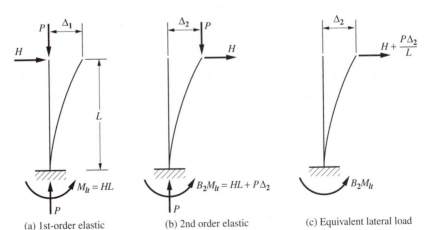

(a) 1st-order elastic   (b) 2nd order elastic   (c) Equivalent lateral load

Figure 8.10 Structure Second-Order Effect: Sway.

Equation 8.9 can now be solved for Δ_2 where

$$\Delta_2 = \frac{\Delta_1}{1 + \dfrac{P\Delta_1}{HL}}$$

and the result substituted into Equation 8.8. Solving the resulting equation for the amplification factor, B_2, and simplifying yields

$$B_2 = \frac{1}{1 - \dfrac{P\Delta_1}{HL}} \tag{8.10a}$$

Considering that the typical beam-column will be part of some larger structure, this equation must be modified to include the effect of the multistory and multibay characteristics of the actual structure. This is easily accomplished by summing the total gravity load on the columns in the story and the total lateral load in the story. Thus, Equation 8.10a becomes

$$B_2 = \frac{1}{1 - \dfrac{\Sigma P\Delta_1}{\Sigma HL}} \tag{8.10b}$$

This amplification factor is essentially that given by the Specification as Equation C2-3 in combination with Equation C2-6b

$$B_2 = \frac{1}{1 - \dfrac{\alpha \Sigma P_{nt}}{\Sigma P_{e2}}} = \frac{1}{1 - \dfrac{\alpha \Sigma P_{nt} \Delta_H}{R_M \Sigma HL}} \tag{8.11}$$

where

$\Sigma P_{nt} = $ total gravity load on the story

$\Sigma P_{e2} = $ measure of lateral stiffness of the structure $= R_M \dfrac{\Sigma HL}{\Delta_H}$

$\Delta_H = $ story drift from a first-order analysis due to the lateral load, H

$\alpha = $ 1.0 for LRFD and 1.6 for ASD to account for the nonlinear behavior of the structure at its ultimate strength

$R_M = $ 0.85 for moment frames to account for the influence of the member effect on the sidesway displacement that could not be accounted for in the simplified derivation above. For braced frames, $R_m = 1.0$

It is often desirable to limit the lateral displacement, or drift, of a structure during the design phase. This limit can be defined using a drift index which is the story drift divided by the story height, Δ_H/L. The design then proceeds by selecting members so that the final structure performs as desired. This is similar to beam design where deflection is the serviceability criterion. Because the drift index can be established without knowing member sizes, it can be used in Equation 8.11. Thus, an analysis with assumed member sizes is unnecessary.

If, however, the column sizes of the structure are known, determination of B_2 is possible using the sidesway buckling resistance given as

$$\Sigma P_{e2} = \Sigma \frac{\pi^2 EI}{(K_2 L)^2} \tag{8.12}$$

where K_2 is the sidesway buckling effective length factor for each column in the story that participates in the lateral load resistance.

With this amplification for sidesway, the moment, M_r, to be used in Equations 8.2 and 8.3, can be evaluated. M_r must include both the member and structure second-order effects. Thus, a first-order analysis without sidesway is carried out, yielding moments, M_{nt}, that is without translation, to be amplified by B_1. Next, a first-order analysis including lateral loads and permitting translation must be carried out. This yields moments, M_{lt}, that is with translation, to be amplified by B_2. The resulting second-order moment is

$$M_r = B_1 M_{nt} + B_2 M_{lt} \qquad (8.13)$$

where

B_1 is given by Equation 8.6

B_2 is given by Equation 8.11

M_{nt} = first-order moments with no translation

M_{lt} = first-order moments that result from lateral translation

M_{lt} could include moments that result from unsymmetrical frame properties or loading as well as from lateral loads. In most real structures, however, moments resulting from this lack of symmetry are usually small and often ignored.

The second-order force is

$$P_r = P_{nt} + B_2 P_{lt}$$

The sum of P_{nt} and P_{lt} should equal the total gravity load on the structure but for an individual column, it is important to amplify the portion of the individual column force that comes from the lateral load.

For situations where there is no lateral load on the structure, it may be necessary to incorporate a minimum lateral load in order to capture the second order effects of the gravity loads. This is covered briefly in Section 8.7 where the three methods provided in the Specification for treating second order effects are discussed.

EXAMPLE 8.2a
Strength Check for Combined Compression and Bending by LRFD

GOAL: Using the LRFD provisions, determine whether the W14×90, A992 column shown in Figure 8.11 is adequate to carry the imposed loading.

GIVEN: An exterior column from an intermediate level of a multi-story moment frame is shown in Figure 8.11. The column is part of a braced frame out of the plane of the figure. Figure 8.11a shows the member to be checked. The same column section will be used for the level above and below the column *AB*. A first-order analysis of the frame for gravity loads plus the minimum lateral load results in the forces shown in Figure 8.11b, whereas the results for gravity plus wind are shown in Figure 8.11c. Assume that the frame drift under service loads is limited to height/300.

SOLUTION

Step 1: Determine the column effective length factor in the plane of bending.

Using the effective length alignment chart introduced in Chapter 5 and given in the Commentary, determine the effective length for buckling in the plane of the moment frame.

$$G_A = G_B = \frac{2\left(\dfrac{999}{12.5}\right)}{\left(\dfrac{2100}{30.0}\right)} = 2.28$$

thus, $K = 1.66$

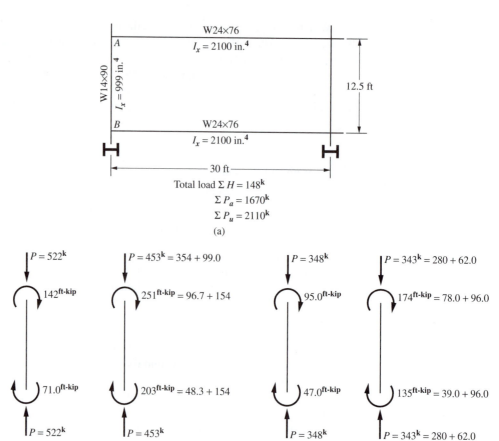

Figure 8.11 Exterior Column From an Intermediate Level of a Multistory Rigid Frame (Example 8.2).

Step 2: Determine the controlling effective length.
With $r_x/r_y = 1.66$ for the W14×90

$$KL_x = 1.66(12.5)/1.66 = 12.5 \text{ ft}$$

$$KL_y = 1.0(12.5) = 12.5 \text{ ft}$$

Step 3: Determine the column design axial strength.
From the column tables, Manual Table 4-1, for $KL = 12.5$ ft

$$\phi P_n = 1060 \text{ kips}$$

Step 4: Determine the first-order moments and forces for the loading case including wind.
The column end moments given in Figure 8.11c are a combination of moments resulting from a nonsway gravity load analysis and a wind analysis. These moments are
Moment for end A

$$M_{nt} = 96.7 \text{ ft-kips}$$

$$M_{lt} = 154 \text{ ft-kips}$$

Moment for end B

$$M_{nt} = 48.3 \text{ ft-kips}$$

$$M_{lt} = 154 \text{ ft-kips}$$

Compression

$$P_{nt} = 354 \text{ kips}$$

$$P_{lt} = 99.0 \text{ kips}$$

Step 5: Determine the second-order moments by amplifying the first-order moments. The no-translation moments must be amplified by B_1, where

$$\frac{M_1}{M_2} = \frac{48.3}{96.7} = 0.50$$

$$C_m = 0.6 - 0.4(0.50) = 0.4$$

$$P_{e1} = \frac{\pi^2(29,000)(999)}{(1.0(12.5)(12))^2} = 12,700 \text{ kips}$$

$$B_1 = \frac{0.4}{1 - \dfrac{354}{12,700}} = 0.411 < 1.0$$

Therefore, $B_1 = 1.0$.

The translation moments must be amplified by B_2. Because the complete design is not known and the design drift limit is known, Equation 8.11 using the drift index formulation is used here.

Additional given information:

The total lateral load on this story is

$$\Sigma H = 148 \text{ kips}$$

The total gravity load for this load combination is

$$\Sigma P_u = 2110 \text{ kips}$$

The drift limit under the lateral load of 148 kips is

$$\Delta_H = L/300 = 12.5(12)/300 = 0.50 \text{ in.}$$

Thus, with $\alpha = 1.0$ for LRFD

$$\Sigma P_{e2} = \frac{0.85(148)(12.5)(12)}{0.50} = 37,700 \text{ kips}$$

$$B_2 = \frac{1.0}{1.0 - \left(\dfrac{(1.0)2110}{37,700}\right)} = 1.06$$

Thus, the second-order compressive force and moments are

$$P_r = 354 + 1.06(99) = 459 \text{ kips}$$

$$M_r = 1.0(96.7) + 1.06(154) = 260 \text{ ft-kips}$$

These represent the required strength for this load combination.

Step 6: Determine whether this shape will provide the required strength based on the appropriate interaction equation.

The unbraced length of the compression flange for pure bending is 12.5 ft, which is less than $L_p = 15.2$ ft for this section, taking into account that its flange is noncompact. Thus, from Manual Table 3-2 the design moment strength of the section is

$$\phi M_n = 573 \text{ ft-kips}$$

Determine the appropriate interaction equation.

$$\frac{P_r}{\phi P_n} = \frac{459}{1060} = 0.433 > 0.2, \quad \text{thus use Equation 8.2 (H1-1a)}$$

which yields

$$0.433 + \frac{8}{9}\left(\frac{260}{573}\right) = 0.836 < 1.0$$

Thus

> the W14×90 is adequate for this load combination

Step 7:   Check the section for the gravity-only load combination, $1.2D + 1.6L$.

Because this is a gravity-only load combination, Specification Section C2.2a requires that the analysis include a minimum lateral load of 0.002 times the gravity load. For this frame the minimum lateral load is $0.002(2110) = 4.2$ kips at this level.

The forces and moments given in Figure 8.11b include the effects of this minimum lateral load. The magnitude of the lateral translation effect is small in this case. Thus, the forces and moments used for this check will be assumed to come from a no-translation case.

A quick review of the determination of B_1 from the first part of this solution indicates that there is no change, thus

$$B_1 = 1.0$$

With the assumption that there is no lateral load

$$M_{lt} = 0.0 \text{ and } B_2 \text{ is unnecessary}$$

Again using Equation 8.2 (H1-1a)

$$\frac{522}{1060} + \frac{8}{9}\left(\frac{142}{573}\right) = 0.713 < 1.0$$

Thus, the

> W14×90 is adequate for both loading conditions considered

EXAMPLE 8.2b
Strength Check for Combined Compression and Bending by ASD

GOAL: Using the ASD provisions, determine whether the W14×90, A992 column shown in Figure 8.11 is adequate to carry the imposed loading.

GIVEN: An exterior column from an intermediate level of a multi-story moment frame is shown in Figure 8.11. The column is part of a braced frame out of the plane of the figure. Figure 8.11a shows the member to be checked. The same column section will be used for the level above and below the column AB. A first-order analysis of the frame for gravity loads plus the minimum lateral load results in the forces shown in Figure 8.11d, whereas the results for gravity plus wind are shown in Figure 8.11e. Assume that the frame drift under service loads is limited to height/300.

SOLUTION

Step 1:   Determine the column effective length factor in the plane of bending.

Using the effective length alignment chart introduced in Chapter 5 and given in the Commentary, determine the effective length for buckling in the plane of the moment frame.

$$G_A = G_B = \frac{2\left(\dfrac{999}{12.5}\right)}{\left(\dfrac{2100}{30.0}\right)} = 2.28$$

thus, $K = 1.66$

Step 2: Determine the controlling effective length.
With $r_x/r_y = 1.66$ for the W14×90

$$KL_x = 1.66(12.5)/1.66 = 12.5 \text{ ft}$$

$$KL_y = 1.0(12.5) = 12.5 \text{ ft}$$

Step 3: Determine the column allowable axial strength.
From the column tables, Manual Table 4-1, for $KL = 12.5$ ft

$$P_n/\Omega = 703 \text{ kips}$$

Step 4: Determine the first-order moments and forces for the loading case including wind.
The column end moments given in Figure 8.11e are a combination of moments resulting from a nonsway gravity load analysis and a wind analysis. These moments are

Moment for end A

$$M_{nt} = 78.0 \text{ ft-kips}$$

$$M_{lt} = 96.0 \text{ ft-kips}$$

Moment for end B

$$M_{nt} = 39.0 \text{ ft-kips}$$

$$M_{lt} = 96.0 \text{ ft-kips}$$

Compression

$$P_{nt} = 280 \text{ kips}$$

$$P_{lt} = 62.0 \text{ kips}$$

Step 5: Determine the second-order moments by amplifying the first-order moments.
The no-translation moments must be amplified by B_1, where

$$\frac{M_1}{M_2} = \frac{39.0}{78.0} = 0.50$$

$$C_m = 0.6 - 0.4(0.50) = 0.4$$

$$P_{e1} = \frac{\pi^2(29,000)(999)}{(1.0(12.5)(12))^2} = 12,700 \text{ kips}$$

$$B_1 = \frac{0.4}{1 - \dfrac{1.6(280)}{12,700}} = 0.415 < 1.0$$

Therefore, $B_1 = 1.0$.
The translation moment must be amplified by B_2. Because the complete design is not known and the design drift limit is known, Equation 8.11 using the drift index formulation is used here.

Additional given information:

The total lateral load on this story is

$$\Sigma H = 148 \text{ kips}$$

The total gravity load for this load combination is

$$\Sigma P_a = 1670 \text{ kips}$$

The drift limit under the lateral load of 148 kips is

$$\Delta_H = L/300 = 12.5(12)/300 = 0.50 \text{ in.}$$

Thus, with $\alpha = 1.6$ for ASD

$$\Sigma P_{e2} = \frac{0.85(148)(12.5)(12)}{0.50} = 37,700 \text{ kips}$$

$$B_2 = \frac{1.0}{1.0 - \left(\dfrac{(1.6)1670}{37,700}\right)} = 1.08$$

Thus, the second-order compressive force and moment are

$$P_r = 280 + 1.08(62.0) = 347 \text{ kips}$$

$$M_r = 1.0(78.0) + 1.08(96.0) = 182 \text{ ft kips}$$

These represent the required strength for this load combination.

Step 6: Determine whether this shape will provide the required strength based on the appropriate interaction equation.

The unbraced length of the compression flange for pure bending is 12.5 ft, which is less than $L_p = 15.2$ ft for this section, taking into account that its flange is noncompact. Thus, from Manual Table 3-2 the allowable moment strength of the section is

$$M_n/\Omega = 382 \text{ ft-kips}$$

Determine the appropriate interaction equation.

$$\frac{P_r}{P_n/\Omega} = \frac{347}{703} = 0.494 > 0.2, \text{ thus use Equation 8.2 (H1-1a)}$$

which yields

$$0.494 + \frac{8}{9}\left(\frac{182}{382}\right) = 0.918 < 1.0$$

Thus

> the W14×90 is adequate for this load combination

Step 7: Check the section for the gravity-only load combination, $D + L$.

Because this is a gravity-only load combination, Specification Section C2.2a requires that the analysis include a minimum lateral load of 0.002 times the gravity load, which for ASD requires the use of $\alpha = 1.6$. For this frame the minimum lateral load is $0.002(1.6)(1670) = 5.34$ kips at this level.

The forces and moments given in Figure 8.11d include the effects of this minimum lateral load. The magnitude of the lateral translation effect is small in this case. Thus, the forces and moments used for this check will be assumed to come from a no-translation case.

A quick review of the determination of B_1 from the first part of this solution indicates that there is no change, thus

$$B_1 = 1.0$$

With the assumption that there is no lateral load

$$M_{lt} = 0.0 \text{ and } B_2 \text{ is unnecessary}$$

Again using Equation 8.2 (H1-1a)

$$\frac{348}{703} + \frac{8}{9}\left(\frac{95.0}{382}\right) = 0.716 < 1.0$$

Thus, the

W14×90 is adequate for both loading conditions considered

The moments in the beams and the beam-to-column connections must also be amplified for the critical case to account for the second-order effects. This is done by considering equilibrium of the beam-to-column joint. The amplified moments in the column above and below the joint are added together and this sum is distributed to the beams which frame into the joint according to their stiffnesses. These moments then establish the connection design moments.

8.7   SPECIFICATION PROVISIONS FOR STABILITY ANALYSIS AND DESIGN

Up to this point, the discussion of the interaction of compression and bending has concentrated on the development of the interaction equations and one approach to incorporate second-order effects. The Specification actually provides three overlapping approaches to deal with these two closely linked issues. As mentioned earlier, the most direct approach is to use the Direct Analysis Method described in Appendix 7.

The Direct Analysis Method yields forces and moments that can be used directly in the interaction equations of Chapter H. The nominal strength of members is determined using the strength provisions already discussed with the additional provision that the effective length of compression members shall be taken as the actual length, that is, $K = 1.0$. The analysis required in this approach can be either a general second-order analysis or the amplified first-order analysis already presented. There are no limitations on the use of the direct analysis method, although the specific provisions do require modifications of member stiffnesses and the application of additional lateral loads, called notional loads. The other two design methods given in the Specification are based on the direct analysis method.

The second method, Design by Second-Order Analysis, given in Specification Section C2.2a, is the approach already described in this chapter for braced and unbraced frames. This approach is valid as long as the ratio of second-order deflection to first-order deflection, Δ_2/Δ_1, is equal to or less than 1.5. Another way to state this requirement is to remember that $\Delta_2/\Delta_1 = B_2$, therefore the method is valid as long as $B_2 \le 1.5$. An additional modification can be applied when $B_2 \le 1.1$. In this case, columns can be designed using $K = 1.0$. The Manual calls this method the *effective length method* because it is essentially the same method used in recent practice with the addition of the requirement of a minimum lateral load to be applied in all load cases.

A third method is given in the Specification, Section C2.2b, and called Design by First-Order Analysis. This approach permits design without direct consideration of second-order effects except through the application of additional lateral loads. This is possible because of the many limits placed on the implementation of this method. For further information on this approach, consult Section C2.2b and the Manual.

8.8   INITIAL BEAM-COLUMN SELECTION

Beam-column design is a trial-and-error process that requires the beam-column section be known before any of the critical parameters can be determined for use in the appropriate interaction equations. There are numerous approaches to determining a preliminary beam-column size. Each incorporates its own level of sophistication and results in its own level of accuracy. Regardless of the approach used to select the trial section, one factor remains: The trial section must ultimately satisfy the appropriate interaction equations.

To establish a simple, yet useful, approach to selecting a trial section, Equation H1-1a is rewritten. Using Equation 8.2 and multiplying each term by P_c yields

$$P_r + \frac{8}{9}\frac{M_{rx}P_c}{M_{cx}} + \frac{8}{9}\frac{M_{ry}P_c}{M_{cy}} \le P_c \tag{8.14}$$

Multiplying the third term by M_{cx}/M_{cx}, letting

$$m = \frac{8P_c}{9M_{cx}} \quad \text{and} \quad U = \frac{M_{cx}}{M_{cy}}$$

and substituting into Equation 8.14 yields

$$P_r + mM_{rx} + mUM_{ry} \le P_c \tag{8.15}$$

Because Equation 8.15 calls for the comparison of the left side of the equation to the column strength, P_c, Equation 8.15 can be thought of as an effective axial load. Thus

$$P_{eff} = P_r + mM_{rx} + mUM_{ry} \le P_c \tag{8.16}$$

The accuracy used in the evaluation of m and U dictates the accuracy with which Equation 8.16 represents the strength of the column being selected. Because at this point in a design the actual column section is not known, exact values of m and U cannot be determined.

Past editions of the AISC Manual have presented numerous approaches to the evaluation of these multipliers. A simpler approach however, is more useful for preliminary design. If the influence of the length, that is, all buckling influence on P_c and M_{cx}, is neglected, the ratio, P_c/M_{cx}, becomes A/Z_x and $m = 8A/9Z_x$. Evaluation of this m for all W6 to W14 shapes with the inclusion of a units correction factor of 12 results in the average m values given in Table 8.2. If the relationship between the area, A, and the plastic section modulus, Z_x, is established using an approximate internal moment arm of $0.89d$, where d is the nominal depth of the member in inches, m reduces to $24/d$. This value is also presented in Table 8.2. This new m is close enough to the average m that it may be readily used for preliminary design.

Table 8.2   Simplified Bending Factors

Shape	m	24/d	U
W6	4.41	4.00	3.01
W8	3.25	3.00	3.11
W10	2.62	2.40	3.62
W12	2.08	2.00	3.47
W14	1.72	1.71	2.86

When bending occurs about the y-axis, U must be evaluated. A review of the same W6 to W14 shapes results in the average U values given in Table 8.2. However, an in-depth review of the U values for these sections shows that only the smallest sections for each nominal depth have U values appreciably larger than 3. Thus, a reasonable value of $U = 3.0$ can be used for the first trial.

More accurate evaluations of these multipliers, including length effects, have been conducted, but there does not appear to be a need for this additional accuracy in a preliminary design. Once the initial section is selected, however, the actual Specification provisions must be satisfied.

EXAMPLE 8.3a
Initial Trial Section
Selection by LRFD

GOAL: Determine the initial trial section for a column.

GIVEN: The loadings of Figure 8.11c are to be used. Assume the column is a W14 and use A992 steel. Also, use the simplified values of Table 8.2.

SOLUTION

Step 1:   Obtain the required strength from Figure 8.11c.

$$P_u = 453 \text{ kips}$$
$$M_u = 251 \text{ ft-kips}$$

Step 2:   Determine the effective load by combining the axial force and the bending moment.
For a W14, $m = 1.71$, thus

$$P_{eff} = 453 + 1.71(251) = 882 \text{ kips}$$

Step 3:   Select a trial column size to carry the required force, P_{eff}.
Using an effective length $KL = 12.5$ ft, from Manual Table 4-1, the lightest W14 to carry this load is

$$\boxed{\text{W14}\times\text{90} \text{ with }   \phi P_n = 1060 \text{ kips}}$$

Example 8.2 showed that this column adequately carries the imposed load. Because the approach used here is expected to be conservative, it would be appropriate to consider the next smaller selection, a W14×82, and check it against the appropriate interaction equations.

EXAMPLE 8.3b
Initial Trial Section
Selection by ASD

SOLUTION

GOAL: Determine the initial trial section for a column.

GIVEN: The loadings of Figure 8.11e are to be used. Assume the column is a W14 and use A992 steel. Also, use the simplified values of Table 8.2.

Step 1:   Obtain the required strength from Figure 8.11e.

$$P_a = 343 \text{ kips}$$

$$M_a = 174 \text{ ft-kips}$$

Step 2:   Determine the effective load by combining the axial force and the bending moment. For a W14, $m = 1.71$, thus

$$P_{eff} = 343 + 1.71(174) = 641 \text{ kips}$$

Step 3:   Select a trial column size to carry the required force, P_{eff}.

Using an effective length $KL = 12.5$ ft, from Manual Table 4-1, the lightest W14 to carry this load is

> W14×90 with $P_n/\Omega = 703$ kips

Example 8.2 showed that this column adequately carries the imposed load. Because the approach used here is expected to be conservative, it would be appropriate to consider the next smaller selection, a W14×82, and check it against the appropriate interaction equations.

Every column section selected must be checked through the appropriate interaction equations. Thus, the process for the initial selection should be quick and reasonable. The experienced designer will rapidly learn to rely on that experience rather than these simplified approaches.

8.9   BEAM-COLUMN DESIGN USING MANUAL PART 6

Perhaps the most useful tables in the Manual are those in Part 6, Design of Members Subject to Combined Loading. Although these tables are presented here as they relate to combined loading, they can also be used for pure compression, pure bending, and pure tension, each with only a slight modification needed.

The Specification interaction equation, H1-1a, is repeated here in a slightly modified form as

$$\left(\frac{1}{P_c}\right)P_r + \left(\frac{8}{9M_{cx}}\right)M_{rx} + \left(\frac{8}{9M_{cy}}\right)M_{ry} \leq 1.0$$

This equation can be rewritten as

$$pP_r + b_xM_{rx} + b_yM_{ry} \leq 1.0 \qquad (8.17)$$

where

$$p = \frac{1}{P_c}$$

$$b_x = \frac{8}{9M_{cx}}$$

$$b_y = \frac{8}{9M_{cy}}$$

Equation H1-1b can then be rewritten as

$$\tfrac{1}{2}\, p P_r + \tfrac{9}{8}\, (b_x M_{rx} + b_y M_{ry}) \leq 1.0 \tag{8.18}$$

It should be clear that p, b_x, and b_y are functions of the strength of the member. In Example 8.2, the column section was checked by determining the axial strength and bending strength from the appropriate beam and column equations or corresponding Manual tables. Using the formulation presented here in Equations 8.17 and 8.18, all the necessary information is obtained from a single table in Part 6 of the Manual.

Figure 8.12 is a portion of Manual Table 6-1. It shows that the compressive strength term, p, for a given section is a function of unbraced length about the weak axis of the member. This table is used in exactly the same way as the column tables in Part 4 of the Manual. The strong axis bending strength, b_x, is a function of the unbraced length of the compression flange of the beam. Previously, this information was available only through the beam curves in Part 3 of the Manual. Weak axis bending is not a function of length so only one value for b_y is found for each shape. Although not used for beam-columns, when tension is combined with bending, the table also provides values for t_y and t_r.

EXAMPLE 8.4a
Combined Strength Check Using Manual Part 6 and LRFD

GOAL: Check the strength of a beam-column using Manual Part 6 and compare to the results of Example 8.2a.

GIVEN: It has already been shown that the W14x90 column of Example 8.2a is adequate by LRFD. Use the required strength values given in Example 8.2a and recheck this shape using the values found in Figure 8.12 or Manual Table 6-1.

SOLUTION

Step 1: Determine the values needed from Manual Table 6-1 (Figure 8.12). The column is required to carry a compressive force with an effective length about the y-axis of 12.5 ft and an x-axis moment with an unbraced length of 12.5 ft. Thus, from Figure 8.12

$$p = 0.000947$$
$$b_x = 0.00155$$

Step 2: Determine which interaction equation to use.

$$p P_r = 0.000947(459) = 0.435 > 0.2$$

Therefore, use Equation 8.17.
$$p P_r + b_x M_r \leq 1.0$$

$$\boxed{0.000947(459) + 0.00155(260) = 0.838 < 1.0}$$

Therefore, as previously determined in Example 8.2a, the shape is adequate for this column and this load combination. The results from Manual Tables 6-2 and 4-1 have slight differences due to rounding. Thus, the results by this approach will not always be exactly the same as those from the approach of Example 8.2a.

| | | Table 6–1 (continued) **Combined Axial and Bending** W Shapes. | | | | | | | F_y = 50 ksi | | | |

Table 6–1 (continued) Combined Axial and Bending W Shapes. F_y = 50 ksi

Shape		W14×										
		90[f]				82				74			
Design		$p \times 10^3$ (kips)$^{-1}$		$b_x \times 10^3$ (kip-ft)$^{-1}$		$p \times 10^3$ (kips)$^{-1}$		$b_x \times 10^3$ (kip-ft)$^{-1}$		$p \times 10^3$ (kips)$^{-1}$		$b_x \times 10^3$ (kip-ft)$^{-1}$	
		ASD	LRFD	ASD	LRFD	ASD	LRFD	ASD	LRFD	ASD	LRFD	ASD	LRFD
	0	1.26	0.840	2.33	1.55	1.39	0.924	2.56	1.71	1.53	1.02	2.83	1.88
	6	1.30	0.863	2.33	1.55	1.48	0.983	2.56	1.71	1.63	1.09	2.83	1.88
	7	1.31	0.872	2.33	1.55	1.51	1.00	2.56	1.71	1.67	1.11	2.83	1.88
	8	1.33	0.882	2.33	1.55	1.55	1.03	2.56	1.71	1.71	1.14	2.83	1.88
	9	1.34	0.894	2.33	1.55	1.60	1.06	2.57	1.71	1.76	1.17	2.84	1.89
	10	1.36	0.907	2.33	1.55	1.65	1.10	2.61	1.74	1.82	1.21	2.89	1.92
	11	1.38	0.921	2.33	1.55	1.71	1.14	2.66	1.77	1.89	1.26	2.94	1.96
	12	1.41	0.938	2.33	1.55	1.78	1.18	2.70	1.80	1.96	1.31	2.99	1.99
	13	1.44	0.956	2.33	1.55	1.85	1.23	2.75	1.83	2.05	1.36	3.05	2.03
	14	1.47	0.976	2.33	1.55	1.94	1.29	2.79	1.86	2.14	1.43	3.10	2.07
	15	1.50	0.998	2.33	1.55	2.04	1.36	2.84	1.89	2.25	1.50	3.16	2.10
	16	1.54	1.02	2.35	1.57	2.15	1.43	2.89	1.92	2.38	1.58	3.22	2.15
	17	1.58	1.05	2.38	1.59	2.28	1.52	2.94	1.96	2.52	1.67	3.29	2.19
	18	1.62	1.08	2.42	1.61	2.42	1.61	3.00	1.99	2.67	1.78	3.35	2.23
	19	1.67	1.11	2.45	1.63	2.58	1.71	3.05	2.03	2.85	1.89	3.42	2.28
	20	1.72	1.14	2.48	1.65	2.75	1.83	3.11	2.07	3.04	2.02	3.50	2.33
	22	1.83	1.22	2.55	1.70	3.18	2.12	3.23	2.15	3.51	2.34	3.65	2.43
	24	1.97	1.31	2.62	1.74	3.73	2.48	3.37	2.24	4.12	2.74	3.82	2.54
	26	2.12	1.41	2.70	1.79	4.38	2.91	3.51	2.34	4.83	3.22	4.00	2.66
	28	2.31	1.53	2.78	1.85	5.08	3.38	3.67	2.44	5.61	3.73	4.20	2.80
	30	2.52	1.68	2.86	1.91	5.83	3.88	3.84	2.55	6.44	4.28	4.42	2.94
	32	2.77	1.85	2.95	1.97	6.63	4.41	4.03	2.68	7.32	4.87	4.73	3.15
	34	3.07	2.04	3.05	2.03	7.49	4.98	4.28	2.85	8.27	5.50	5.09	3.38
	36	3.43	2.28	3.16	2.10	8.39	5.59	4.57	3.04	9.27	6.17	5.44	3.62
	38	3.82	2.54	3.27	2.17	9.35	6.22	4.86	3.24	10.3	6.87	5.80	3.86
	40	4.23	2.81	3.39	2.25	10.4	6.90	5.15	3.43	11.4	7.61	6.15	4.09
Other Constants and Properties													
$b_y \times 10^3$ (kip-ft)$^{-1}$		4.90		3.26		7.95		5.29		8.80		5.85	
$t_y \times 10^3$ (kips)$^{-1}$		1.26		0.840		1.39		0.924		1.53		1.02	
$t_r \times 10^3$ (kips)$^{-1}$		1.55		1.03		1.71		1.14		1.88		1.26	
r_x/r_y		1.66				2.44				2.44			

Effective length KL (ft) with respect to least radius of gyration r_y or Unbraced Length L_b (ft) for X-X axis bending

[f] Shape does not meet compact limit for flexure with F_y = 50 ksi.

Figure 8.12 Combined Axial and Bending Strength for W-Shapes. Copyright © American Institute of Steel Construction, Inc. Reprinted with Permission. All rights reserved.

EXAMPLE 8.4b
Combined Strength
Check Using Manual
Part 6 and ASD

GOAL: Check the strength of a beam-column using Manual Part 6 and compare to the results of Example 8.2b.

GIVEN: It has already been shown that the W14×90 column of Example 8.2b is adequate by ASD. Use the required strength values given in Example 8.2b and recheck this shape using the values found in Figure 8.12 or Manual Table 6-1.

SOLUTION

Step 1:   Determine the values needed from Manual Table 6-1 (Figure 8.12). The column is required to carry a compressive force with an effective length about the y-axis of 12.5 ft and an x-axis moment with an unbraced length of 12.5 ft. Thus, from Figure 8.12

$$p = 0.00143$$
$$b_x = 0.00233$$

Step 2:   Determine which interaction equation to use.

$$pP_r = 0.00143(347) = 0.496 > 0.2$$

Therefore, use Equation 8.17.

$$pP_r + b_xM_r \leq 1.0$$

$$\boxed{0.00143(347) + 0.00233(182) = 0.920 < 1.0}$$

Therefore, as previously determined in Example 8.2b, the shape is adequate for this column and this load combination. The results from Manual Tables 6-2 and 4-1 have slight differences due to rounding. Thus, the results by this approach will not always be exactly the same as those from the approach of Example 8.2b.

8.10   COMBINED SIMPLE AND RIGID FRAMES

The practical design of steel structures often results in frames that combine segments of rigidly connected elements with segments that are pin connected. If these structures rely on the moment frame to resist lateral load and to provide the overall stability of the structure, the rigidly connected columns are called upon to carry more load than what appears to be directly applied to them. In these combined simple and moment frames, the simple columns "lean" on the moment frames in order to maintain their stability. Thus, they are often called *leaning columns*. They are also called *gravity columns* because they participate only in carrying gravity loads. These columns can be designed with an effective length factor, $K = 1.0$. Because these leaning columns have no lateral stability of their own, the moment frame columns must be designed to provide the lateral stability for the full frame. Although this combination of framing types makes design of a structure more complicated, it can also be economically advantageous, because the combination can reduce the number of moment connections for the full structure.

Numerous design approaches have been proposed for consideration of the *leaning column* and associated moment frame design.[3,4] Yura proposes to design columns that

---

[3]Yura, J. A., "The Effective Length of Columns in Unbraced Frames," *Engineering Journal*, AISC, Vol. 8, No. 2, 1971, pp. 37–42.

[4]LeMessurier, W. J., "A Practical Method of Second Order Analysis," *Engineering Journal*, AISC, Vol. 14, No. 2, 1977, pp. 49–67.

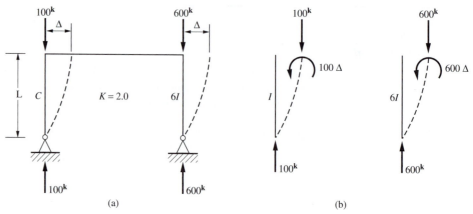

Figure 8.13 Pinned Base Unbraced Frame.

provide lateral stability for the total load on the frame at the story in question, whereas LeMessurier presents a modified effective length factor that accounts for the full frame stability. Perhaps the most straightforward approach is that presented by Yura, as follows.

The two-column frame shown in Figure 8.13a is a moment frame with pinned base columns and a rigidly connected beam. The column sizes are selected so that, under the loads shown, they buckle in a sidesway mode simultaneously because their load is directly proportional to the stiffness of the members. Equilibrium in the displaced position is shown in Figure 8.13b. The lateral displacement of the frame, Δ, results in a moment at the top of each column equal to the load applied on the column times the displacement, as shown. These are the second-order effects discussed in Section 8.6. The total load on the frame is 700 kips and the total $P\Delta$ moment is $700\,\Delta$, divided between the two columns based on the load that each carries.

If the load on the right-hand column is reduced to 500 kips, the column does not buckle sideways because the moment at the top is now less than $600\,\Delta$. To reach the buckling condition, a horizontal force must be applied at the top of the column, as shown in Figure 8.14b. This force can result only from action on the left column that is transmitted through

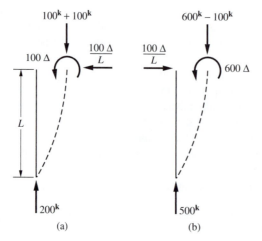

Figure 8.14 Columns From Unbraced Frame with Revised Loading.

the beam. Equilibrium of the left column, shown in Figure 8.14a, requires that an additional column load of 100 kips be applied to that column in order for the load on the frame to be in equilibrium in this displaced position. The total frame capacity is still 700 kips and the total second order moment is still 700Δ.

The maximum load that an individual column can resist is limited to that permitted for the column in a braced frame for which $K = 1.0$. In this example case, the left column could resist 400 kips and the right column 2400 kips. This is an increase of 4 times the load originally on the column because the effective length factor for each column would be reduced from 2.0 to 1.0. The additional capacity of the left column is only with respect to the bending axis. The column would have the same capacity about the other axis as it did prior to reducing the load on the right column.

The ability of one column to carry increased load when another column in the frame is called upon to carry less than its critical load for lateral buckling is an important characteristic. This allows a pin-ended column to lean on a moment frame column, provided that the total gravity load on the frame can be carried by the rigid frame.

EXAMPLE 8.5a
Moment Frame Strength and Stability by LRFD

GOAL: Determine whether the structure shown in Figure 8.15 has sufficient strength and stability to carry the imposed loads.

GIVEN: The frame shown in Figure 8.15 is similar to that in Example 8.1 except that the in-plane stability and lateral load resistance is provided by the moment frame action at the four corners. The exterior columns are W8×40 and the roof girder is assumed to be rigid. Out-of-plane stability and lateral load resistance is provided by X-bracing along column lines 1 and 4.

The loading is the same as that for Example 8.1: Dead Load = 50 psf, Snow Load = 20 psf, Roof Live Load = 10 psf, and Wind Load = 20 psf horizontal. Use A992 steel.

SOLUTION

Step 1: The analysis of the frame for gravity loads as given for Example 8.1 will be used. Because different load combinations may be critical, however, the analysis results for nominal Snow and nominal Dead Load are given in Figure 8.16b. The analysis results for nominal Wind Load acting to the left are given in Figure 8.16c.

Step 2: Determine the first-order forces and moments.
For ASCE 7 load case 3

$$P_u = 1.2(15.8) + 1.6(6.33) + 0.8(0.710) = 29.1 + 0.568 = 29.7 \text{ kips}$$

$$M_u = 1.2(20.5) + 1.6(8.20) + 0.8(32.0) = 37.7 + 25.6 = 63.3 \text{ ft-kips}$$

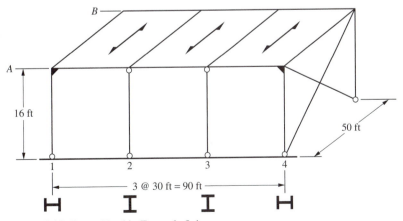

Figure 8.15 Frame Used in Example 8.4.

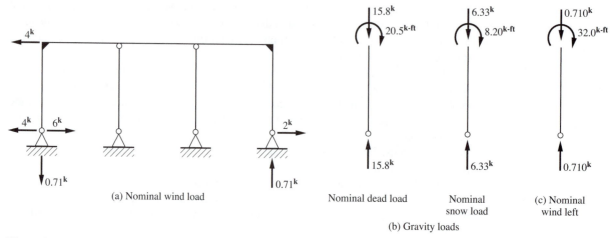

(a) Nominal wind load

(b) Gravity loads

Nominal dead load   Nominal   (c) Nominal
                           snow load   wind left

Figure 8.16 Nominal Wind Load, Snow Load, and Dead Load (Example 8.4).

For ASCE 7 load case 4

$$P_u = 1.2(15.8) + 0.5(6.33) + 1.6(0.710) = 22.1 + 1.34 = 23.4 \text{ kips}$$
$$M_u = 1.2(20.5) + 0.5(8.20) + 1.6(32.0) = 28.7 + 51.2 = 79.9 \text{ ft-kips}$$

Step 3: Determine the total story gravity load acting on one frame.

$$\text{Dead} = 0.05 \text{ ksf } (90 \text{ ft})(50 \text{ ft})/2 \text{ frames} = 113 \text{ kips}$$
$$\text{Snow} = 0.02 \text{ ksf } (90 \text{ ft})(50 \text{ ft})/2 \text{ frames} = 45.0 \text{ kips}$$

Step 4: Determine the second-order forces and moments for loading case 3.
From Step 2

$$P_u = 29.7 \text{ kips}, \quad M_{nt} = 37.7 \text{ ft-kips}, \quad M_{lt} = 25.6 \text{ ft-kips}$$

For the W8×40

$$A = 11.7 \text{ in.}^2, \quad I_x = 146 \text{ in.}^4, \quad r_x = 3.53 \text{ in.}, \quad r_x/r_y = 1.73$$

In the plane of the frame

$$C_m = 0.6 - 0.4\left(\frac{0}{37.7}\right) = 0.6$$

$$P_{e1} = \frac{\pi^2 EI_x}{(1.0L)^2} = \frac{\pi^2(29,000)(146)}{(16.0(12))^2} = 1130 \text{ kips}$$

and

$$B_1 = \frac{0.6}{1 - \dfrac{29.7}{1130}} = 0.616 < 1.0$$

Therefore, use $B_1 = 1.0$.

To determine the sway amplification, the total gravity load on the frame is

$$P_u = 1.2(113) + 1.6(45.0) = 208 \text{ kips}$$

A serviceability drift index of 0.003 is maintained under the actual wind loads. Therefore, $\Sigma H = 4.0$ kips and $\Delta/L = 0.003$ is used to determine the sway amplification factor.

$$B_2 = \frac{1}{1 - \left(\dfrac{\Sigma P_u}{\Sigma H}\left(\dfrac{\Delta}{L}\right)\right)} = \frac{1}{1 - \dfrac{208}{4.0}(0.003)} = 1.18$$

Thus, the second-order force and moment are

$$M_r = 1.0(37.7) + 1.18(25.6) = 67.9 \text{ ft-kips}$$

$$P_r = 29.1 + 1.18(0.568) = 29.8 \text{ kips}$$

Step 5:   Determine whether the column satisfies the interaction equation.

Because the roof beam is assumed to be rigid in this example, use the recommended design value of $K = 2.0$ from Figure 5.17 case f in the plane of the frame, $KL_x = 2.0(16.0) = 32.0$ ft. Out of the plane of the frame, this is a braced frame where $K = 1.0$; thus, $KL_y = 16.0$ ft.

Determine the critical buckling axis.

$$KL_{eff} = \frac{KL_x}{r_x/r_y} = \frac{32.0}{1.73} = 18.5 \text{ ft} > KL_y = 16.0 \text{ ft}$$

Thus, from Manual Table 4-1, using $KL_{eff} = 18.5$ ft

$$\phi P_n = 222 \text{ kips}$$

and from Manual Table 3-10 with an unbraced length of $L_b = 16$ ft

$$\phi M_{nx} = 128 \text{ ft-kips}$$

Determine the appropriate interaction equation to use.

$$\frac{P_r}{\phi P_n} = \frac{29.8}{222} = 0.134 < 0.2$$

Therefore, use Equation 8.3 (H1-1b).

$$\boxed{\frac{29.8}{2(222)} + \frac{67.9}{128} = 0.598 < 1.0}$$

Thus, the column is adequate for this load combination.

Step 6:   Determine the first-order forces and moments for loading case 4 from Step 2.

$$P_u = 23.4 \text{ kips}, \quad M_{nt} = 28.7 \text{ ft-kips}, \quad M_{lt} = 51.2 \text{ ft-kips}$$

Step 7:   Determine the second-order forces and moments.

In the plane of the frame, as in Step 4

$$C_m = 0.6 - 0.4\left(\frac{0}{28.7}\right) = 0.6$$

$$P_{e1} = \frac{\pi^2 EI_x}{(1.0L)^2} = \frac{\pi^2(29,000)(146)}{(16(12))^2} = 1130 \text{ kips}$$

and

$$B_1 = \frac{0.6}{1 - \dfrac{23.4}{1130}} = 0.613 < 1.0$$

Therefore, use $B_1 = 1.0$.

To determine the sway amplification, the total gravity load on the frame is

$$P_u = 1.2(113) + 0.5(45.0) = 158 \text{ kips}$$

Again, a serviceability drift index of 0.003 is maintained under the actual wind loads. Therefore, $\Sigma H = 4.0$ kips and $\Delta/L = 0.003$ is used to determine the sway amplification factor.

$$B_2 = \frac{1}{1 - \left(\dfrac{\Sigma P_u}{\Sigma H}\left(\dfrac{\Delta}{L}\right)\right)} = \frac{1}{1 - \dfrac{158}{4.0}(0.003)} = 1.13$$

Thus, the second-order moments are

$$M_r = 1.0(28.7) + 1.13(51.2) = 86.6 \text{ ft-kips}$$

and adding in the effect of lateral load (1.6W) amplified by B_2

$$P_r = 22.1 + 1.13(1.34) = 23.6 \text{ kips}$$

Step 8:   Determine whether the column satisfies the interaction equation.
Using the same strength values found in Step 5, determine the appropriate interaction equation

$$\frac{P_r}{\phi P_n} = \frac{23.6}{222} = 0.106 < 0.2$$

Therefore, use Equation 8.3 (H1-1b).

$$\boxed{\frac{23.6}{2(222)} + \frac{86.6}{128} = 0.730 < 1.0}$$

Thus, the column is adequate for this load combination also.

Step 9:   The W8×40 is shown to be adequate for gravity and wind loads in combination. Now, check to see that these columns have sufficient capacity to brace the interior pinned columns for gravity load only.

Step 10:   For stability in the plane of the frame, the total load on the structure is to be resisted by the four corner columns; thus

$$\text{Dead Load} = 0.05 \text{ ksf } (50 \text{ ft})(90 \text{ ft})/4 \text{ columns} = 56.3 \text{ kips}$$
$$\text{Snow Load} = 0.02 \text{ ksf } (50 \text{ ft})(90 \text{ ft})/4 \text{ columns} = 22.5 \text{ kips}$$

Thus, for load combination 3

$$P_u = 1.2(56.3) + 1.6(22.5) = 104 \text{ kips}$$
$$M_u = 1.2(20.5) + 1.6(8.20) = 37.7 \text{ ft-kips}$$

As determined in Step 5 for in plane buckling

$$\phi P_{nx} = 222 \text{ kips}$$
$$\phi M_{nx} = 128 \text{ ft-kips}$$

Checking for the appropriate interaction equation

$$\frac{P_u}{\phi P_n} = \frac{104}{222} = 0.468 > 0.2$$

Thus, use Equation 8.2 (H1-1a).
For an effective length $Kl_x = 16.0 \text{ ft}$, $P_{e1} = 1130 \text{ kips}$.
As before, $C_m = 0.6$. Thus

$$B_1 = \frac{0.6}{1 - \dfrac{104}{1130}} = 0.66 < 1.0$$

Therefore, use $B_1 = 1.0$
and

$$\boxed{\frac{P_u}{\phi P_n} + \frac{8}{9}\left(\frac{M_{ux}}{\phi M_{nx}}\right) = \frac{104}{222} + \frac{8}{9}\left(\frac{37.7}{128}\right) = 0.730 < 1.0}$$

Thus, the W8×40 is adequate for both strength under combined load and stability for supporting the leaning columns.

EXAMPLE 8.5b
Moment Frame Strength and Stability by ASD

GOAL: Determine whether the structure shown in Figure 8.15 has sufficient strength and stability to carry the imposed loads.

GIVEN: The frame shown in Figure 8.15 is similar to that in Example 8.1 except that the in-plane stability and lateral load resistance is provided by the rigid frame action at the four corners. The exterior columns are W8×40 and the roof girder is assumed to be rigid. Out-of-plane stability and lateral load resistance is provided by X-bracing along column lines 1 and 4.

The loading is the same as that for Example 8.1: Dead Load = 50 psf, Snow Load = 20 psf, Roof Live Load = 10 psf, and Wind Load = 20 psf horizontal. Use A992 steel.

SOLUTION

Step 1:   The analysis of the frame for gravity loads as given for Example 8.1 will be used. Because different load combinations may be critical, however, the analysis results for nominal Snow and nominal Dead Load are given in Figure 8.16b. The analysis results for nominal Wind Load acting to the left are given in Figure 8.16c.

Step 2:   Determine the first-order forces and moments.
For ASCE 7 load case 3

$$P_a = (15.8) + (6.33) = 22.1 \text{ kips}$$

$$M_a = (20.5) + (8.20) = 28.7 \text{ ft-kips}$$

For ASCE 7 load case 6

$$P_a = (15.8) + 0.75(6.33) + 0.75(0.710) = 20.5 + 0.533 = 21.0 \text{ kips}$$

$$M_a = (20.5) + 0.75(8.20) + 0.75(32.0) = 26.7 + 24.0 = 50.7 \text{ ft-kips}$$

Step 3:   Determine the total story gravity load acting on one frame.

$$\text{Dead} = 0.05 \text{ ksf } (90 \text{ ft})(50 \text{ ft})/2 \text{ frames} = 113 \text{ kips}$$

$$\text{Snow} = 0.02 \text{ ksf } (90 \text{ ft})(50 \text{ ft})/2 \text{ frames} = 45.0 \text{ kips}$$

Step 4:   Determine the second-order forces and moments for loading case 3.
From Step 2

$$P_a = 22.1 \text{ kips}, \quad M_{nt} = 28.7 \text{ ft-kips}, \quad M_{lt} = 0 \text{ ft-kips}$$

For the W8×40

$$A = 11.7 \text{ in.}^2, \quad I_x = 146 \text{ in.}^4, \quad r_x = 3.53 \text{ in.}, \quad r_x/r_y = 1.73$$

In the plane of the frame

$$C_m = 0.6 - 0.4\left(\frac{0}{28.7}\right) = 0.6$$

$$P_{e1} = \frac{\pi^2 EI_x}{(1.0L)^2} = \frac{\pi^2(29,000)(146)}{(16.0(12))^2} = 1130 \text{ kips}$$

and

$$B_1 = \frac{0.6}{1 - \dfrac{1.6(22.1)}{1130}} = 0.619 < 1.0$$

Therefore, use $B_1 = 1.0$.

To determine the sway amplification, the total gravity load on the frame is

$$P_a = (113) + (45.0) = 158 \text{ kips}$$

A serviceability drift index of 0.003 is maintained under the actual wind loads. Therefore, $\Sigma H = 4.0$ kips and $\Delta/L = 0.003$ is used to determine the sway amplification factor. Again, for ASD, $\alpha = 1.6$.

$$B_2 = \frac{1}{1 - \left(\dfrac{\alpha \Sigma P_a}{\Sigma H}\left(\dfrac{\Delta}{L}\right)\right)} = \frac{1}{1 - \dfrac{1.6(158)}{4.0}(0.003)} = 1.23$$

Thus, the second-order force and moment are

$$M_r = 1.0(28.7) + 1.23(0) = 28.7 \text{ ft-kips}$$

$$P_r = 22.1 + 1.23(0) = 2.1 \text{ kips}$$

Step 5: Determine whether the column satisfies the interaction equation.

Because the roof beam is assumed to be rigid in this example, use the recommended design value of $K = 2.0$ from Figure 5.17 case f in the plane of the frame, $KL_x = 2(16.0) = 32.0$ ft. Out of the plane of the frame, this is a braced frame where $K = 1.0$; thus, $KL_y = 16.0$ ft.

Determining the critical buckling axis.

$$KL_{eff} = \frac{KL_x}{r_x/r_y} = \frac{32}{1.73} = 18.5 \text{ ft} > KL_y = 16.0 \text{ ft}$$

Thus, from Manual Table 4-1, using $KL_{eff} = 18.5$ ft

$$P_n/\Omega = 148 \text{ kips}$$

and from Manual Table 3-10 with an unbraced length of $L_b = 16$ ft

$$M_{nx}/\Omega = 85.0 \text{ ft-kips}$$

Determine the appropriate interaction equation to use.

$$\frac{P_r}{P_n/\Omega} = \frac{22.1}{148} = 0.149 < 0.2$$

Therefore, use Equation 8.3 (H1-1b).

$$\boxed{\frac{22.1}{2(148)} + \frac{28.7}{85.0} = 0.412 < 1.0}$$

Thus, the column is adequate for this load combination.

Step 6: Determine the first-order forces and moments for loading case 6 from Step 2.

$$P_a = 21.0 \text{ kips}, \quad M_{nt} = 26.7 \text{ ft-kips}, \quad M_{lt} = 24.0 \text{ ft-kips}$$

Step 7:   Determine the second-order forces and moments.
In the plane of the frame, as in Step 4

$$C_m = 0.6 - 0.4\left(\frac{0}{26.7}\right) = 0.6$$

$$P_{e1} = \frac{\pi^2 EI_x}{(1.0L)^2} = \frac{\pi^2 (29,000)(146)}{(16(12))^2} = 1130 \text{ kips}$$

and

$$B_1 = \frac{0.6}{1 - \dfrac{1.6(21.0)}{1130}} = 0.618 < 1.0$$

Therefore, use $B_1 = 1.0$.
To determine the sway amplification, the total gravity load on the frame is

$$P_a = (113) + 0.75(45.0) = 147 \text{ kips}$$

Again, a serviceability drift index of 0.003 is maintained under the actual wind loads. Therefore, $\Sigma H = 4.0$ kips and $\Delta/L = 0.003$ is used to determine the sway amplification factor.

$$B_2 = \frac{1}{1 - \left(\dfrac{\alpha \Sigma P_a}{\Sigma H}\left(\dfrac{\Delta}{L}\right)\right)} = \frac{1}{1 - \dfrac{1.6(147)}{4.0}(0.003)} = 1.21$$

Thus, the second-order force and moment are

$$M_r = 1.0(26.7) + 1.21(24.0) = 55.7 \text{ ft-kips}$$

and adding in the lateral load effect amplified by B_2

$$P_r = 20.5 + 1.21(0.533) = 21.1 \text{ kips}$$

Step 8:   Determine whether the column satisfies the interaction equation.
Using the same values found in Step 5, determine the appropriate interaction equation

$$\frac{P_r}{P_n/\Omega} = \frac{21.1}{148} = 0.143 < 0.2$$

Therefore, use Equation 8.3 (H1-1b).

$$\boxed{\frac{21.1}{2(148)} + \frac{55.7}{85.0} = 0.727 < 1.0}$$

Thus, the column is adequate for this load combination also.

Step 9:   The W8×40 is shown to be adequate for gravity and wind loads in combination. Now, check to see that these columns have sufficient capacity to brace the interior pinned columns for gravity load only.

Step 10:   For stability in the plane of the frame, the total load on the structure is to be resisted by the four corner columns; thus

$$\text{Dead Load} = 0.05 \text{ ksf } (50 \text{ ft})(90 \text{ ft})/4 \text{ columns} = 56.3 \text{ kips}$$

$$\text{Snow Load} = 0.02 \text{ ksf } (50 \text{ ft})(90 \text{ ft})/4 \text{ columns} = 22.5 \text{ kips}$$

Thus, for load combination 3

$$P_a = (56.3) + (22.5) = 78.8 \text{ kips}$$

$$M_a = (20.5) + (8.20) = 28.7 \text{ ft-kips}$$

As determined in Step 5 for in plane buckling

$$P_n/\Omega = 148 \text{ kips}$$

$$M_{nx}/\Omega = 85.0 \text{ ft-kips}$$

Checking for the appropriate interaction equation

$$\frac{P_u}{P_n/\Omega} = \frac{78.8}{148} = 0.532 > 0.2$$

Thus, use Equation 8.2 (H1-1a).
For an effective length $Kl_x = 16.0$ ft, $P_{ex} = 1130$ kips.
As before, $C_m = 0.6$. Thus

$$B_1 = \frac{0.6}{1 - \dfrac{1.6(78.8)}{1130}} = 0.675 < 1.0$$

Therefore, use $B_1 = 1.0$
and

$$\boxed{\frac{P_a}{P_n/\Omega} + \frac{8}{9}\left(\frac{M_{ax}}{M_{nx}/\Omega}\right) = \frac{78.8}{148} + \frac{8}{9}\left(\frac{28.7}{85.0}\right) = 0.833 < 1.0}$$

Thus, the W8×40 is adequate for both strength under combined load and stability for supporting the leaning columns.

8.11 PARTIALLY RESTRAINED (PR) FRAMES

The beams and columns in the frames considered up to this point have all been connected with moment-resisting fully restrained (FR) connections or simple pinned connections. These latter connections had been considered a special case of the more general partially restrained (PR) connections provided for in the Specification. However, these simple connections are now separately defined in Specification Section B3.6a. Partially restrained connections, defined in Specification Section B3.6b along with FR conections, have historically been referred to as *semirigid connections*. When these PR connections are included as the connecting elements in a structural frame, they influence both the strength and stability of the structure.

Before considering the partially restrained frame, it will be helpful to look at the partially restrained beam. The relationship between the end moment and end rotation for a symmetric, uniformly loaded prismatic beam can be obtained from the well-known slope deflection equation as

$$M = -2\frac{EI\theta}{L} + \frac{WL}{12} \tag{8.19}$$

This equation is plotted in Figure 8.17a and labeled as the *beam line*.

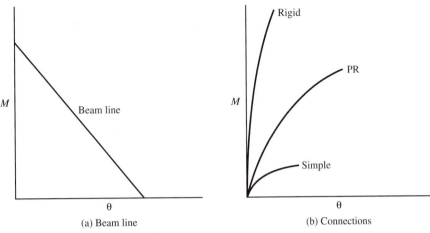

(a) Beam line   (b) Connections

Figure 8.17 Moment Rotation Curves for Uniformly Loaded Beam and Typical Connections.

All PR connections exhibit some rotation as a result of an applied moment. The moment-rotation characteristics of these connections are the key to determining the type of connection and thus the behavior of the structure. Moment-rotation curves for three generic connections are shown in Figure 8.17b and are labeled rigid, simple, and PR. A lot of research has been conducted in an effort to identify the moment-rotation curves for real connections. Two compilations of these curves have been published.[5,6]

The relationship between the moment-rotation characteristics of a connection and a beam can be seen by plotting the *beam line* and *connection curve* together, as shown in Figure 8.18. Normal engineering practice treats connections capable of resisting at least 90% of the fixed-end moment as rigid and those capable of resisting no more than 20% of the fixed-end moment as simple. All connections that exhibit an ability to resist moment between these limits must be treated as partially restrained connections, accounting for their true moment-rotation characteristics.

The influence of the PR connection on the maximum positive and negative moments on the beam is seen in Figure 8.19. Here, the ratio of positive or negative moment to the fixed-end moment is plotted against the ratio of beam stiffness, EI/L, to a linear connection stiffness, M/θ. The moment for which the beam must be designed ranges from 0.75 times the fixed-end moment to 1.5 times the fixed-end moment, depending on the stiffness of the connection.

When PR connections are used to connect beams and columns to form PR frames, the analysis becomes much more complex. The results of numerous studies dealing with this issue have been reported. Although some practical designs have been carried out, widespread practical design of PR frames is still some time off. In addition to the problems associated with modeling a particular connection, the question of loading sequence arises. Because real, partially restrained connections behave nonlinearly, the sequence of applied loads influences the structural response. The approach to load application may have more significance than the accuracy of the connection model used in the analysis.

---

[5]Goverdhan, A. V., *A Collection of Experimental Moment Rotation Curves and Evaluation of Prediction Equations for Semi-Rigid Connections*, Master of Science Thesis, Vanderbilt University, Nashville, TN, 1983.

[6]Kishi, N., and Chen, W. F,. *Data Base of Steel Beam-to-Column Connections*, CE-STR-86-26, West Lafayette, IN: Purdue University, School of Engineering, 1986.

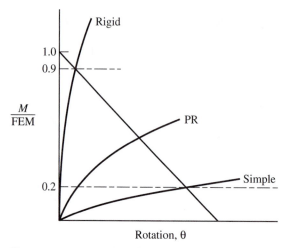

Figure 8.18 Beam Line and Connection Curves.

Although a complete, theoretical analysis of a partially restrained frame may currently be beyond the scope of normal engineering practice, a simplified approach exists that is not only well within the scope of practice, but is commonly carried out in everyday design. This approach can be referred to as *Flexible Moment Connections*. It has historically been called *Type 2 with Wind*. The Flexible Moment Connection approach relies heavily on the nonlinear moment-rotation behavior of the PR connection although the actual curve is not used. In addition, it relies on a phenomenon called *shake down*, which shows that the

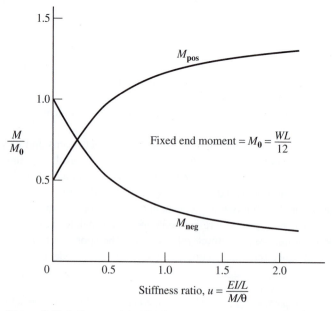

Figure 8.19 Influence of the PR Connection on the Maximum Positive and Negative Moments of a Beam.

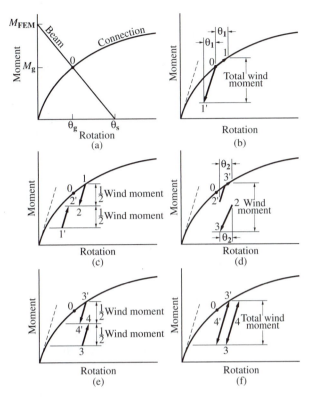

Figure 8.20 Moment-Rotation Curves Showing Shake-Down. Courtesy American Iron and Steel Institute.

connection, although exhibiting nonlinear behavior initially, behaves linearly after a limited number of applications of lateral load.[7]

The moment-rotation curve for a typical PR connection is shown in Figure 8.20a along with the beam line for a uniformly loaded beam. The point labeled 0 represents equilibrium for the applied gravity loads. The application of wind load produces moments at the beam ends that add to the gravity moment at the leeward end of the beam and subtract from the windward end. Because moment at the windward end is being removed, the connection behaves elastically with a stiffness close to the original connection stiffness, whereas at the leeward end, the connection continues to move along the nonlinear connection curve. Points labeled 1 and 1′ in Figure 8.20b represent equilibrium under the first application of wind to the frame.

When the wind load is removed, the connection moves from points 1 and 1′ to points 2 and 2′, as shown in Figure 8.20c. The next application of a wind load that is larger than the first and in the opposite direction will see the connection behavior move to points 3 and 3′. Note that on the windward side, the magnitude of this applied wind moment dictates whether the connection behaves linearly or follows the nonlinear curve, as shown in Figure 8.20d. Removal of this wind load causes the connection on one end to unload and on the other end to load, both linearly. Any further application of wind load, less than the maximum already applied, will see the connection behave linearly. In addition, the maximum moment on the connection is still close to that applied originally from the gravity load. Thus, the condition described in Figure 8.20f shows that shake-down has taken place and the connection now behaves linearly for both loading and unloading.

[7]Geschwindner, L. F., and Disque, R. O., "Flexible Moment Connections for Unbraced Frames Subject to Lateral Forces—A Return to Simplicity." *Engineering Journal*, AISC, Vol. 42, No. 2, 2005, pp 99–112.

The design procedure used to account for this shake-down is straight forward. All beams are designed as simple beams using the appropriate load combinations. This assures that the beams are adequate, regardless of the actual connection stiffness, as was seen in Figure 8.19. Wind load moments are determined through a modified portal analysis where the leeward column is assumed not to participate in the lateral load resistance. Connections are sized to resist the resulting moments, again for the appropriate load combinations. In addition, it is particularly important to provide connections that have sufficient ductility to accommodate the large rotations that will occur, without overloading the bolts or welds under combined gravity and wind.

Columns must be designed to provide frame stability under gravity loads as well as gravity plus wind. The columns may be designed using the approach that was presented for columns in moment frames, but with two essential differences from the conventional rigid frame design:

1. Because the gravity load is likely to load the connection to its plastic moment capacity, the column can be restrained only by a girder on one side and this girder will act as if it is pinned at its far end. Therefore, in computing the girder stiffness rotation factor, I_g/L_g, for use in the effective length alignment chart, the girder length should be doubled.

2. One of the external columns, the leeward column for the wind loading case, cannot participate in frame stability because it will be attached to a connection that is at its plastic moment capacity. The stability of the frame may be assured, however, by designing the remaining columns to support the total frame load.

For the exterior column, the moment in the beam to column joint is equal to the capacity of the connection. It is sufficiently accurate to assume that this moment is distributed one-half to the upper column and one-half to the lower column. For interior columns, the greatest, realistically possible difference in moments resulting from the girders framing into the column should be distributed equally to the columns above and below the joint.

EXAMPLE 8.6a
Column Design with
Flexible Wind
Connections by LRFD

GOAL: Design the girders and columns of a building with flexible wind connections and determine the moments for which the connections must be designed.

GIVEN: An intermediate story of a three-story building is given in Figure 8.21. Story height is 12 ft. The frame is braced in the direction normal to that shown. Use the LRFD provisions and A992 steel.

SOLUTION

Step 1:   Determine the required forces and moments.
The loads shown in Figure 8.21 are the code-specified nominal loads. The required forces are calculated using tributary areas as follows. Gravity loads on exterior columns.

$$1.2DL = 1.2(25 \text{ kips} + 0.75 \text{ kips/ft}(15 \text{ ft})) = 43.5 \text{ kips}$$
$$1.6LL = 1.6(75 \text{ kips} + 2.25 \text{ kips/ft}(15 \text{ ft})) = \underline{174 \text{ kips}}$$

Total 218 kips

Gravity loads on interior columns

$$1.2DL = 1.2(50 \text{ kips} + 0.75 \text{ kips/ft}(30 \text{ ft})) = 87.0 \text{ kips}$$
$$1.6LL = 1.6(150 \text{ kips} + 2.25 \text{ kips/ft}(30 \text{ ft})) = \underline{348 \text{ kips}}$$

Total 435 kips

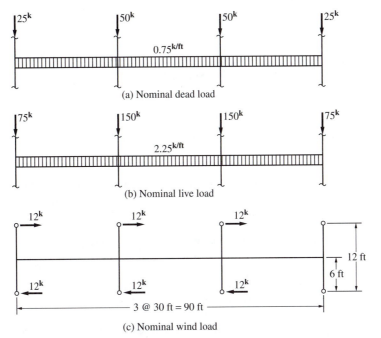

Figure 8.21 Intermediate Story of a Three-Story Building (Example 8.5).

Gravity load on girders:

$$1.2DL = 1.2(0.75 \text{ kips/ft}(30 \text{ ft})) = 27.0 \text{ kips}$$

$$1.6LL = 1.6(2.25 \text{ kips/ft}(30 \text{ ft})) = \underline{108 \text{ kips}}$$

Total 135 kips

Step 2:   Design the girder for the simple beam moment using Table 3-2.

$$M_u = 135(30.0)/8 = 506 \text{ ft-kips}$$

Use W21×62 ($\phi M_n = 540$ ft-kips, $I_x = 1330$ in.⁴)

Step 3:   Design the columns for the gravity load on the interior column using Table 4-1. For buckling out of the plane in a braced frame

$$K = 1.0 \text{ and } KL_y = 12.0$$

Thus, with $P_u = 435$ kips

try W14×53, ($\phi P_n = 465$ kips)

Step 4:   To check the column for stability in the plane, determine the effective length factor from the alignment chart with

$$G_{top} = G_{bottom} = \frac{2\left(\dfrac{541}{12.0}\right)}{\left(\dfrac{1330}{2(30.0)}\right)} = 4.07$$

Note that only one beam is capable of restraining the column and that beam is pinned at its far end, thus the effective beam length is taken as twice its actual length.

Considering the stress in the column under load, the stiffness reduction factor can be determined.

$$f_u = 435/15.6 = 27.9 \text{ ksi}$$

Thus, from the Manual Table 4-21, the stiffness reduction factor, $\tau_a = 0.807$. The inelastic stiffness ratio then becomes

$$G_{top} = G_{bottom} = 0.807(4.07) = 3.28$$

which yields, from the alignment chart

$$K = 1.87$$

Step 5: Determine the effective length in the plane of bending.

$$\frac{KL}{r_x/r_y} = \frac{1.87(12.0)}{3.07} = 7.31 \text{ ft}$$

Step 6: Determine the column compressive strength from Manual Table 4-1.

$$\phi P_n = 602 \text{ kips}$$

Step 7: Determine the second-order moment.

The applied wind moment is $M_u = 1.6(6.0)(12.0) = 115$ ft-kips and the applied force is $P_u = 435$ kips.

Considering all the moment as a translation moment

$$P_{e2} = \frac{\pi^2(29,000)(541)}{(1.87(12.0)(12))^2} = 2140 \text{ kips}$$

Therefore

$$B_2 = \frac{1}{1 - \dfrac{3(435)}{3(2140)}} = 1.26$$

and $M_r = 1.26(115) = 145$ ft-kips

Step 8: Determine whether the column satisfies the interaction equation

$$\frac{P_r}{\phi P_n} = \frac{435}{602} = 0.721 > 0.2$$

Therefore, use Equation 8.2 (H1-1a), $\phi M_n = 285$, from Manual Table 3-10, which results in

$$\boxed{0.721 + \frac{8}{9}\left(\frac{145}{285}\right) = 1.17 > 1.0}$$

This indicates that the W14×53 is not adequate for stability. The next larger column should be considered.

Step 9: Determine the required moment strength for the connections.

All beam-to-column connections must be designed to resist the amplified wind moments.

Thus

$$\boxed{M_{u\ conn} = 290 \text{ ft-kips}}$$

EXAMPLE 8.6b
Column Design with Flexible Wind Connections by ASD

SOLUTION

GOAL: Design the girders and columns of a building with flexible wind connections and determine the moments for which the connections must be designed.

GIVEN: An intermediate story of a three-story building is given in Figure 8.21. Story height is 12 ft. The frame is braced in the direction normal to that shown. Use the ASD provisions and A992 steel.

Step 1: Determine the required forces and moments.

The loads shown in Figure 8.21 are the code-specified nominal loads. The required forces are calculated using tributary areas as follows. Gravity loads on exterior columns

$$DL = (25\,k + 0.75\,k/ft(15\,ft)) = 36.3 \text{ kips}$$

$$LL = (75\,k + 2.25\,k/ft(15\,ft)) = \underline{108 \text{ kips}}$$

Total 144 kips

Gravity loads on interior columns

$$DL = (50\,k + 0.75\,k/ft(30\,ft)) = 72.5 \text{ kips}$$

$$LL = (150\,k + 2.25\,k/ft(30\,ft)) = \underline{218 \text{ kips}}$$

Total 291 kips

Gravity load on girders

$$DL = (0.75\,k/ft(30\,ft)) = 22.5 \text{ kips}$$

$$LL = (2.25\,k/ft(30\,ft)) = \underline{67.5 \text{ kips}}$$

Total 90.0 kips

Step 2: Design the girder for the simple beam moment.

$$M_a = 90.0(30.0)/8 = 338 \text{ ft-kips}$$

Use W21×62 ($M_n/\Omega = 359$ ft-kips, $I_x = 1330$ in.[4])

Step 3: Design the columns for gravity load on the interior column.
For buckling out of the plane in a braced frame

$$K = 1.0 \quad \text{and} \quad KL_y = 12.0$$

Thus, with $P_a = 290$ kips

$$\text{try W14×53}, \quad (P_n/\Omega = 310 \text{ kips})$$

Step 4: To check the column for stability in the plane, determine the effective length factor from the alignment chart with

$$G_{top} = G_{bottom} = \frac{2\left(\dfrac{541}{12.0}\right)}{\left(\dfrac{1330}{2(30.0)}\right)} = 4.07$$

Note that only one beam is capable of restraining the column and that beam is pinned at its far end, thus the effective beam length is taken as twice its actual length.

Considering the stress in the column under load, the stiffness reduction factor can be determined.

$$f_a = 290/15.6 = 18.6 \text{ ksi}$$

Thus, from the Manual Table 4-21, the stiffness reduction factor, $\tau_a = 0.805$. The inelastic stiffness ratio then becomes

$$G_{top} = G_{bottom} = 0.805(4.07) = 3.28$$

which yields, from the alignment chart

$$K = 1.87$$

Step 5: Determine the effective length in the plane of bending.

$$\frac{KL}{r_x/r_y} = \frac{1.87(12.0)}{3.07} = 7.31 \text{ ft}$$

Step 6: Determine the column compressive strength from Manual Table 4-1.

$$P_n/\Omega = 401 \text{ kips}$$

Step 7: Determine the second-order moment.

The applied wind moment is $M_a = 6.0(12.0) = 72.0$ ft-kips and the applied force is $P_a = 290$ kips.

Considering all the moment as a translation moment

$$P_{e2} = \frac{\pi^2(29,000)(541)}{(1.87(12)(12))^2} = 2140 \text{ kips}$$

$$\alpha P_a = 1.6(290) = 464 \text{ kips}$$

Therefore

$$B_2 = \frac{1}{1 - \dfrac{3(464)}{3(2140)}} = 1.28$$

and $M_r = 1.28(72.0) = 92.2$ ft-kips

Step 8: Determine whether the column satisfies the interaction equation

$$\frac{P_r}{P_n/\Omega} = \frac{290}{401} = 0.723 > 0.2$$

Therefore, use Equation 8.2 (H1-1a), $M_n/\Omega = 190$, from Manual Table 3-10, which results in

$$\boxed{0.723 + \frac{8}{9}\left(\frac{92.2}{190}\right) = 1.15 > 1.0}$$

This indicates that the W14×53 is not adequate for stability. The next larger column should be considered.

Step 9: Determine the required moment strength for the connections.

All beam-to-column connections must be designed to resist the amplified wind moments. Thus

$$\boxed{M_{a\ conn} = 184 \text{ ft-kips}}$$

After an acceptable column is selected, the lateral displacement of the structure must be checked. Coverage of drift in wind moment frames is beyond the treatment intended here but is covered in Geschwindner and Disque.

8.12   BRACING DESIGN

Braces in steel structures are used to reduce the effective length of columns, reduce the unbraced length of beams, and provide overall structural stability. The discussion of columns in Chapter 5 showed how braces could be effective in reducing effective length and thereby increasing column strength. Chapter 6 demonstrated how the unbraced length of a beam influenced its strength and earlier in this chapter the influence of sway on the stability of a structure was discussed. Every case assumed that the given bracing requirements were satisfied; however, nothing was said about the strength or stiffness of the required braces. Appendix 6 of the Specification treats bracing for columns and beams similarly, although the specific requirements are different. Two types of braces are defined, nodal braces and relative braces.

Nodal braces control the movement of a point on the member without interaction with any adjacent braced points. These braces would be attached to the member and then to a fixed support, such as the abutment shown in Figure 8.22b.

Relative braces rely on other braced points of the structure to provide support. A diagonal brace within a frame would be a relative brace, as shown in Figure 8.22a. In this case, the diagonal brace and the horizontal strut together compose the relative brace. Because the horizontal strut is usually a part of a very stiff floor system that has significant strength in its plane, the strength and stiffness of the diagonal element usually controls the overall behavior of this braced system.

The brace requirements of the Specification are intended to enable the members being designed to reach their maximum strength based on the length between bracing points and an effective length factor, $K = 1.0$. A brace has two requirements: strength and stiffness. A brace that is inadequate in either of these respects is not sufficient to enable the member it is bracing to perform as it was designed.

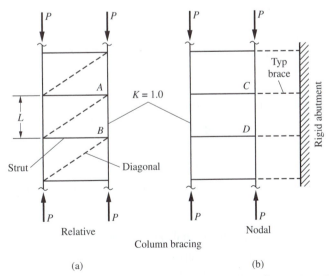

Figure 8.22 Definitions of Bracing Types. Copyright © American Institute of Steel Construction, Inc. Reprinted with Permission. All rights reserved.

8.12.1   Column Bracing

For a column relative brace, the required brace strength is

$$P_{br} = 0.004 P_r \qquad (8.20)$$

and the required brace stiffness is

$$\beta_{br} = \frac{1}{\phi}\left(\frac{2P_r}{L_b}\right) \text{ (LRFD)} \qquad \beta_{br} = \Omega\left(\frac{2P_r}{L_b}\right) \text{ (ASD)} \qquad (8.21)$$

$$\phi = 0.75 \text{ (LRFD)} \qquad \Omega = 2.00 \text{ (ASD)}$$

where

$L_b =$ distance between braces

$P_r =$ required strength for ASD or LRFD as appropriate for the design method being used.

For a column nodal brace, the required brace strength is

$$P_{br} = 0.01 P_r \qquad (8.22)$$

and the required brace stiffness is

$$\beta_{br} = \frac{1}{\phi}\left(\frac{8P_r}{L_b}\right) \text{ (LRFD)} \qquad \beta_{br} = \Omega\left(\frac{8P_r}{L_b}\right) \text{ (ASD)} \qquad (8.23)$$

$$\phi = 0.75 \text{ (LRFD)} \qquad \Omega = 2.00 \text{ (ASD)}$$

where

$L_b =$ distance between braces

$P_r =$ required strength for ASD or LRFD as appropriate for the design method being used.

8.12.2   Beam Bracing

For a beam relative brace, the required strength of the brace is

$$P_{br} = 0.008 M_r C_d / h_o \qquad (8.24)$$

and the required brace stiffness is

$$\beta_{br} = \frac{1}{\phi}\left(\frac{4M_r C_d}{L_b h_o}\right) \text{ (LRFD)} \qquad \beta_{br} = \Omega\left(\frac{4M_r C_d}{L_b h_o}\right) \text{ (ASD)} \qquad (8.25)$$

$$\phi = 0.75 \text{ (LRFD)} \qquad \Omega = 2.00 \text{ (ASD)}$$

where

$h_o =$ distance between flange centroids

$C_d =$ 1.0 for single curvature and 2.0 for double curvature

$L_b =$ laterally unbraced length

$M_r =$ required flexural strength

For a beam nodal brace, the required strength of the brace is

$$P_{br} = 0.02 M_r C_d / h_o \tag{8.26}$$

and the required brace stiffness is

$$\beta_{br} = \frac{1}{\phi}\left(\frac{10 M_r C_d}{L_b h_o}\right) \text{(LRFD)} \qquad \beta_{br} = \Omega\left(\frac{10 M_r C_d}{L_b h_o}\right) \text{(ASD)} \tag{8.27}$$

$$\phi = 0.75 \text{ (LRFD)} \qquad \Omega = 2.00 \text{ (ASD)}$$

where

$h_o =$ distance between flange centroids
$C_d =$ 1.0 for single curvature and 2.0 for double curvature
$L_b =$ laterally unbraced length
$M_r =$ required flexural strength

8.12.3   Frame Bracing

Frame bracing and column bracing are accomplished by the same relative and nodal braces and use the same stiffness and strength equations.

EXAMPLE 8.7a	**GOAL:** Determine the required bracing for a braced frame to resist lateral load.
Bracing Design by LRFD	
	GIVEN: Using the LRFD requirements, select a rod to provide the nodal bracing shown in the three-bay panel of Figure 8.9a to resist a wind load of 4 kips and provide stability for a gravity live load of 113 kips and dead load of 45 kips.
SOLUTION	**Step 1:** Determine the required brace stiffness for gravity load. For the gravity load, the required brace stiffness is based on $1.2D + 1.6L$.

$$P_r = 1.2(113) + 1.6(45.0) = 208 \text{ kips}$$

$$\beta_{br} = \frac{1}{\phi}\left(\frac{8P_r}{L_b}\right) = \frac{1}{0.75}\left(\frac{8(208)}{16.0}\right) = 139 \text{ kips/ft}$$

Step 2:   Determine the required brace area accounting for the angle of the brace.
Based on the geometry of the brace from Figure 8.9, where θ is the angle of the brace with the horizontal

$$\beta_{br} = \frac{A_{br}E}{L_r}\cos^2\theta = 139 \text{ kips/ft}$$

This results in a required brace area

$$A_{br} = \frac{139(34.0)}{29,000\left(\dfrac{30}{34}\right)^2} = 0.209 \text{ in.}^2$$

Step 3:   Determine the required brace force for gravity load. The required horizontal brace force for a nodal brace is

$$P_{br} = 0.01 P_r = 0.01(208) = 2.08 \text{ kips}$$

which gives a force in the member of

$$P_{br(angle)} = 2.08(34/30) = 2.36 \text{ kips}$$

and a required area, assuming $F_y = 36$ ksi for a rod, of

$$A_{br} = \frac{2.36}{0.9(36)} = 0.0728 \text{ in.}^2$$

Step 4: For gravity plus wind, determine the stiffness and strength requirements. The stiffness requirement will be the same.

The strength must be sufficient to resist the 4.0 kip wind load, thus for the wind portion

$$A_{br} = \frac{1.6(4.0)\left(\dfrac{34}{30}\right)}{0.9(36)} = 0.224 \text{ in.}^2$$

Step 5: Determine the required area for the combined wind and gravity loading.

Combining the required area from the gravity force and that from the wind force yields

$$A_{req} = 0.0728 + 0.224 = 0.296 \text{ in.}^2$$

Step 6: Select a rod to meet the required area for stiffness and strength.

$$\boxed{\text{use a 5/8-in. rod with } A = 0.307 \text{ in.}^2}$$

EXAMPLE 8.7b
Bracing Design by ASD

GOAL: Determine the required bracing for a braced frame to resist lateral load.

GIVEN: Using the ASD requirements, select a rod to provide the nodal bracing shown in the three-bay panel of Figure 8.9a to resist a wind load of 4 kips and provide stability for a gravity live load of 113 kips and dead load of 45 kips.

SOLUTION

Step 1: Determine the required brace stiffness for gravity load.

For the gravity load, the required brace stiffness is based on $D + L$.

$$P_r = 113 + 45.0 = 158 \text{ kips}$$

$$\beta_{br} = \Omega\left(\frac{8P_r}{L_b}\right) = 2.00\left(\frac{8(158)}{16.0}\right) = 158 \text{ kips/ft}$$

Step 2: Determine the required brace area accounting for the angle of the brace.

Based on the geometry of the brace from Figure 8.9, where θ is the angle of the brace with the horizontal

$$\beta_{br} = \frac{A_{br}E}{L_r}\cos^2\theta = 158 \text{ kips/ft}$$

This results in a required brace area

$$A_{br} = \frac{158(34.0)}{29,000\left(\dfrac{30}{34}\right)^2} = 0.238 \text{ in.}^2$$

Step 3: Determine the required brace force for gravity load.
The horizontal brace force is

$$P_{br} = 0.01P_r = 0.01(158) = 1.58 \text{ kips}$$

which gives a force in the member of

$$P_{br(angle)} = 1.58(34/30) = 1.79 \text{ kips}$$

and a required area, assuming $F_y = 36$ ksi for a rod, of

$$A_{br} = \frac{1.79}{0.6(36)} = 0.0829 \text{ in.}^2$$

Step 4: For gravity plus wind, determine the stiffness and strength requirements. The stiffness requirement will be the same. The strength must be sufficient to resist the 4.0 kip wind load, thus for the wind portion

$$A_{br} = \frac{4.0\left(\dfrac{34}{30}\right)}{0.6(36)} = 0.210 \text{ in.}^2$$

Step 5: Determine the required area for the combined wind and gravity loading.
Combining the required area from the gravity force and that from the wind force yields

$$A_{req} = 0.0829 + 0.210 = 0.293 \text{ in.}^2$$

Step 6: Select a rod to meet the required area for stiffness and strength.

$$\boxed{\text{use a 5/8-in. rod with } A = 0.307 \text{ in.}^2}$$

8.13 PROBLEMS

Unless noted otherwise, all columns should be considered pinned in a braced frame out of the plane being considered in the problem with bending about the strong axis.

1. Determine whether a W14×90, A992 column with a length of 12.5 ft is adequate in a braced frame to carry the following loads: a compressive dead load of 100 kips and live load of 300 kips, a dead load moment of 30 ft-kips and live load moment of 70 ft-kips at one end, and a dead load moment of 15 ft-kips and a live load moment of 35 ft-kips at the other. The member is bending in reverse curvature. Determine by (a) LRFD and (b) ASD.

2. A W12×58, A992 is used as a 14-ft column in a braced frame to carry a compressive dead load of 50 kips and live load of 150 kips. Will this column be adequate to carry a dead load moment of 20 ft-kips and live load moment of 50 ft-kips at each end, bending the column in single curvature by (a) LRFD and (b) ASD?

3. Given a W14×120, A992 16-ft column in a braced frame with a compressive dead load of 90 kips and live load of 270 kips. Maintaining a live load to dead load ratio of 3, determine the maximum live and dead load moments that can be applied about the strong axis on the upper end when the lower end is pinned by (a) LRFD and (b) ASD.

4. Reconsider the column and loadings in Problem 1 if that column were bent in single curvature by (a) LRFD and (b) ASD.

5. Reconsider the column and loadings in Problem 2 if that column were bent in reverse curvature by (a) LRFD and (b) ASD.

6. A pin-ended column in a braced frame must carry a compressive dead load of 85 kips and live load of 280 kips, along with a uniformly distributed transverse dead load of 0.4 kips/ft and live load of 1.3 kips/ft. Will a W14×74, A992 member be adequate if the transverse load is applied to put bending about the strong axis? Determine by (a) LRFD and (b) ASD.

For Problems 7 through 9, assume that the ratio of total gravity load on the story to the Euler buckling load, $(\alpha \Sigma P_{nt}/\Sigma P_{e2})$, is the same as the ratio of the column load to the column Euler buckling load for the specific column, $(\alpha P_{nt}/P_{e2})$.

7. An unbraced frame includes a 12-ft column that is called upon to carry a compressive dead load of 100 kips and live load of 300 kips. The top of the column is loaded with a no-translation dead load moment of 25 ft-kips and a no-translation live load moment of 80 ft-kips. The translation moments applied to that column end are a dead load moment of 35 ft-kips and a live load moment of 100 ft-kips. The lower end of the column feels half of these moments and the column is bending in reverse curvature. Will a W14×109, A992 member be adequate to carry this loading? Assume that the effective length factor in the plane of bending is 1.66. Determine by (a) LRFD and (b) ASD.

8. A W14×176, A992 member is proposed for use as a 12.5-ft column in an unbraced frame. Will this member be adequate to carry a compressive dead load of 160 kips and live load of 490 kips? The top of the column is loaded with a no-translation dead load moment of 15 ft-kips and a no-translation live load moment of 30 ft-kips. The translation moments applied to that column end are a dead load moment of 80 ft-kips and a live load moment of 250 ft-kips. The lower end of the column is considered pinned and the effective length factor is taken as 1.5. Determine by (a) LRFD and (b) ASD.

9. Will a W14×38 be adequate as a 13-ft column in an unbraced frame with a compressive dead load of 25 kips and live load of 80 kips? The top and bottom of the column are loaded with a no-translation dead load moment of 20 ft-kips and a no-translation live load moment of 55 ft-kips. The translation moments applied to the column ends are a dead load moment of 10 ft-kips and a live load moment of 50 ft-kips. The column is bent in reverse curvature and $K_x = 1.3$. Determine by (a) LRFD and (b) ASD.

10. Determine whether a 10-ft long braced frame W14×43, A992 column can carry a compressive dead load of 35 kips and live load of 80 kips along with a dead load moment of 20 ft-kips and live load moment of 40 ft-kips. One half of these moments are applied at the other end, bending it in single curvature.

11. A two-story single bay frame is shown below. The uniform live and dead loads are indicated along with the wind load. A first-order elastic analysis has yielded the results shown in the figure for the given loads and the appropriate notional loads. Assuming that the story drift is limited to height/300 under the given wind loads, determine whether the first- and second-story columns are adequate. The members are shown and are all A992 steel. Determine by (a) LRFD and (b) ASD.

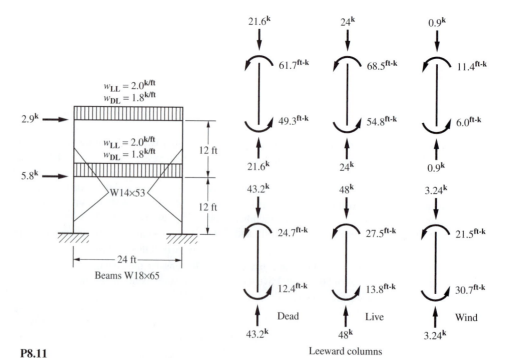

P8.11

12. Determine whether the columns of the two-bay unbraced frame shown below are adequate to support the given loading. Results for the first-order analysis are provided. Because the structure drift is unknown, use the ratio of total applied load to total Euler load in calculating the second-order amplifications. All members are A992 steel and the sizes are as shown. Determine by (a) LRFD and (b) ASD.

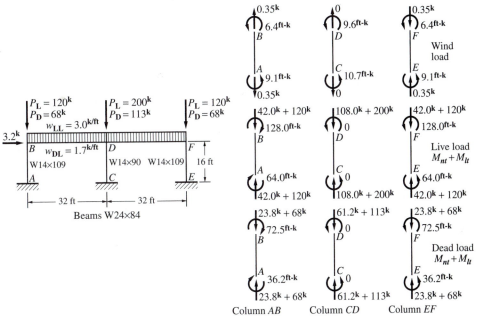

P8.12

13. A nonsymmetric two-bay unbraced frame is required to support the live and dead loads given in the figure below. Using the results from the first-order elastic analysis provided, determine whether each column will be adequate. All members are A992 steel and the sizes are as shown. Determine by (a) LRFD and (b) ASD.

$w_{LL} = 3.2^{k/ft}$
$w_{DL} = 1.9^{k/ft}$

16 ft   All columns W12×96
        All beams W21×73

30 ft ——— 20 ft

25.4^k 55.1^k 14.4^k

77.2 − 11.3 = 65.7^{ft-k} 23.9 + 11.3 = 35.2^{ft-k} 14.5 + 13.7 = 28.2^{ft-k}

38.6 − 16.6 = 22.0^{ft-k} 0   7.3 + 17.8 = 25.1^{ft-k}

25.4^k 55.1^k 14.4^k

Dead load $M_{nt} + M_{lt}$

43^k 92.7^k 24.3^k

130.0 − 19.0 = 111^{ft-k} 42.0 + 19 = 59.2^{ft-k} 24.4 + 23.0 = 47.4^{ft-k}

65.0 − 28.0 = 37.0^{ft-k} 0   12.2 + 30.0 = 42.2^{ft-k}

43^k 92.7^k 24.3^k

Live load $M_{nt} + M_{lt}$

P8.13

Assume that the columns described in Problems 14 through 17 are members of moment frames for which a second-order direct analysis has been performed. Use the approach described in Section 8.8 to select trial sections to carry the indicated loads and check the appropriate interaction equations for column strength. Assume all members are A992 and that the given forces and moments are from either LRFD load combinations or 1.6(ASD) load combinations.

14. Select a W-shape for a column with a length of 18 ft to carry a force of 700 kips and a moment of 350 ft-kips by (a) LRFD and (b) ASD.

15. Select a W-shape for a column with a length of 28 ft to carry a force of 1100 kips and a moment of 170 ft-kips by (a) LRFD and (b) ASD.

16. Select a W-shape for a column with a length of 14 ft to carry a force of 350 kips and a moment of 470 ft-kips by (a) LRFD and (b) ASD.

17. Select a W-shape for a column with a length of 16 ft to carry a force of 1250 kips and a moment of 450 ft-kips by (a) LRFD and (b) ASD.

18. The two-bay moment frame shown below contains a single leaning column. The results of a first-order elastic analysis for each load are given. Determine whether the exterior columns are adequate to provide stability for the frame. All W-shapes are given and the steel is A992. Determine by (a) LRFD and (b) ASD.

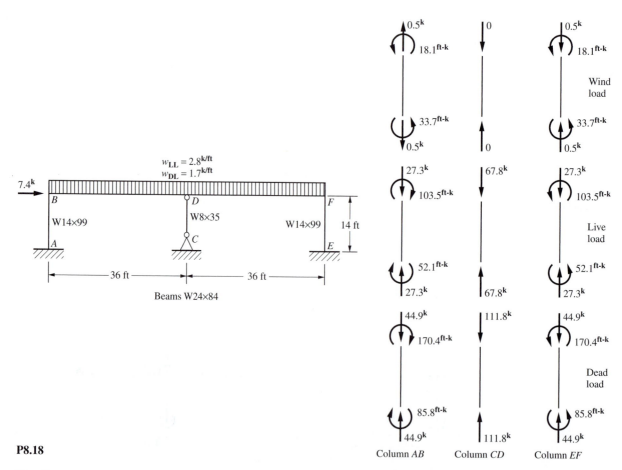

P8.18

19. The two-story frame shown on page 263 relies on the left-hand columns to provide stability. Using the first-order analysis results shown, determine whether the given structure is adequate if the steel is A992.

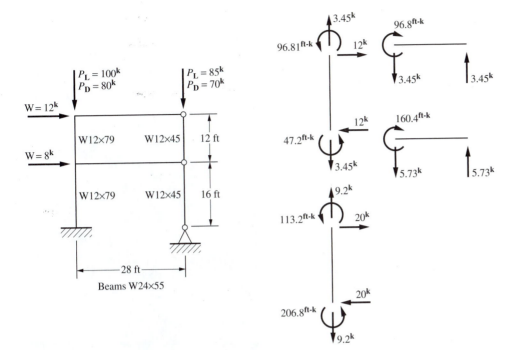

P8.19

20. The two-bay, two-story frame shown below is to be designed. Using the Live, Dead, Snow, and Wind Loads given in the figure, design the columns and beams to provide the required strength and stability by (a) LRFD and (b) ASD.

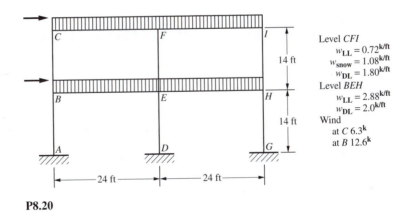

Level *CFI*
$w_{LL} = 0.72^{k/ft}$
$w_{snow} = 1.08^{k/ft}$
$w_{DL} = 1.80^{k/ft}$
Level *BEH*
$w_{LL} = 2.88^{k/ft}$
$w_{DL} = 2.0^{k/ft}$
Wind
   at *C* 6.3^k
   at *B* 12.6^k

P8.20

Chapter 9

WaMu Center, Seattle.
Photo courtesy Michael Dickter/Magnusson Klemencic
Associates.

Composite Construction

9.1 INTRODUCTION

Any structural member in which two or more materials having different stress-strain relationships are combined and called upon to work as a single member may be considered a *composite member*.

Many different types of members have been used that could be called composite. Such members as shown in Figure 9.1 are (a) a reinforced concrete beam, (b) a precast concrete beam and cast-in-place slab, (c) a "flitch" girder combining wood side members and a steel plate, (d) a stressed skin panel where plywood is combined with solid wood members, and (e) a steel shape combined with concrete.

This last type of member, and those similar members, are normally thought of as composite members in building applications. The Specification, Chapter I, provides rules for design of the composite members illustrated in Figure 9.2. These members are (a) steel beams fully encased in concrete, (b) steel beams with flat soffit concrete slabs, (c) steel beams combined with formed steel deck, (d) steel columns fully encased with concrete, and (e) hollow steel shapes filled with concrete.

Encased beams and filled columns, as shown in Figures 9.2a and e, do not specifically require mechanical anchorage between the steel and concrete, other than the natural

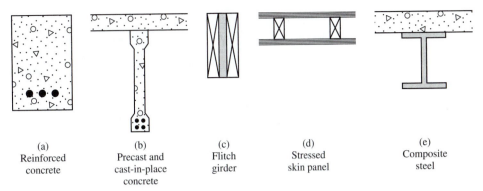

<table>
<tr><td>(a)</td><td>(b)</td><td>(c)</td><td>(d)</td><td>(e)</td></tr>
<tr><td>Reinforced
concrete</td><td>Precast and
cast-in-place
concrete</td><td>Flitch
girder</td><td>Stressed
skin panel</td><td>Composite
steel</td></tr>
</table>

Figure 9.1 Composite Members.

bond that exists between the two materials; however, the other flexural and compression members shown in Figures 9.2b, c, and d always require some form of mechanical shear connection.

Regardless of the type of mechanical shear device provided, it must connect the steel and concrete to form a unit and permit them to work together to resist the load. This considerably increases the strength of the bare steel shape. Composite beams were first used in bridge design in the United States in about 1935. Until the invention of the shear stud, the concrete floor slab was connected to the stringer beams by means of wire spirals or channels welded to the top flange of the beam, as shown in Figure 9.3.

In the l940s the Nelson Stud Company invented the shear stud, a headed rod welded to the steel beam by means of a special device or gun, as shown in Figure 9.4. The company did not enforce its patent but instead encouraged nonproprietary use of the system, assuming correctly that the company would get its share of the business if it became popular. In a very short time, studs replaced spirals and channels, so that today, studs are used almost exclusively in composite beam construction.

In 1952, AISC adopted composite design rules for encased beams in its specification for building design and later, in l956, extended them to beams with flat soffits. Although the design procedure was based on the ultimate strength of the composite section, the rules were written in the form of an allowable stress procedure as was common for the time. As a result, allowable stress design for composite beams has often been criticized as being convoluted and difficult to understand.

In the current specification, whether for ASD or LRFD, the rules for the design of composite beams are straightforward and surprisingly simple. The ultimate flexural strength

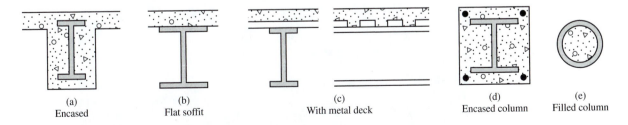

<table>
<tr><td>(a)</td><td>(b)</td><td>(c)</td><td>(d)</td><td>(e)</td></tr>
<tr><td>Encased</td><td>Flat soffit</td><td>With metal deck</td><td>Encased column</td><td>Filled column</td></tr>
</table>

Figure 9.2 Composite Steel Beams and Columns.

Figure 9.3 Composite Beam Using a Spiral Shear Connector.

Figure 9.4 Installation of a Shear Stud with a Stud Gun.
Photo Courtesy W. Samuel Easterling.

Table 9.1 Sections of Specification and Parts of Manual Found in this Chapter

	Specification
I1	General Provisions
I2	Axial Members
I3	Flexural Members
I4	Combined Axial Force and Flexure
	Manual
Part 3	Design of Flexural Members
Part 4	Design of Compression Members

of the composite member is based on plastic stress distribution with the ductile shear connector transferring shear between the steel section and the concrete slab.

Specification Section I1 gives limitations on material properties for use in composite concrete members. Concrete is limited to f'_c between 3 ksi and 10 ksi for normal weight concrete and between 3 ksi and 6 ksi for lightweight concrete. The specified minimum yield strength of the structural steel and reinforcing steel is up to 75 ksi.

This chapter discusses the design of both composite beams and composite columns.

Table 9.1 lists the sections of the Specification and parts of the Manual discussed in this chapter.

9.2 ADVANTAGES AND DISADVANTAGES OF COMPOSITE BEAM CONSTRUCTION

One feature of composite construction makes it particularly advantageous for use in building structures. The typical building floor system is composed of two main parts: a floor structure that carries load to supporting members, usually a concrete slab or slab on metal deck; and the supporting members that span between girders, usually steel beams or joists. The advantage of a composite floor system stems from the "double counting" of the already existing concrete slab. All other factors that could be identified as advantages of this type of construction can be traced back to this single feature. A composite beam takes the already existing concrete slab and makes it work with the steel beam to carry the load to the girders. Thus, the resulting system has a greater strength than would have been available from the bare steel beam alone. The composite beam is stronger and stiffer than the noncomposite beam.

This factor manifests itself in reduced weight and/or shallower depths of members to carry the same loads when compared to the bare steel beam. Because the concrete slab is in compression and the majority of the steel is in tension, both materials are working to their best advantage. In addition, the effective beam depth has been increased from just the depth of the steel to the total distance from the top of the slab to the bottom of the steel, thus increasing the overall efficiency of the member.

With regard to stiffness, the composite section also has an increased elastic moment of inertia when compared to the bare steel beam. Although actual calculations for the stiffness of the composite section may be somewhat approximate in many cases, the impact of the increased stiffness profoundly effects the static deflection.

The only disadvantage with composite construction is the added cost of the required shear connectors. Because the increased strength, or reduction in required steel weight, is normally sufficient to offset the added cost of the shear connectors, this increased cost is usually not a disadvantage.

9.3 SHORED VERSUS UNSHORED CONSTRUCTION

Two methods of construction are available for composite beams: *shored* and *unshored construction*. Each has advantages and disadvantages, which will be discussed briefly. The difference between these two approaches to the construction of a composite beam is how the self-weight of the wet concrete is carried.

When the steel shape alone is called upon to carry the concrete weight, the beam is considered to be unshored. In this case, the steel is stressed and it deflects. This is the simplest approach to constructing the composite beam because the formwork and/or decking is supported directly on the steel beam. Unshored construction may, however, lead to a deflection problem during the construction phase because, as the wet concrete is placed, the steel beam deflects. To obtain a level slab, more concrete is placed where the beam deflection is greatest. This means that the contractor must place more concrete than the initial slab thickness called for and that the designer needs to provide more strength than would have been needed if the slab had remained of uniform thickness.

For shored construction, temporary supports called *shores* are placed under the steel beam to carry the wet concrete weight. In this case, the composite section carries the entire load after the shores are removed. No load is carried by the bare steel beam alone and thus, no deflection occurs during concrete placement. Two factors must be considered in the selection of shored construction (1) the additional cost, both in time and money, of placing and removing the temporary shoring; and (2) the potential increase in long-term, dead load, deflection due to creep in the concrete, which now is called upon to participate in carrying the permanent weight of the slab.

Although elastic stress distribution and deflection under service load conditions are influenced by whether the composite beam is shored or unshored, research has shown that the ultimate strength of the composite section is independent of the shoring situation. Thus, the use of shoring is entirely a serviceability and constructability question that must be considered by both the designer and constructor. Whether using ASD or LRFD provisions, the nominal strength of the shored and unshored system is the same and is given in Specification Section I3.

9.4 EFFECTIVE FLANGE

A cross section taken through a series of typical composite beams is shown in Figure 9.5. Because the concrete slab is normally part of the transverse spanning floor system, its thickness and the spacing of the steel beams are usually established prior to the design of the composite beams. Because the ability of the slab to participate in load carrying decreases as the distance from the beam centerline increases, some limit must be established to determine the portion of the slab that can be used in the calculations to determine the strength of the composite beam. The Specification provides two criteria for determining the effective width of the concrete slab for an interior beam and an additional criterion for an edge beam in Section I3.1a. As shown in Figure 9.5, the effective width, b_{eff}, is the sum of b' on each side

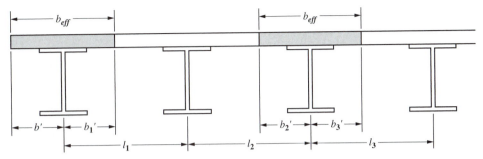

Figure 9.5 Effective Flange Width.

of the centerline of the steel section. For an interior beam, b' is the least of

$$b' \leq \frac{\text{span}}{8}$$

$$b' \leq \frac{1}{2} \text{ distance to the adjacent beam}$$

For an edge beam, the additional criteria is

$$b' \leq \text{ distance to the edge of the slab}$$

The entire thickness of the concrete slab is available to carry a compressive force. However, the depth of the concrete used in calculations is that required to provide sufficient area in compression to balance the force transferred by the shear connectors to the steel shape. The slab thickness does not influence the effective width of the slab.

9.5 STRENGTH OF COMPOSITE BEAMS AND SLAB

Flat soffit composite beams, Figure 9.2b, are constructed using formwork which is set at the same elevation as the top of the steel section. The concrete slab is placed directly on the steel section, resulting in a flat surface at the level of the top of the steel. Composite beams with formed steel deck, Figure 9.2c, are constructed with the steel deck resting on top of the steel beam or girder. The concrete is placed on top of the deck so that the concrete ribs and voids alternate. Provided that the portion of concrete required to balance the tension force in the steel is available above the tops of the ribs, the ultimate strength of both types of composite beams are determined in a similar fashion. Although the steel member may be either shored or unshored, the strength of the composite member is independent of shores and the design rules are independent of the method of construction.

The flexural strength of a composite beam under positive moment where concrete is in compression is presented in Section I3.2a of the Specification. In this section, strength is developed for flat soffit beams. The required modifications to account for the use of metal deck are presented in the next section. For steel sections with a web slenderness ratio, $h/t_w \leq 3.76\sqrt{E/F_y}$, the case for all rolled W-, S-, and HP-shapes, the nominal moment, M_n, is determined from the plastic distribution of stress on the composite section and

$$\phi_b = 0.90 \text{ (LRFD)} \qquad \Omega_b = 1.67 \text{ (ASD)}$$

A composite beam cross section is shown in Figure 9.6 with three possible plastic stress distributions. Regardless of the stress distribution considered, equilibrium requires that the total tension force must equal the total compression force, $T = C$. In Figure 9.6a, the plastic neutral axis (PNA) is located at the top of the steel shape. The compression force

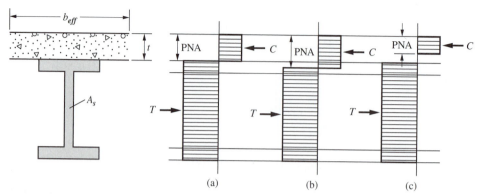

Figure 9.6 Plastic Stress Distribution.

developed using all of the concrete is exactly equal to the tension force developed using all of the steel. For the distribution of Figure 9.6b, the PNA is located within the steel shape. In this case, all of the concrete is taking compression but this is not sufficient to balance the tension force that the full steel shape could provide. Thus, some of the steel shape is in compression in order to satisfy $T = C$. The plastic stress distribution shown in Figure 9.6c is what occurs when less than the full amount of concrete is needed to balance the tensile force developed in the steel shape. Here the PNA is located within the concrete and that portion of the concrete below the PNA is not used because it would be in tension.

In all three cases, equilibrium of the cross section requires that the shear connectors be capable of transferring the force carried by the concrete into the steel. For the cases in Figures 9.6a and b, this is the full strength of the concrete. For the case in Figure 9.6c, this is the strength of the steel shape. Because the shear connectors are carrying the full amount of shear force required to provide equilibrium using the maximum capacity of one of the elements, this is called a *fully composite* beam. It is also possible to design a composite beam when the shear force that can be transferred by the shear connectors is less that this amount. In this case, the beam is called a *partially composite* beam. Although it has less strength than the fully composite member, it is often the most economical solution.

The Specification indicates that the plastic stress distribution in the concrete shall be taken as a uniform stress at a magnitude of $0.85 f_c'$. This is the same distribution specified by ACI 318, the specification for reinforced concrete. In addition, the distribution of stress in the steel is taken as a uniform F_y, as was the case for determining the plastic moment strength of a steel shape.

The Specification also provides for the use of a strain compatibility method in determining the strength of a composite section. This approach should be considered when a section is of unusual geometry or the steel does not have a compact web.

9.5.1 Fully Composite Beams

Establishing which stress distribution is in effect for a particular combination of steel and concrete requires calculating the minimum compressive force as controlled by the three components of the composite beam: concrete, steel, and shear connectors.

If all of the concrete were working in compression

$$V_c' = 0.85 f_c' b_{eff} t \tag{9.1}$$

If all of the steel shape were working in tension

$$V'_s = A_s F_y \tag{9.2}$$

If the shear studs were carrying their full capacity

$$V'_q = \Sigma Q_n \tag{9.3}$$

Because full composite action is assumed at this time, V'_q does not control and is not considered further. If $V'_s \leq V'_c$, the steel is fully stressed and only a portion of the concrete is stressed. This is the distribution given in either Figure 9.6a or c. If $V'_c \leq V'_s$, the concrete is fully stressed and the steel is called upon to carry both tension and compression in order to assure equilibrium. This results in the distribution shown in Figure 9.6b. Once the proper stress distribution is known, the corresponding forces can be determined and their point of application found. With this information, the nominal moment, M_n, can be found by taking moments about some reference point. Because the internal forces are equivalent to a force couple, any point of reference can be used for taking moments; however, it is convenient to use a consistent reference point. These calculations use the top of the steel as the point about which moments are taken.

Determination of the PNA for the cases in Figures 9.6a and c is quite straightforward. In both cases the steel is fully stressed in tension so it is referred to as the *steel controls* and it is known that the concrete must carry a compressive force equal to V'_s. Only that portion of the concrete required to resist this force will be used so that $C_c = 0.85 f'_c b_{eff} a$ where a defines the depth of the concrete stressed to its ultimate. Setting $V'_s = C_c$, and solving for a yields

$$a = \frac{V'_s}{0.85 f'_c b_{eff}} = \frac{A_s F_y}{0.85 f'_c b_{eff}} \tag{9.4}$$

For the special case where V'_s was exactly equal to V'_c, the value of a thus obtained is equal to the actual slab thickness, t. This is the case shown in Figure 9.6a. For all other values of a, the distribution of Figure 9.6c results. The nominal flexural strength can then be obtained by taking moments about the top of the steel so that

$$M_n = T_s(d/2) + C_c(t - a/2) \tag{9.5}$$

When the concrete controls, $V'_c < V'_s$, the determination of the PNA is a bit more complex. It is best to consider this case as two separate subcases: (1) the PNA occurring within the steel flange, and (2) the PNA occurring within the web. Once it is determined that V'_c controls, the next step is to determine the force in the steel flange and web from

$$T_f = F_y(b_f t_f) \tag{9.6}$$

$$T_w = T_s - 2T_f \tag{9.7}$$

A comparison between the force in the concrete and the force in the bottom flange plus the web shows whether the PNA is in the top flange or web. Thus, if $C_c > T_w + T_f$, more tension is needed for equilibrium and the PNA must be in the top flange. If $C_c < T_w + T_f$, less tension is needed for equilibrium and the PNA is in the web. In either case, the difference between the concrete force, C_c, and the available steel force, T_s, must be divided evenly between tension and compression in order to obtain equilibrium. This allows determination of the PNA location and the nominal moment strength. Thus, with

$$A_{s-c} = \text{area of steel in compression}$$

and

$$A_s = \text{total area of steel}$$

equilibrium is given by

$$C_c + F_y A_{s-c} = T_s - F_y A_{s-c} \qquad (9.8)$$

Solving for the area of steel in compression yields

$$A_{s-c} = \frac{T_s - C_c}{2F_y} \qquad (9.9)$$

For the case where the PNA is in the flange, the distance from the top of the flange to the PNA is given by x, where

$$x = \frac{A_{s-c}}{b_f} \qquad (9.10)$$

and for the case where the PNA is in the web

$$x = \frac{A_{s-c} - b_f t_f}{t_w} + t_f \qquad (9.11)$$

Equation 9.11 can be more easily understood if it is related to the areas being considered. The area of the web in compression is the area of steel in compression less the flange area. This web compression area is divided by the web thickness and the result is the location of the PNA measured from the underside of the flange. Thus, x is simply the thickness of the flange plus the depth of the web in compression.

EXAMPLE 9.1
Fully Composite Beam
Strength

GOAL: Determine the nominal moment strength for the interior composite beam shown as Beam A in Figure 9.7. Also determine the design moment and the allowable moment.

GIVEN: The section is a W21×44 and supports a 4.5-in. concrete slab. The dimensions are as shown. $F_y = 50$ ksi. $f'_c = 4$ ksi. Assume full composite action.

SOLUTION

Step 1:   Determine the effective flange width, the minimum of

$$b_{eff} = 30.0(12 \text{ in./ft})/4 = 90.0 \text{ in.}$$

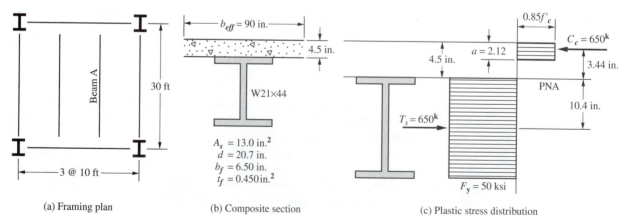

(a) Framing plan   (b) Composite section   (c) Plastic stress distribution

Figure 9.7 Interior Composite Beam (Example 9.1).

or

$$b_{eff} = (10.0 + 10.0)(12\ \text{in./ft})/2 = 120\ \text{in.}$$

Therefore use

$$b_{eff} = 90.0\ \text{in.}$$

Step 2:   Determine the controlling compression force.

$$V_c' = 0.85(4.0)(90.0)(4.5) = 1380\ \text{kips}$$
$$V_s' = 13.0(50) = 650\ \text{kips}$$

Assuming full composite action, the shear connectors must carry the smallest of V_c' and V_s', thus

$$V_q' = 650\ \text{kips}$$

Because V_s' is less than V_c', the PNA is in the concrete.

Step 3:   Determine the PNA location using Equation 9.4.

$$a = \frac{650}{0.85(4)(90.0)} = 2.12\ \text{in.}$$

The resulting plastic stress distribution is shown in Figure 9.7c.

Step 4:   Determine the nominal moment strength using Equation 9.5.

$$M_n = 650\left(\frac{20.7}{2}\right) + 650\left(4.50 - \frac{2.12}{2}\right) = 9000\ \text{in. kips}$$

$$\boxed{M_n = \left(\frac{9000}{12}\right) = 750\ \text{ft kips}}$$

Step 5:   For LRFD, the design moment is

$$\boxed{\phi M_n = 0.9(750) = 675\ \text{ft kips}}$$

Step 3:   For ASD, the allowable moment is

$$\boxed{\frac{M_n}{\Omega} = \frac{750}{1.67} = 449\ \text{ft kips}}$$

EXAMPLE 9.2
Fully Composite Beam Strength

GOAL: Determine the nominal moment strength for the interior composite beam shown as Beam A in Figure 9.7 using a larger W-shape. Also determine the design moment and the allowable moment.

GIVEN: Use a W21×111 for the steel member and the same materials as in Example 9.1. Again, assume full composite action.

SOLUTION

Step 1: Determine the effective flange width.
The effective flange width will remain the same, thus

$$b_{eff} = 90.0 \text{ in.}$$

Step 2: Determine the controlling compression force.

$$V_c' = 0.85(4.0)(4.5)(90.0) = 1380 \text{ kips}$$

$$V_s' = 32.7(50) = 1640 \text{ kips}$$

Assuming full composite action

$$V_q' = 1380 \text{ kips}$$

Because V_c' is less than V_s', the PNA is in the steel.

Step 3: Determine whether the PNA is in the steel flange or web.

$$T_f = 12.3(0.875)(50) = 538 \text{ kips}$$

$$T_w = 1640 - 2(538) = 564 \text{ kips}$$

Thus

$$C_c = 1380 > T_f + T_w = 538 + 564 = 1100$$

Because additional tension is required to balance the compression in the concrete, the PNA is in the flange.

Step 4: Determine the area of steel in compression.
Use Equation 9.9.

$$A_{s-c} = \frac{1640 - 1380}{2(50)} = 2.60 \text{ in.}^2$$

Step 5: Determine the location of the PNA in the flange.
The PNA is located down from the top of the steel x as given by Equation 9.10.

$$x = 2.60/12.3 = 0.211 \text{ in.}$$

The stress distribution for this PNA location is shown in Figure 9.8b.

Step 6: Determine the nominal moment strength of the composite beam.
Moments could be taken about any point to determine the nominal moment; however, a simplified mathematical model is shown in Figure 9.8c that makes the analysis quicker. In this case, the full area of steel is shown in tension and the portion in compression is first removed (130 kips on the compression side) and then added in compression (another 130 kips on the compression side), shown by the 2(130) = 260 kips. This results in only three

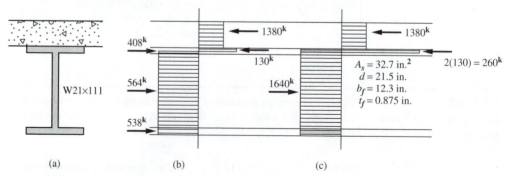

(a)   (b)   (c)

Figure 9.8 Interior Composite Beam (Example 9.2).

forces and moment arms entering the moment equation. Thus

$$M_n = T_s \left(\frac{d}{2}\right) + C_c \left(\frac{t}{2}\right) - 2A_{sc}F_y \left(\frac{x}{2}\right)$$

$$M_n = 1640\left(\frac{21.5}{2}\right) + 1380\left(\frac{4.5}{2}\right) - 2(130)\left(\frac{0.211}{2}\right) = 20,700 \text{ in.-kips}$$

$$M_n = \frac{20,700}{12} = 1730 \text{ ft-kips}$$

Step 7:   For LRFD, the design moment is

$$\phi M_n = 0.9(1730) = 1560 \text{ ft-kips}$$

Step 7:   For ASD, the allowable moment is

$$\frac{M_n}{\Omega} = \frac{1730}{1.67} = 1040 \text{ ft-kips}$$

9.5.2   Partially Composite Beams

The composite members considered thus far have been fully composite. This means that the shear connectors were assumed to be capable of transferring whatever force was required for equilibrium when either the concrete or steel were fully stressed. There are many conditions where the required strength of the composite section is less than what would result from full composite action. In particular, these are cases where the size of the steel member is dictated by factors other than the strength of the composite section. Because shear connectors compose a significant part of the cost of a composite beam, economies can result if the lower required flexural strength can be reflected in a reduced number of shear connectors when the steel section and concrete geometry are already given.

If the composite section is viewed under elastic stress distributions, partial composite action can be more easily understood. Figure 9.9 shows elastic stress distributions for three

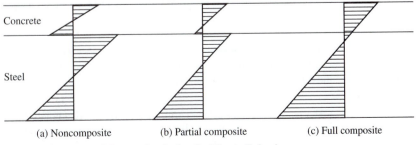

Concrete

Steel

(a) Noncomposite   (b) Partial composite   (c) Full composite

Figure 9.9 Levels of Composite Action for Elastic Behavior.

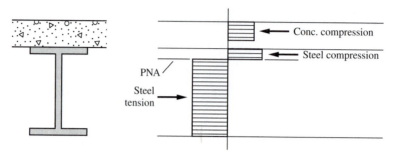

Figure 9.10 Plastic Stress Distribution for Partial Composite Action.

cases of combined steel and concrete. The first case, Figure 9.9a, is what results when the concrete simply rests on the steel with no shear transfer between the two materials. The result is two independent members that slip past each other at the interface. If the two materials are fully connected, the elastic stress distribution is as shown in Figure 9.9c and the materials are not permitted to slip at all. If some limited amount of slip is permitted between the steel and the concrete, the resulting elastic stress distribution is similar to that shown in Figure 9.9b. This is how the partially composite beam would behave in the elastic region.

The plastic moment strength for a partially composite member is the result of a stress distribution similar to that shown in Figure 9.10. The PNA will be in the steel and the magnitude of the compression force in the concrete will be controlled by the strength of the shear connectors.

$$V'_q = \Sigma Q_n$$

Regardless of the final location of the PNA, the force in the concrete is limited by the strength of the shear studs. Thus, an approach combining those taken for the three cases of fully composite sections is used for the partially composite member. By the definition of partially composite members

$$C_q = V'_q = \Sigma Q_n$$

and the depth of the concrete acting in compression is given by

$$a = \frac{\Sigma Q_n}{0.85 f'_c b_{eff}} \tag{9.12}$$

Equations 9.6 through 9.11 can then be used to determine the location of the PNA within the steel and the nominal moment can be obtained as before.

EXAMPLE 9.3
Partially Composite Beam Strength

GOAL: Determine the nominal moment strength of a partially composite beam. Also determine the design moment and the allowable moment.

GIVEN: Consider the concrete and steel given in Example 9.1 and shown in Figure 9.7. In this case, however, assume that the shear connectors are capable of transferring only $C_q = 500$ kips.

SOLUTION

Step 1:   Determine the effective flange width.
This is the same as determined for Example 9.1.

$$b_{eff} = 90.0 \text{ in.}$$

Step 2:   Determine the controlling compression force.
From Example 9.1

$$V_c' = 1380 \text{ kips}$$

$$V_s' = 650 \text{ kips}$$

From the given data

$$V_q' = C_q = 500 \text{ kips}$$

Because the lowest value of the compressive force is given by V_q', this is a partially composite member.

Step 3:   Determine the depth of the concrete working in compression from Equation 9.12.

$$a = \frac{500}{0.85(4)(90.0)} = 1.63 \text{ in.}$$

Step 4:   Determine the area of steel in compression from Equation 9.9.

$$A_{s-c} = \frac{650 - 500}{2(50)} = 1.50 \text{ in.}^2$$

Because this is less than the area of the flange, $6.50(0.450) = 2.93 \text{ in.}^2$, the PNA is in the flange.

Step 5:   Determine the location of the PNA from Equation 9.10.

$$x = 1.50/6.50 = 0.231 \text{ in.}$$

Step 6:   Determine the nominal moment strength using the three forces shown in Figure 9.11.

$$M_n = 650\left(\frac{20.7}{2}\right) + 500\left(4.50 - \frac{1.63}{2}\right) - 2(75.0)\left(\frac{0.231}{2}\right) = 8590 \text{ in.-kips}$$

$$M_n = \frac{8590}{12} = 716 \text{ ft-kips}$$

Step 7:   For LRFD, the design moment is

$$\phi M_n = 0.9(716) = 644 \text{ ft-kips}$$

Step 8:   For ASD, the allowable moment is

$$\frac{M_n}{\Omega} = \frac{716}{1.67} = 429 \text{ ft-kips}$$

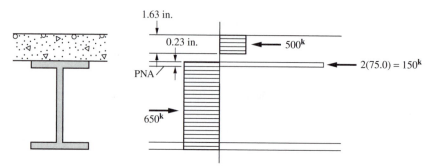

Figure 9.11 Stress Distribution and Forces Used in Example 9.3.

The nominal moment strength decreased from 750 ft-kips for the full composite action of Example 9.1 to 716 ft-kips for the level of partial composite action given in Example 9.3. This is approximately a 5% reduction in strength corresponding to more than a 23% reduction in shear connector strength. In both cases, the strength of the composite beam is significantly greater than that of the bare steel beam where the plastic moment strength of the bare steel beam is $M_p = 398$ ft kips. It is acceptable to make comparisons at the nominal strength level because for the bare steel beam and the composite beam, the resistance factors and safety factors are the same.

9.5.3 Composite Beam Design Tables

The force transferred between the steel and concrete governs the strength of the composite beams. The shear studs transfer that force to the concrete so design can be linked to the total shear force, ΣQ_n. Design tables have been developed that use the shear stud strength in combination with an infinite variety of concrete areas and strengths to determine the flexural strength of the composite beam. These are given in Manual Table 3-19, an example of which is shown here as Figure 9.12.

The variables used in Manual Table 3-19 are defined in Figure 9.13. The beam is divided into seven PNA locations; five are in the flange and two are in the web. When the PNA is at the top of the flange, position 1, the entire steel section is in tension. This is a fully composite beam. When the PNA is in the web at location 7, 25% of the potential steel section force is transferred to the concrete through the studs. As shown in Figure 9.13, the flange has five PNA locations and the stud strength for location 6 is one-half the difference between that at locations 5 and 7. These seven PNA locations establish corresponding stud strengths, ΣQ_n, which are also given in the tables.

The contribution of the concrete to the beam strength requires knowledge of the location of the concrete compressive force. As already discussed, the force in the concrete is equal to the force in the studs, ΣQ_n. The moment arm for that force is defined as Y2 in Figure 9.13. It is a function of the concrete strength and concrete geometry. These tables are quite flexible and accommodate any permitted concrete strength and effective slab width. The thickness of the slab is limited only by the maximum moment arm given in the table.

Although these tables are of most value in the design of a composite beam, they can also be used to check a particular combination. Selection of a composite beam is illustrated in Section 9.9.

Table 3–19 (continued)
Composite W Shapes
Available Strength in Flexure, kip-ft

$F_y = 50$ ksi

W16–W14

Shape	M_p/Ω_b ASD	$\phi_b M_p$ LRFD	PNA[c]	Y1[a] in.	ΣQ_n kip	Y2=2 ASD	Y2=2 LRFD	Y2=2.5 ASD	Y2=2.5 LRFD	Y2=3 ASD	Y2=3 LRFD	Y2=3.5 ASD	Y2=3.5 LRFD
W16×26	110	166	TFL	0	384	189	284	198	298	208	312	217	327
			2	0.0863	337	184	276	192	289	201	302	209	314
			3	0.173	289	179	269	186	280	193	291	201	301
			4	0.259	242	174	261	180	270	186	279	192	288
			BFL	0.345	194	168	253	173	260	178	267	183	275
			6	2.04	145	161	241	164	247	168	252	171	258
			7	4.00	96.0	148	223	151	227	153	230	156	234
W14×38	153	231	TFL	0	558	252	379	266	400	280	421	294	442
			2	0.129	471	243	365	255	383	267	401	278	418
			3	0.258	384	234	351	243	365	253	380	262	394
			4	0.386	297	223	336	231	347	238	358	246	369
			BFL	0.515	209	213	320	218	328	223	335	228	343
			6	1.42	174	208	312	212	319	216	325	221	332
			7	2.55	140	201	302	204	307	208	312	211	318
W14×34	136	205	TFL	0	500	224	337	237	356	249	375	262	393
			2	0.114	423	216	325	227	341	238	357	248	373
			3	0.228	347	208	313	217	326	225	339	234	352
			4	0.341	270	199	300	206	310	213	320	220	330
			BFL	0.455	193	190	286	195	293	200	300	205	308
			6	1.41	159	185	279	189	285	193	291	197	297
			7	2.60	125	179	268	182	273	185	278	188	283
W14×30	118	177	TFL	0	442	197	296	208	313	219	329	230	346
			2	0.0963	378	190	286	200	300	209	314	219	329
			3	0.193	313	183	276	191	287	199	299	207	311
			4	0.289	248	176	265	182	274	189	283	195	293
			BFL	0.385	183	169	253	173	260	178	267	182	274
			6	1.48	147	163	246	167	251	171	257	174	262
			7	2.82	111	156	234	159	239	162	243	164	247
W14×26	100	151	TFL	0	385	172	258	181	273	191	287	201	302
			2	0.105	332	166	250	175	263	183	275	191	287
			3	0.210	279	161	242	168	252	175	263	182	273
			4	0.315	226	155	233	160	241	166	250	172	258
			BFL	0.420	174	149	223	153	230	157	236	162	243
			6	1.67	135	143	215	146	220	150	225	153	230
			7	3.18	96.1	134	202	137	206	139	209	142	213

ASD	LRFD
$\Omega_b = 1.67$	$\phi_b = 0.90$

[a] Y1 = distance from top of the steel beam to plastic neutral axis.
[b] Y2 = distance from top of the steel beam to concrete flange force.
[c] See Figure 3–3c for PNA locations.

Figure 9.12 Composite W Shapes. Available Strength in Flexure. Copyright © American Institute of Steel Construction, Inc. Reprinted with Permission. All rights reserved.

Table 3–19 (continued)
Composite W Shapes
Available Strength in Flexure, kip-ft.

F_y = 50 ksi

W16–W14

Shape	$Y2^b$, in.													
	4		4.5		5		5.5		6		6.5		7	
	ASD	LRFD	ASD	LRFD	ASD	LRFD	ASD	LRFD	ASD	LRFD	ASD	LRFD	ASD	LRFD
W16×26	227	341	237	356	246	370	256	385	265	399	275	413	285	428
	218	327	226	340	234	352	243	365	251	377	260	390	268	403
	208	312	215	323	222	334	229	345	237	356	244	366	251	377
	198	297	204	306	210	315	216	324	222	334	228	343	234	352
	188	282	192	289	197	296	202	304	207	311	212	318	217	326
	175	263	179	269	182	274	186	279	190	285	193	290	197	296
	158	237	160	241	163	245	165	248	168	252	170	255	172	259
W14×38	308	463	322	483	336	504	350	525	363	546	377	567	391	588
	290	436	302	454	314	471	325	489	337	507	349	524	361	542
	272	409	281	423	291	437	301	452	310	466	320	481	329	495
	253	380	260	391	268	403	275	414	283	425	290	436	297	447
	234	351	239	359	244	367	249	375	255	383	260	390	265	398
	225	338	229	345	234	351	238	358	243	365	247	371	251	378
	215	323	218	328	222	333	225	339	229	344	232	349	236	354
W14×34	274	412	287	431	299	450	312	468	324	487	337	506	349	525
	259	389	269	405	280	421	290	436	301	452	311	468	322	484
	243	365	251	378	260	391	269	404	277	417	286	430	295	443
	226	340	233	350	240	360	247	371	253	381	260	391	267	401
	209	315	214	322	219	329	224	337	229	344	234	351	238	358
	201	303	205	308	209	314	213	320	217	326	221	332	225	338
	191	287	194	292	197	297	200	301	204	306	207	311	210	315
W14×30	241	362	252	379	263	396	274	412	285	429	296	445	307	462
	228	343	237	357	247	371	256	385	266	399	275	413	285	428
	215	323	222	334	230	346	238	358	246	369	254	381	261	393
	201	302	207	311	213	321	219	330	226	339	232	348	238	358
	187	281	191	288	196	295	201	301	205	308	210	315	214	322
	178	268	182	273	185	279	189	284	193	290	196	295	200	301
	167	251	170	255	173	259	175	264	178	268	181	272	184	276
W14×26	210	316	220	330	229	345	239	359	249	374	258	388	268	402
	200	300	208	312	216	325	224	337	233	350	241	362	249	375
	189	283	196	294	203	304	209	315	216	325	223	336	230	346
	177	267	183	275	189	284	194	292	200	301	206	309	211	317
	166	249	170	256	175	262	179	269	183	275	188	282	192	288
	156	235	160	240	163	245	166	250	170	255	173	260	177	265
	144	216	146	220	149	224	151	227	154	231	156	234	158	238

ASD	LRFD	
Ω_b = 1.67	ϕ_b = 0.90	[a] $Y1$ = distance from top of the steel beam to plastic neutral axis. [b] $Y2$ = distance from top of the steel beam to concrete flange force. [c] See Figure 3–3c for PNA locations.

Figure 9.12 (*Continued*)

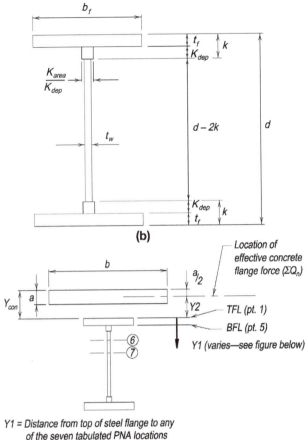

(b)

Y1 = Distance from top of steel flange to any
    of the seven tabulated PNA locations

$$\Sigma Q_n \; (@\; point \textcircled{6}) = \frac{\Sigma Q_n \; (@\; pt.\; 5) + \Sigma Q_n \; (@\; pt.\; 7)}{2}$$

$$\Sigma Q_n \; (@\; point \textcircled{7}) = 0.25 F_y A_s$$

PNA FLANGE LOCATIONS

Figure 9.13 Definition of Variables for Use with Composite Beam Design Tables. Copyright ©
American Institute of Steel Construction, Inc. Reprinted with Permission. All rights reserved.

EXAMPLE 9.4	**GOAL:** Determine the design flexural strength and allowable flexural strength for the fully

EXAMPLE 9.4
Composite Beam
Strength Using Tables

GOAL: Determine the design flexural strength and allowable flexural strength for the fully
composite W16×26.

GIVEN: The W16×26 beam is used with the metal deck and slab shown in Figure 9.14. The
effective flange width is given, $b_{eff} = 60.0$ in. $f_c' = 4$ ksi. Use Table 3-19 from Figure 9.12.

SOLUTION   **Step 1:**   Determine the controlling compression force.

$$V_c' = 0.85(4)(3.0)(60.0) = 612 \text{ kips}$$

$$V_s' = 7.68(50) = 384 \text{ kips}$$

where

t_f = thickness of channel flange, in.

t_w = thickness of channel web, in.

L_c = length of channel, in.

The strength of the channel shear connector must be developed by welding the channel to the beam flange for the force Q_n with appropriate consideration of the eccentricity of the force on the connector.

9.6.1   Number and Placement of Shear Studs

Although a shear stud serves to transfer load between the steel beam and the concrete slab, it is not necessary to place the studs in accordance with the shear diagram of the loaded beam. Tests have demonstrated that the studs have sufficient ductility to redistribute the shear load under the ultimate load condition. Therefore, in design, it is assumed that the studs share the load equally. Thus, the total shear force determined according to Section 9.5 must be transferred over the distance between maximum moment and zero moment. For a uniform load this results in V'_q/Q_n connectors on each side of the maximum moment at the center line of the beam span. In the case of concentrated loads placed at the third points of the beam, the same number of studs would be on each side of the beam between the load and the support, and a minimum number of studs would be required between the loads. This is shown in Figure 9.15.

Shear studs must be placed so that they have a minimum of 1 in. of lateral concrete cover, unless placed in ribs of formed steel deck. The diameter of the studs must be no greater than 2.5 times the flange thickness of the beam to which they are welded if they are not located directly over the beam web. For studs not placed in steel deck ribs, the minimum center-to-center spacing of studs is six stud diameters along the member and four stud diameters transverse to the member. When placed in metal deck ribs, the spacing is to be no less than four diameters in any direction and the maximum stud spacing is eight times the total slab thickness or 36 in.

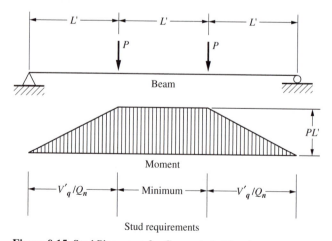

Figure 9.15 Stud Placement for Concentrated Load.

EXAMPLE 9.5
Shear Stud
Determination

GOAL: Determine the required number of $^3/_4$-in. shear studs required over the complete beam span.

GIVEN: Use the fully composite beam of Example 9.1. Assume normal weight concrete and the values of Example 9.1.

SOLUTION

Step 1:   Determine the strength of a single shear stud.
From Table 9.2, based on the concrete

$$Q_n = 26.1 \text{ kips}$$

and based on the stud

$$Q_n = 28.7 \text{ kips}$$

Use the least Q_n, so

$$Q_n = 26.1 \text{ kips}$$

Step 2:   Determine the number of studs required.
From Example 9.1

$$V_q' = \Sigma Q_n = 650 \text{ kips}$$

Thus

$$\# \text{studs} = 650/26.1 = 24.9 \text{ studs}$$

Step 3:   Determine the total number of studs required for the beam.
Place 25 $^3/_4$-in. shear studs on each side of the beam between the maximum moment and the zero moment. Thus,

> use 50 studs for the entire beam span

Note that these calculations are independent of ASD or LRFD because the calculations are carried out at the nominal strength level.

9.7   COMPOSITE BEAMS WITH FORMED METAL DECK

The combination of formed steel deck and composite design is considered today to be one of the most economical methods of floor construction. The steel deck is a stay-in-place formwork for the concrete slab. Cells, which can be formed by enclosing the space below the deck and between the ribs, can then be used to distribute the electrical and electronic systems of the building, contributing greatly to the overall economy of the system.

The Specification provides rules for steel decks with nominal rib heights of up to 3 in. and average rib widths of 2 in. or more. For a deck that has ribs narrower at the top than at the interface with the beam, the width of the rib used in calculations must be taken as no more than the width at the narrow portion. Deck with this profile is shown in Figure 9.16 along with other common deck profiles. Studs must be $^3/_4$ in. in diameter or less and extend at least $1^1/_2$ inches above the top of the steel deck. The concrete slab thickness must be sufficient to provide $^1/_2$ in. of cover over the top of the installed stud. The deck must be anchored to the supporting beam by a combination of puddle welds and studs at a spacing not to exceed 18 in.

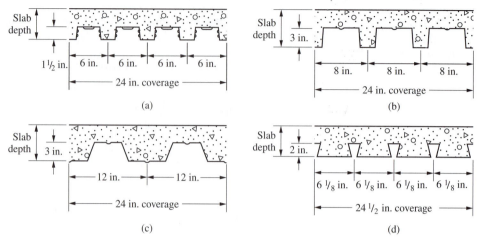

Figure 9.16 Common Steel Deck Profiles.

9.7.1 Deck Ribs Perpendicular to Steel Beam

Beams supporting the steel deck have the ribs running perpendicular to the beam, as shown in Figure 9.17. The space below the top of the rib contains concrete only in the alternating spaces so there is no opportunity to transfer force at this level. Thus, the only concrete available for calculating the full concrete force is above the top of the deck.

The right-hand side of the inequality of Equation 9.13, without R_p or R_g, accounts for the tensile strength of the stud material. This strength must be reduced to account for the fact that the force exerted on the stud is applied at a higher point in this application than in a flat soffit beam. It must also be reduced to account for the location of the stud within the concrete rib, because there is a difference in strength if the stud is placed closer to the rib wall in the direction of force or closer to the rib wall away from the force. The stud strength value specified on the right-hand side of the inequality of Equation 9.13 includes two multipliers, R_g and R_p. A simplified table of the values for these adjustment factors is given in Table 9.3 and the strength of the stud as controlled by F_u is given in Table 9.2.

R_g is used to account for the number of studs in a given concrete rib. If a rib contains a single stud, $R_g = 1.0$; if the rib contains two studs, $R_g = 0.85$; and if a rib contains 3 or more studs, $R_g = 0.7$.

R_p is used to account for the location of the stud in the rib in either the strong or weak position. Figure 9.18 shows the strong and weak location of a stud in relation to the applied force. Because it is difficult to insure that the studs are located in the strong position, it is

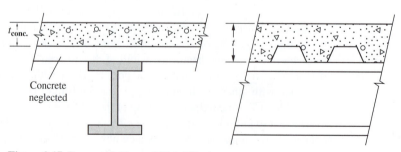

Figure 9.17 Beam with Formed Metal Deck.

Table 9.3   Shear Stud Strength Adjustment Factors
Copyright © American Institute of Steel Construction, Inc. Reprinted with
Permission. All rights reserved.

Condition	R_g	R_p
No decking*	1.0	1.0
Decking oriented parallel to the steel shape		
$\quad \dfrac{w_r}{h_r} \geq 1.5$	1.0	0.75
$\quad \dfrac{w_r}{h_r} < 1.5$	0.85**	0.75
Decking oriented perpendicular to the steel shape		
Number of studs occupying the same decking rib		
$\quad$ 1	1.0	0.6+
$\quad$ 2	0.85	0.6+
$\quad$ 3 or more	0.7	0.6+

h_r = nominal rib height, in. (mm)

w_r = average width of concrete rib or haunch (as defined in Section I3.2c), in. (mm)

*to qualify as "no decking," stud shear connectors shall be welded directly to the steel shape and no more than 50 percent of the top flange of the steel shape may be covered by decking or sheet steel, such as girder filler material.

**for a single stud

+this value may be increased to 0.75 when $e_{mid-ht} \geq 2$ in. (51 mm)

recommended that $R_p = 0.6$ be used unless it is critical enough to warrant the extra effort to insure that studs are placed in the strong position. When studs are placed in the strong position, which is when they are at least 2.0 in. from the loaded side of the rib edge at mid-height to the stud, as shown in Figure 9.18, R_p can be increased to 0.75.

The maximum stud spacing is specified as 36 in., which is convenient because many decks have a rib spacing of 6 in.

9.7.2   Deck Ribs Parallel to Steel Beam

For girders supporting beams that carry steel deck, the ribs run parallel to the steel section, as shown in Figure 9.19. Concrete below the top of the deck can be used in calculating the composite section properties and must be used in shear stud calculations. For calculation purposes, concrete below the top of the steel deck can be neglected unless it is needed to balance the shear stud strength. The design procedure described for flat soffit beams applies here as well, provided sufficient concrete is available above the top of the steel deck as determined through Equations 9.4 or 9.12. If the concrete below the top of the steel deck is

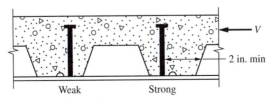

Figure 9.18 Strong and Weak Shear Stud Locations.

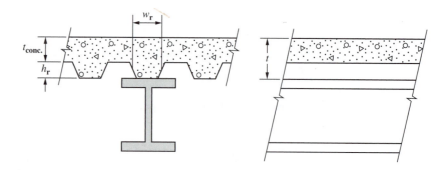

Figure 9.19 Girder with Formed Metal Deck.

needed to balance the stud or steel beam strength, the only difference in the determination of strength is related to the changed geometry when including a portion of the ribs.

When the depth of the steel deck is $1\frac{1}{2}$ in. or greater, the average width, w_r, of the haunch or rib is not to be less than 2 in. for the first stud in the transverse row plus four stud diameters for each additional stud. If the deck rib is too narrow, the deck can be split over the beam and spaced in such a way as to allow for the necessary rib width without adversely effecting member strength.

For this deck orientation, R_g is used to account for the width-to-height ratio of the deck rib. When $w_r/h_r \geq 1.5$, $R_g = 1.0$. When $w_r/h_r < 1.5$, $R_g = 0.85$.

R_p is taken as 0.75 in all cases where the deck ribs are parallel to the supporting member.

EXAMPLE 9.6
Composite Beam
Design

GOAL: Calculate the design moment and allowable moment and determine the stud requirements for a composite section.

Carry out the calculations for the following three cases.

(a) Full composite action. (Figure 9.20)

(b) Partial composite action with $V_q' = 387$ kips, which results in the PNA at the center of the top flange of the steel beam. (Figure 9.21)

(c) Partial composite action with $V_q' = 260$ kips, which results in the PNA at the bottom of the top flange of the steel beam. (Figure 9.22)

GIVEN: Use a W18×35 with a 6-in. slab on a 3-in. metal deck perpendicular to the beam with the profile shown in Figure 9.16c. The beam spacing is 12 ft and the beam span is 40 ft. $f_c' = 4$ ksi. $F_y = 50$ ksi.

SOLUTION

Step 1:   Determine the effective flange width.

$$b_{eff} = 40.0/4 = 10.0 \text{ ft (governs)}$$
$$b_{eff} = \text{beam spacing} = 12.0 \text{ ft}$$

Step 2:   Determine the compression force using the full concrete and full steel areas.

$$V_c' = 0.85(4)(120)(3.0) = 1220 \text{ kips}$$
$$V_s' = 10.3(50) = 515 \text{ kips}$$

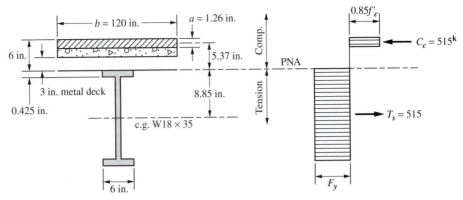

Figure 9.20 Example 9.6, case (a).

Part (a) Full composite action, Figure 9.20

Step 3: Determine the controlling concrete force.

For full composite action, V'_q is the smallest of V'_c and V'_s, thus

$$V'_q = 515 \text{ kips}$$

Step 4: Calculate the effective depth of the concrete.

$$a = \frac{515}{0.85(4)(120)} = 1.26 \text{ in.}$$

Because $a < 3$ in. available in the concrete above the deck, the procedures for a flat soffit beam can be used.

Step 5: Determine the nominal moment strength.

$$M_n = 515\left(\frac{17.7}{2}\right) + 515\left(6.0 - \frac{1.26}{2}\right) = 7320 \text{ in.-kips}$$

$$M_n = \frac{7320}{12} = 610 \text{ ft-kips}$$

Step 6: For LRFD, the design moment is

$$\phi M_n = 0.9(610) = 549 \text{ ft kips}$$

Step 6: For ASD, the allowable moment is

$$\frac{M_n}{\Omega} = \frac{610}{1.67} = 365 \text{ ft kips}$$

Step 7:   Determine the strength of a single stud.

From Table 9.2, the value of a single $^3/_4$-in. stud, with normal weight concrete $f'_c = 4$ ksi, is 26.1 kips.

However, because the studs are used in conjunction with the metal deck, a check for any required reduction must be made. Assuming two studs per rib with the studs placed in the weakest location, from Tables 9.2 and 9.3

$$R_p = 0.85$$
$$R_g = 0.6$$
$$A_{sc}F_u = 28.7 \text{ kips}$$
$$R_p R_g A_{sc}F_u = 0.85(0.6)(28.7) = 14.6 \leq 26.1 \text{ kips}$$

The stud strength is the lower value, based on the stud placement in the metal deck.

Step 8:   Determine the number of studs required on each side of the maximum moment.

The shear that is to be transferred is 515 kips. Therefore

$$\text{Number of studs} = 515/14.6 = 35.3$$

Step 9:   Determine the total number of studs required for the beam.

Use 36 studs on each half span or

> 72 studs for the full beam span

With this deck profile, studs can be placed every 12 in. This will nicely accommodate the 72 studs on the 40-ft span with two studs placed in each rib.

Part (b) Partial composite action, Figure 9.21

Step 10:   Determine the controlling concrete force.

Because the value of $V'_q = 387$ kips given is less than V'_c and V'_s, V'_q controls and this is a partially composite beam.

Step 11:   Calculate the effective depth of the concrete.

$$a = \frac{387}{0.85(4)(120)} = 0.949 \text{ in.}$$

Because $a < 3.0$ in. sufficient concrete is available above the metal deck as in Part (a).

Step 12:   Determine the area of steel in compression.

$$A_{s-c} = \frac{515 - 387}{2(50)} = 1.28 \text{ in.}^2$$

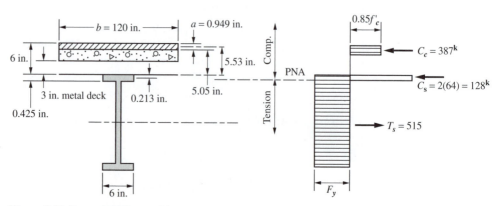

Figure 9.21 Example 9.6, case (b).

Step 13: Determine the location of the PNA in the steel.
Assume the PNA is in the flange.

$$x = \frac{A_{s-c}}{b_f} = \frac{1.28}{6.00} = 0.213 \text{ in.} < t_f = 0.425 \text{ in.}$$

Therefore, the PNA is in the flange.

Step 14: Determine the nominal moment strength.

$$M_n = 515\left(\frac{17.7}{2}\right) + 387\left(6.0 - \frac{0.949}{2}\right) - 128\left(\frac{0.213}{2}\right) = 6680 \text{ in kips}$$

$$M_n = \frac{6680}{12} = 557 \text{ ft-kips}$$

Step 15: For LRFD, the design moment is

$$\phi M_n = 0.9(557) = 501 \text{ ft-kips}$$

Step 15: For ASD, the allowable moment is

$$\frac{M_n}{\Omega} = \left(\frac{557}{1.67}\right) = 334 \text{ ft-kips}$$

Step 16: Determine the stud requirements.
The shear to be transferred is 387 kips.
The value of stud strength previously determined in Part (a)

$$Q_n = 14.6 \text{ kips}$$

Step 17: Determine the required number of shear studs.

$$\# \text{studs} = 387/14.6 = 26.5 \text{ studs}$$

Step 18: Determine the total number of studs required for the beam.
Use 27 studs in each half span or

$$\boxed{54 \text{ studs for the full beam span}}$$

Part (c) Partial composite action, Figure 9.22

Step 19: Determine the controlling concrete force.
Because the value of $V'_q = 260$ kips given is less than V'_c and V'_s, V'_q controls and this is a partially composite beam.

Step 20: Calculate the effective depth of concrete.

$$a = \frac{260}{0.85(4)(120)} = 0.637 \text{ in.}$$

Step 21: Determine the area of steel in compression.

$$A_{s-c} = \frac{515 - 260}{2(50)} = 2.55 \text{ in.}^2$$

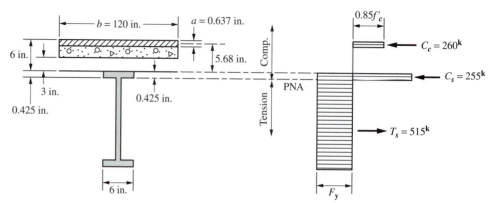

Figure 9.22 Example 9.6, case (c).

Step 22:   Determine the location of the PNA in the steel.
Assume the PNA is in the flange

$$x = \frac{A_{s-c}}{b_f} = \frac{2.55}{6.00} = 0.425 \text{ in.}$$

which is the flange thickness, as expected.

Step 23:   Determine the nominal moment strength.

$$M_n = 515\left(\frac{17.7}{2}\right) + 260\left(6.0 - \frac{0.637}{2}\right) - 255\left(\frac{0.425}{2}\right) = 5980 \text{ in.-kips}$$

$$M_n = \frac{5980}{12} = 498 \text{ ft-kips}$$

Step 24:   For LRFD, the design moment is

$$\phi M_n = 0.9(498) = 448 \text{ ft-kips}$$

Step 24:   For ASD, the allowable moment is

$$\frac{M_n}{\Omega} = \left(\frac{498}{1.67}\right) = 298 \text{ ft kips}$$

Step 25:   Determine the stud requirements.
The shear to be transferred is 260 kips.
As before, the single stud strength is

$$Q_n = 14.6 \text{ kips}$$

Step 26:   Determine the required number of shear studs.

$$\# \text{ studs} = 260/14.6 = 17.8 \text{ studs.}$$

Step 27:   Determine the total number of studs required for the beam.
Use 18 studs in each half span or 36 studs for the full beam span.

However, this requires only a single stud in each rib so the $R_g = 0.85$ does not need to be applied. Therefore,

$$Q_n = 1.0(0.6)(28.7) = 17.2 < 26.1 \text{ kips}$$

and

$$\# \text{studs} = 260/17.2 = 15.1 \text{ studs}$$

Thus,

use a total of 32 studs

which will also be accommodated with one stud per rib.

9.8   FULLY ENCASED STEEL BEAMS

Steel beams fully encased in concrete that contributes to the strength of the final member are called *encased beams*. Such beams can be designed by one of two procedures given in Specification Section I3.3. The flexural strength can be calculated from the superposition of elastic stresses, considering the effects of shoring, with $\phi_b = 0.9$ and $\Omega_b = 1.67$. Or, when shear connectors are provided, the flexural strength can be based on the plastic stress distribution or strain compatibility approach for the composite section with $\phi_b = 0.85$ and $\Omega_b = 1.76$. Alternatively, the strength can be calculated from the plastic stress distribution on the steel section alone, with $\phi_b = 0.9$ and $\Omega_b = 1.67$.

9.9   SELECTING A SECTION

The design of a composite beam is somewhat of a trial-and-error procedure, as are numerous other design situations. The material presented thus far in this chapter has been directed toward the determination of section strength when the cross-section and concrete dimensions are known. This section addresses the preliminary selection of the steel shape to go along with an already known concrete slab. This procedure is followed by a discussion of the design tables found in the Manual.

With an estimate of beam depth, the weight of the beam can be estimated. This is based on the assumption that the PNA is within the concrete so that the full steel section is at yield. The resulting dimensions are given in Figure 9.23. The moment arm between the tension force in the steel and the compression force in the concrete is given by

$$\text{moment arm} = \frac{d}{2} + \left(t - \frac{a}{2}\right) \tag{9.15}$$

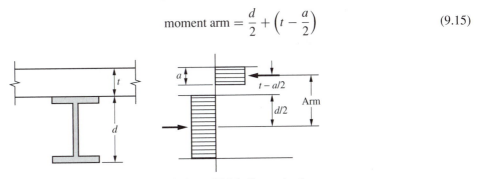

Figure 9.23 Moment Arm for Preliminary Weight Determination.

If the nominal moment strength is divided by the moment arm, the required tension force will be determined. If that force is divided by the steel yield stress, the required area is determined. Multiplying the required area by weight of steel, 3.4 lb./ft for each in.^{2}, yields an estimate of the beam weight. Thus

$$\text{beam weight} = \frac{M_n}{\left(\dfrac{d}{2} + t - \dfrac{a}{2}\right)F_y} \quad (3.4) \tag{9.16}$$

To determine the beam weight by this approach, the depth of the beam must be estimated. Several approaches have been suggested for this. One simple approach would be to take the span in feet and divide it by 24 to get the depth in inches. Another approach to determine the depth of the total composite section $(d + t)$ is to take the span in feet and divide by 16 to obtain the depth in inches. Any reasonable approach gives a starting point. Because the thickness of the slab is determined from the design in the transverse direction, only the effective depth of the concrete is left to be determined. It is generally sufficient to assume that the effective depth of the concrete is 1 in.; therefore, $a/2 = 0.5$ in.

Although this approach to finding a starting point for composite beam design might be helpful, the design tables in the Manual are such that this much effort is not needed to establish a starting point. Manual Table 3-19 (Figure 9.12) has already been discussed in the context of determining the strength of a given combination of steel and concrete. It will now be approached from the perspective of selecting a section through the use of an example.

EXAMPLE 9.6a
Composite Beam Design by LRFD

GOAL: Select the most economical W-shape to be used as a composite beam.

GIVEN: The composite beam spans 30.0 ft and is spaced at 10.0 ft from adjacent beams. It supports a 5-in. slab on a 2-in. formed steel deck with a profile similar to that shown in Figure 9.16a. The beam must carry a dead load moment of 50.0 ft-kips and a live load moment of 150 ft-kips. In addition, the bare steel beam must be checked for the dead load plus construction live load, which produces a moment of 40.0 ft-kips. $F_y = 50$ ksi and $f'_c = 3$ ksi.

SOLUTION

Step 1:   Determine the required moment for the composite beam.

$$M_u = 1.2(50.0) + 1.6(150) = 300 \text{ ft-kips}$$

Step 2:   Determine the starting moment arm for the concrete from the top of the steel.
    Manual Tables 3-19 are most effective when entered with a value for Y2. In order to start the design process, the moment arm for the compressive force in the concrete must be estimated. It is almost always adequate to assume, as a starting point, that $a = 1.0$ in. Thus

$$Y2 = 5 - \frac{1.0}{2} = 4.5 \text{ in.}$$

Step 3:   Select potential W-shapes from Figure 9.12.
    Enter the column in Figure 9.12 with Y2 = 4.5 and proceed down to identify the potential section that will carry the design moment of 300 ft-kips.

W16×26 with $M_u = 306$ ft kips and $\Sigma Q_n = 242$ kips

Using additional portions of Table 3.19 from the Manual yields additional possibilities

W16×31 with $M_u = 319$ ft kips and $\Sigma Q_n = 164$ kips
W14×34 with $M_u = 308$ ft kips and $\Sigma Q_n = 159$ kips
W14×30 with $M_u = 311$ ft kips and $\Sigma Q_n = 248$ kips
W14×26 with $M_u = 312$ ft kips and $\Sigma Q_n = 332$ kips

Step 4:   Determine the effective flange width.

$$b_{eff} \leq \frac{30.0(12)}{4} = 90.0 \text{ in.}$$

$$b_{eff} \leq 10.0(12) = 120 \text{ in.}$$

Therefore $b_{eff} = 90.0$ in.

Step 5:   Determine the depth of concrete needed to balance the force in the shear studs.
   For the beam span and spacing given, considering further the W16×31, using the value of ΣQ_n and the effective flange width already determined.

$$a = \frac{164}{0.85(3)(90.0)} = 0.715 \text{ in.}$$

Because this is less than $a = 1.0$ in. that was assumed to start the problem, the assumption was conservative. Design could continue with the determination of a more accurate required stud strength or this conservative solution could be used. The required number of studs would be determined as before, accounting for the presence of any formed steel deck and its influence on the individual stud strength.

Step 6:   Determine the required strength of the bare steel beam under dead load plus construction live load.

$$M_u = 1.2(50.0) + 1.6(40.0) = 124 \text{ ft-kips}$$

Step 7:   Check to verify that the bare steel beam will support the required strength.
   From Manual Table 3-19, the W16×31 has a design strength of,

$$\phi M_p = 203 \text{ ft-kips} > 124 \text{ ft-kips}$$

Therefore, the W16 × 31 is an acceptable selection for strength.

Step 8:   To show what happens when the assumption for a is not quite as good, the W14 × 26 is considered. Again using the ΣQ_n determined from the table and the effective flange width

$$a = \frac{332}{0.85(3)(90.0)} = 1.45 \text{ in.}$$

This is significantly greater than the assumed value. To consider this section further, determine a new Y2 such that

$$Y2 = 5 - \frac{1.45}{2} = 4.28 \text{ in.}$$

Entering Manual Table 3-19 with Y2 = 4.0 as a conservative number, $M_u = 300$ ft-kips is determined and it corresponds to the same required shear stud strength. Thus, this section also meets the strength requirements and the design can proceed with stud selection.

EXAMPLE 9.6b
Composite Beam Design by ASD

GOAL: Select the most economical W-shape to be used as a composite beam.

GIVEN: The composite beam spans 30.0 ft and is spaced at 10.0 ft from adjacent beams. It supports a 5-in. slab on a 2-in. formed steel deck with a profile similar to that shown in Figure 9.16a. The beam must carry a dead load moment of 50.0 ft-kips and a live load moment of 150 ft-kips. In addition, the bare steel beam must be checked for the dead load plus construction live load, which produces a moment of 40.0 ft-kips. $F_y = 50$ ksi and $f_c' = 3$ ksi.

SOLUTION

Step 1: Determine the required moment for the composite beam.

$$M_a = 50.0 + 150 = 200 \text{ ft-kips}$$

Step 2: Determine the starting moment arm for the concrete from the top of the steel.

Manual Tables 3-19 are most effective when entered with a value for Y2. In order to start the design process, the moment arm for the compressive force in the concrete must be estimated. It is almost always adequate to assume, as a starting point, that $a = 1.0$ in. Thus

$$Y2 = 5 - \frac{1.0}{2} = 4.5 \text{ in.}$$

Step 3: Select potential W-shapes from Figure 9.12.

Enter the column in Figure 9.12 with Y2 = 4.5 and proceed down to identify the potential section that will carry the design moment of 200 ft-kips.

$$\text{W16×26 with } M_a = 204 \text{ ft kips and } \Sigma Q_n = 242 \text{ kips}$$

Using additional portions of Table 3-19 from the Manual yields additional possibilities

$$\text{W16×31 with } M_a = 212 \text{ ft kips and } \Sigma Q_n = 164 \text{ kips}$$

$$\text{W14×34 with } M_a = 205 \text{ ft kips and } \Sigma Q_n = 159 \text{ kips}$$

$$\text{W14×30 with } M_a = 207 \text{ ft kips and } \Sigma Q_n = 248 \text{ kips}$$

$$\text{W14×26 with } M_a = 208 \text{ ft kips and } \Sigma Q_n = 332 \text{ kips}$$

Step 4: Determine the effective flange width.

$$b_{eff} \leq \frac{30.0(12)}{4} = 90.0 \text{ in.}$$

$$b_{eff} \leq 10.0(12) = 120 \text{ in.}$$

Therefore $b_{eff} = 90.0$ in.

Step 5: Determine the depth of concrete needed to balance the force in the shear studs.

For the beam span and spacing given, consider further the W16×31, using the value of ΣQ_n and the effective flange width already determined.

$$a = \frac{164}{0.85(3)(90.0)} = 0.715 \text{ in.}$$

Because this is less than $a = 1.0$ in. that was assumed to start the problem, the assumption was conservative. Design could continue with the determination of a more accurate required stud strength or this conservative solution could be used. The required number of studs would be determined as before, accounting for the presence of any formed steel deck and its influence on the individual stud strength.

Step 6: Determine the required strength of the bare steel beam under dead load plus construction live load.

$$M_a = 50.0 + 40.0 = 90.0 \text{ ft-kips}$$

Step 7: Check to verify that the bare steel beam will support the required strength.

From Manual Table 3-19, the W16×31 has an allowable strength of

$$\frac{M_p}{\Omega} = 135 \text{ ft-kips} > 90.0 \text{ ft-kips}$$

Therefore, the W16×31 is an acceptable selection for strength.

Step 8:   To show what happens when the assumption for a is not quite as good, the W14×26 is considered. Again using the ΣQ_n determined from the table and the effective flange width

$$a = \frac{332}{0.85(3)(90.0)} = 1.45 \text{ in.}$$

This is significantly greater than the assumed value. To consider this section further, determine a new Y2 such that

$$Y2 = 5 - \frac{1.45}{2} = 4.28 \text{ in.}$$

Entering Manual Table 3-19 with Y2 = 4.0 as a conservative number, $M_a = 200$ ft kips is determined and it corresponds to the same required shear stud strength. Thus, this section also meets the strength requirements and the design can proceed with stud selection.

Which of the many possible sections should be chosen as the final design depends on the overall economics of the situation. One way to compare several choices is to look at the total weight of the steel sections combined with the total quantity of studs required. To make this comparison it is often effective to assume that an installed single shear stud has the equivalent cost of 10 pounds of steel. To make this type of comparison, the five potential sections found initially are presented in Table 9.4. Here, it was assumed that $Q_n = 21.0$ kips. This means that no consideration was taken for metal deck reduction. In addition, no check was made for the assumed versus actual a dimension. This table is simply to help determine which of the potential shapes should be considered further. Based on this table, it could be said that the W16×26 with an equivalent weight of 1020 lbs. should be investigated further.

9.10   SERVICEABILITY CONSIDERATIONS

Three important serviceability considerations are associated with the design of composite floor systems: deflection during construction, vibration under service loads, and live load deflection under service loads.

9.10.1   Deflection During Construction

As discussed in Section 9.3, the Specification permits either shored or unshored construction. With unshored construction, the Specification requires that the steel section alone have

Table 9.4   Shapes Selected for Example 9.6

Shape	M_u	ΣQ_n	Weight of steel	# studs	Equivalent weight of studs	Equivalent total weight of beam
W16×31	319	164	930	16	160	1090
W16×26	306	242	780	24	240	1020
W14×34	308	159	1020	16	160	1180
W14×30	311	248	900	24	240	1140
W14×26	312	332	780	32	320	1100

adequate strength to support all loads applied prior to the concrete attaining 75% of its specified strength. The bare steel beam, under the weight of these loads, deflects as an elastic member. Because of this deflection of the beam under the weight of the wet concrete, cambering of the steel beam is often specified. Cambering is the imposition of a permanent upward deflection of the beam in its unloaded state so that, under load, the downward deflection results in a beam without excessive deflection. Predicting the necessary camber is difficult because of the varying methods and sequences of concrete placement used by different contractors as well as such factors as the end restraint provided by the beam connection. Even with camber it is often prudent for the designer to add a little extra concrete load into the design dead load and for the contractor to allow for a little extra concrete in the quantity estimate.

In the case of shored construction, deflection during construction is usually not a concern, because the shores are not removed until the concrete has achieved some strength and composite action can be counted upon. The deflection under the wet concrete for shored construction is at a minimum. On the other hand, long-term deflection due to creep of the concrete may have to be investigated because the concrete is stressed under the self-weight as a permanent load along with the sustained service loads.

9.10.2 Vibration Under Service Loads

Composite construction usually is shallower than comparable noncomposite construction and, therefore, may be more susceptible to perceived vibrations. Because vibration calculations assume that the beam behaves compositely, even when it is not a composite beam, the additional stiffness of a composite beam does not improve its vibration characteristics. If problems occur, they usually occur in applications with long spans and little damping. For instance, a large area of a department store containing only a light jewelry display might exhibit vibrations that would be perceptible to some customers. On the other hand, an office building, constructed with the same floor system, could contain full or partial height partitions that would provide sufficient damping to obviate any perceived vibration. Because of wide differences in human perception of vibration and many other factors, vibration problems do not lend themselves to simple solutions. *AISC Design Guide 11 Floor Vibrations Due to Human Activity* provides more information and gives the designer an approach to vibration-acceptance criteria, damping, and rational design techniques.

9.10.3 Live Load Deflections

Live load deflections can be a critical design consideration for many applications. Excessive deflection could cause problems with the proper fit of partitions, doors, and equipment and may also result in an unacceptable appearance, including cracking of finishes and other visible evidence of distress. Therefore, a live load deflection calculation should be carried out for most situations. As discussed in Chapter 6, this calculation is made with the service loads for which deflection is of interest, usually the nominal live loads.

Because beam deflections are a function of the stiffness of the beam, the modulus of elasticity and moment of inertia of the composite section must be determined. The true moment of inertia at the service load level for which deflections are to be calculated are not easily determined. In addition, the modulus of elasticity of the composite section must account for the interaction of steel and concrete. The normal approach is to transform the concrete into a material that behaves like steel, with the same modulus of elasticity. The moment of inertia of the new transformed section can then be determined. Transformation of the concrete into steel is accomplished by dividing the concrete area by the modular ratio, $n = E_s/E_c$, and maintaining the same thickness. Although this seems like a fairly straightforward process, the problem is determining the thickness of the concrete that is actually participating in resisting the deflection.

One approach is to assume that the only concrete participating in resisting deflection is what is also providing strength. Thus, whether it is a fully composite or partially composite member, a moment of inertia can be determined using the known value of a from the strength calculations. Because the nominal strength is calculated at the ultimate load level, the amount of concrete actually participating for service loads could be significantly more than that used in the strength calculations. Thus, a moment of inertia determined by this approach is less than might actually be available and is called a *lower bound moment of inertia*, I_{Lb}. Figure 9.24 is an example of Manual Table 3.24, which gives the lower bound moment of inertia in a format that parallels the strength tables already discussed. Use of these I_{Lb} values results in a conservative estimate of service load deflections.

EXAMPLE 9.7
Deflection

GOAL: Determine the construction load deflection of the bare steel beam and the service load deflection for the composite beam of Example 9.6.

GIVEN: Consider the W16×26 as the beam designed in Example 9.6. Check the beam for a dead load of 0.45 kip/ft, a construction live load of 0.36 kip/ft and an in-service live load of 1.35 kip/ft. Compare the construction load deflection to span/360. For the live load deflection, use the lower bound moment of inertia from Figure 9.24 and compare the calculated deflection to span/360 as a design limit.

SOLUTION

Step 1:   Determine the total construction load for deflection calculations.

$$w = w_D + w_{L-const.} = 0.45 + 0.36 = 0.81 \text{ kip/ft}$$

Step 2:   Determine the moment of inertia of the W16×26 from Manual Table 1-1 or Manual Table 3-20.

$$I_x = 301 \text{ in.}^4$$

Step 3:   Calculate the construction load deflection and compare to span/360.

$$\Delta_{LL} = \frac{5(0.81)(30.0)^4(1728)}{384(29,000)(301)} = 1.69 \text{ in.} > \frac{30(12)}{360} = 1.0 \text{ in.}$$

Because the deflection exceeds our limit, cambering of the beam or shoring during construction would be required. Shoring has a significant cost impact as well as a scheduling impact, therefore, it is likely that the beam would be cambered or a larger section would be used.

Step 4:   Assuming that the construction load deflection issue would be resolved, determine the live load deflection under the in-service live load.
From Example 9.6, the W16×26 was selected using Y2 = 4.5 and the resulting shear stud force was $\Sigma Q_n = 242$ kips at PNA location 4.

Step 3:   Determine the lower bound moment of inertia.
Using Manual Table 3-20, with the values given in Step 4 and select

$$I_{LB} = 754 \text{ in.}^4$$

Step 4:   Determine the live load deflection.

$$\Delta_{LL} = \frac{5(1.35)(30.0)^4(1728)}{384(29,000)(754)} = 1.13 \text{ in.}$$

Table 3–24 (continued)
Lower Bound Elastic Moment of Inertia, I_{LB}, for Plastic Composite Sections

W16–W14

Shape[d]	PNA[c]	Y1[a] in.	ΣQ_n kip	Y2[b], in.										
				2	2.5	3	3.5	4	4.5	5	5.5	6	6.5	7
W16×26	TFL	0	384	673	712	753	795	840	886	935	985	1040	1090	1150
(301)	2	0.0863	337	649	685	723	763	805	848	893	940	989	1040	1090
	3	0.173	289	621	654	689	726	764	804	845	888	933	980	1030
	4	0.259	242	589	619	650	683	718	754	791	830	870	912	955
	BFL	0.345	194	551	577	604	633	663	694	726	760	795	832	869
	6	2.04	145	505	526	549	572	596	622	648	676	705	734	765
	7	4.00	96	450	465	482	499	517	535	554	575	595	617	640
W14×38	TFL	0	558	842	894	949	1010	1070	1130	1200	1260	1340	1410	1490
(385)	2	0.129	471	803	851	901	954	1010	1070	1130	1190	1260	1320	1390
	3	0.258	384	758	800	845	891	941	992	1050	1100	1160	1220	1280
	4	0.386	297	703	739	777	817	858	902	948	996	1050	1100	1150
	BFL	0.515	209	634	662	692	723	756	791	827	864	903	943	985
	6	1.42	174	602	627	653	680	709	739	770	803	837	872	909
	7	2.55	140	568	589	611	634	658	684	710	738	766	796	827
W14×34	TFL	0	500	744	790	839	890	944	1000	1060	1120	1180	1250	1320
(340)	2	0.114	423	710	753	797	844	894	945	999	1050	1110	1170	1240
	3	0.228	347	671	709	749	791	835	881	929	979	1030	1090	1140
	4	0.341	270	623	656	690	726	764	803	844	887	932	978	1030
	BFL	0.455	193	565	591	618	646	676	708	740	774	810	847	885
	6	1.41	159	535	557	581	605	631	659	687	716	747	779	812
	7	2.60	125	502	520	540	560	582	604	628	652	677	704	731
W14×30	TFL	0	442	643	683	726	771	818	868	919	973	1030	1090	1150
(291)	2	0.0963	378	615	653	692	734	777	823	870	920	972	1030	1080
	3	0.193	313	583	616	652	689	728	769	812	857	903	951	1000
	4	0.289	248	544	573	604	636	670	706	743	781	822	863	907
	BFL	0.385	183	497	521	546	572	600	629	659	690	723	757	793
	6	1.48	147	467	487	508	531	554	579	605	631	659	688	719
	7	2.82	111	432	448	466	484	503	522	543	565	587	611	635
W14×26	TFL	0	385	554	589	626	666	707	750	795	842	891	942	994
(245)	2	0.105	332	531	564	598	635	673	713	754	798	843	890	939
	3	0.210	279	504	534	565	598	633	669	707	747	788	830	875
	4	0.315	226	473	500	527	556	587	619	652	687	723	760	799
	BFL	0.420	174	437	459	482	507	533	559	587	617	647	679	712
	6	1.67	135	405	424	443	463	485	507	531	555	580	607	634
	7	3.18	96.1	368	382	397	413	430	447	465	484	503	523	544

[a] Y1 = distance from top of the steel beam to plastic neutral axis.
[b] Y2 = distance from top of the steel beam to concrete flange force.
[c] See Figure 3–3c for PNA locations.
[d] Value in parentheses is I_x (in.^{4}) of non-composite steel shape.

Figure 9.24 Lower Bound Elastic Moment of Inertia, I_{LB}, for Plastic Composite Sections. Copyright ⓒ American Institute of Steel Construction, Inc. Reprinted with Permission. All rights reserved.

Step 5:   Compare the calculated deflection with the given limit.

$$\Delta_{LL} = 1.13 > \frac{\text{span}}{360} = \frac{30(12)}{360} = 1.0 \text{ in.}$$

Because the calculated deflection is greater than the limiting value, the live load deflection is not acceptable based on the given criteria. This result, combined with the construction load deflection issue, would likely lead the designer to select a larger section for this situation as was actually done in Example 9.6.

Deflection calculations are carried out under service loads and are independent of design by LRFD or ASD.

9.11   COMPOSITE COLUMNS

Composite columns in building construction have been much slower to gain acceptance than composite beams. Specification provisions were first provided in the 1986 LRFD Specification and, until the 2005 Specification, were never available for ASD. Although the use of composite columns in buildings is still quite limited, the attention to hardening of structures against blast forces will likely bring them more to the forefront.

Specification Section I2 provides two types of composite columns: open shapes encased by concrete and hollow shapes filled with concrete. Composite columns exist at the interface between specification provisions for steel and those for concrete. The 2005 Specification closes the gap between the AISC and ACI material-specific requirements. For a member to qualify as a composite column under the Specification, it must meet the following limitations:

1. The cross-sectional area of the steel member must comprise at least 1% of the gross area.

2. The concrete encasement must be reinforced with longitudinal steel as well as lateral ties or spirals. The longitudinal steel area must be at least 0.004 times the gross area and the tie area must be at least 0.009 in.^{2} per in. of tie spacing.

3. The concrete strength, f_c', must be between 3 ksi and 10 ksi for normal weight concrete and 3 ksi and 6 ksi for lightweight concrete.

4. The maximum value of F_y to be used in calculations is 75 ksi.

5. Hollow sections must have a minimum wall thickness such that $\dfrac{b}{t} \leq 2.26\sqrt{\dfrac{E}{F_y}}$ for rectangular HSS and $\dfrac{D}{t} \leq 0.15\dfrac{E}{F_y}$ for round HSS.

Although these requirements are usually readily satisfied, for situations where they are not, ACI 318 should also be consulted.

To account for the effects of slenderness on the nominal strength of a composite column, the equations found in Chapter E for steel columns are used with slight modification. Because of the combination of two dissimilar materials and the general uncertainties of composite column behavior, the resistance and safety factors are taken as

$$\phi_c = 0.75 \text{ (LRFD)} \qquad \Omega_c = 2.00 \text{ (ASD)}$$

To convert the column equations, Equations 5.10 and 5.11, for use with a composite column, the yield stress is replaced by a nominal axial strength and the elastic critical buckling stress is replaced by the elastic critical buckling strength of the composite column using an effective stiffness. This is presented in Specification Section I2.1 for encased columns where

For $P_e \geq 0.44P_o$

$$P_n = P_o\left[0.658^{\left(\frac{P_o}{P_e}\right)}\right]$$

(9.17)

and for $P_e < 0.44P_o$

$$P_n = 0.877P_e$$

(9.18)

where

$$P_o = A_sF_y + A_{sr}F_{yr} + 0.85A_cf_c'$$

$$P_e = \pi^2\frac{EI_{eff}}{(KL)^2}$$

$$EI_{eff} = E_sI_s + 0.5E_sI_{sr} + C_1E_cI_c$$

$$C_1 = 0.1 + 2\left(\frac{A_s}{A_c + A_s}\right) \leq 0.3$$

In these equations, the s subscript refers to the steel section, the sr subscript refers to the longitudinal reinforcing steel, and the c subscript refers to the concrete.

For filled columns, the following are to be used

$$P_o = A_sF_y + A_{sr}F_{yr} + C_2A_cf_c'$$

$$EI_{eff} = E_sI_s + E_sI_{sr} + C_3E_cI_c$$

$$C_2 = 0.85 \text{ for rectangular sections}$$

$$= 0.95 \text{ for round sections}$$

$$C_3 = 0.6 + 2\left(\frac{A_s}{A_c + A_s}\right) \leq 0.9$$

These two separate cases can easily be combined through the use of the already defined constant, C_2, and the new constants, C_4 and C_5.

$$P_o = A_sF_y + A_{sr}F_{yr} + C_2A_cf_c'$$

(9.19)

$$EI_{eff} = E_sI_s + C_4E_sI_{sr} + C_5E_cI_c$$

(9.20)

$$C_2 = 0.85 \text{ for encased sections and rectangular HSS}$$

$$= 0.95 \text{ for round HSS}$$

$$C_4 = 0.5 \text{ for encased sections}$$

$$= 1.0 \text{ for filled HSS}$$

$$C_5 = 0.1 + 2\left(\frac{A_s}{A_c + A_s}\right) \leq 0.3 \text{ for encased sections}$$

$$= 0.6 + 2\left(\frac{A_s}{A_c + A_s}\right) \leq 0.9 \text{ for filled HSS}$$

EXAMPLE 9.8
Composite Column
Strength

GOAL: Determine the nominal strength of a composite column. Then determine the design strength and the allowable strength.

GIVEN: The column is composed of a W14×53 encased in 18 in. × 22 in. of concrete as shown in Figure 9.25. Additional given information is as follows:

Column effective length = 15 ft.

Steel shape: $F_y = 50$ ksi

Reinforcing: four #9 bars, Gr. 60, $F_y = 60$ ksi

Concrete strength: $f'_c = 5$ ksi

$E_c = 145^{1.5}\sqrt{5} = 3900$ ksi

SOLUTION

Step 1: Determine the areas of the components.

$$A_s = 15.6 \text{ in.}^2$$
$$A_{sr} = 4(1.0) = 4.0 \text{ in.}^2$$
$$A_c = 18.0(22.0) - 15.6 - 4.0 = 376 \text{ in.}^2$$

Step 2: Check the minimum steel ratios.

$$\rho_s = \frac{A_s}{A_g} = \frac{15.6}{22.0(18.0)} = 0.0394 > 0.001$$

$$\rho_{sr} = \frac{A_{sr}}{A_g} = \frac{4.0}{22.0(18.0)} = 0.0101 > 0.004$$

So the specified minimums are satisfied.

Step 3: Determine P_o and P_e.

$$C_2 = 0.85$$
$$C_4 = 0.5$$
$$C_5 = 0.1 + 2\left(\frac{15.6}{376 + 15.6}\right) = 0.180 < 0.3$$

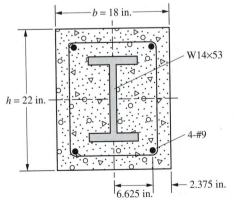

Figure 9.25 Composite Column (Example 9.8).

By inspection, the y-axis will be the critical buckling axis.

$$I_s = I_y = 57.7 \text{ in.}^4$$

$$I_{sr} = 4(1.0)(6.625)^2 = 176 \text{ in.}^4$$

$$I_c = \frac{22.0(18.0)^3}{12} - 57.7 - 176 = 10{,}500 \text{ in.}^4$$

$$P_o = 15.6(50) + 4.0(60) + 0.85(5)(376) = 2620 \text{ kips}$$

$$EI_{eff} = 29{,}000(57.7) + 0.5(29{,}000)(176) + 0.180(3900)(10{,}500) = 11.6 \times 10^6 \text{ in.-kips}$$

$$P_e = \frac{\pi^2 EI_{eff}}{(KL)^2} = \frac{\pi^2(11.6 \times 10^6)}{(15.0(12))^2} = 3530 \text{ kips}$$

Step 4:   Determine the controlling column strength equation.

$$\frac{P_e}{P_o} = \frac{3530}{2620} = 1.35 > 0.44$$

Therefore, because $P_e > 0.44 P_o$, use Equation 9.17.

Step 5:   Determine the nominal compressive strength.

$$P_n = P_o(0.658)^{\frac{P_o}{P_e}} = 2620(0.658)^{\frac{1}{1.35}} = 1920 \text{ kips}$$

Step 6:   For LRFD, the design compressive strength is

$$\phi P_n = 0.75(1920) = 1440 \text{ kips}$$

Step 6:   For ASD, the allowable compressing strength is

$$\frac{P_n}{\Omega} = \frac{1920}{2.00} = 960 \text{ kips}$$

The strength of filled HSS is determined in the same fashion as just discussed for the encased shapes, with the use of the appropriate C constants. Primarily because of the unlimited possible combinations of steel section and concrete size, the Manual does not provide tables for composite encased columns. However, because filled HSS represent a limited set of possible geometries, Manual Part 4 does give strength tables for these shapes. These tables are used exactly like the column tables for the bare steel column previously discussed.

9.12   COMPOSITE BEAM-COLUMNS

Composite beam-columns have the same potential to occur as bare steel beam-columns. Any application where bending moment and axial force are applied simultaneously needs to be addressed according to the provisions of Specification Section I4.

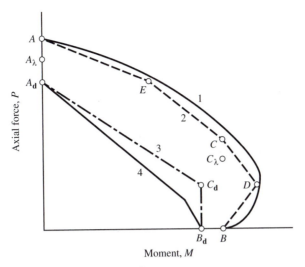

Figure 9.26 Composite Column Interaction Diagrams.

For doubly symmetric composite beam-columns, the most common composite beam-columns found in building construction, the interaction equations of Specification Chapter H can be used conservatively. For a more accurate approach to determining the available strength, the interaction surface can be developed based on plastic stress distributions and the length modification from Specification Section I2, as discussed in Section 9.11.

Figure 9.26 provides several potentially useful interaction diagrams for a composite beam-column. Curve 1 is the interaction curve based on a strain-compatibility approach similar to that used for developing similar diagrams for reinforced concrete columns, without consideration of length effects. Curve 2 represents a segmented straight line approximation based on plastic stress distributions, again without incorporating any length effects. Curve 3 is a further simplification of Curve 2, incorporating resistance or safety factors and the length effects. Curve 4 is the result of applying the equations of Chapter H.

Only Curves 3 and 4 in Figure 9.26 account for the effects of length on beam-column strength. The conservatism of the Chapter H approach is not that great when compared to the Curve 3 approach. Thus, a detailed discussion of beam-column behavior will not be undertaken.

9.13   PROBLEMS

1. Determine the location of the plastic neutral axis and the available moment strength for a flat soffit, fully composite beam composed of a W16×26 spanning 20 ft and spaced 8 ft on center, supporting a 6-in. concrete slab. Use $f_c' = 4$ ksi and A992 steel. Determine (a) design strength by LRFD and (b) allowable strength by ASD.

2. Determine the location of the plastic neutral axis and the available moment strength for a flat soffit, fully composite beam composed of a W16×45 spanning 22 ft and spaced 8 ft on center, supporting a 5-in. concrete slab. Use $f_c' = 5$ ksi and A992 steel. Determine (a) design strength by LRFD and (b) allowable strength by ASD.

3. Determine the location of the plastic neutral axis and the available moment strength for a flat soffit, fully composite beam

composed of a W18×50 spanning 20 ft and spaced 6 ft on center, supporting a 5-in. concrete slab. Use $f_c' = 5$ ksi and A992 steel. Determine (a) design strength by LRFD and (b) allowable strength by ASD.

4. Determine the location of the plastic neutral axis and the available moment strength for a flat soffit, fully composite beam composed of a W18×71 spanning 18 ft and spaced 5 ft on center, supporting a 4-in. concrete slab. Use $f_c' = 4$ ksi and A992 steel. Determine (a) design strength by LRFD and (b) allowable strength by ASD.

5. Determine the location of the plastic neutral axis and the available moment strength for a flat soffit, fully composite beam composed of a W14×43 spanning 20 ft and spaced 5 ft on center, supporting a 4-in. concrete slab. Use $f_c' = 3$ ksi and A992 steel. Determine (a) design strength by LRFD and (b) allowable strength by ASD.

6. Determine the location of the plastic neutral axis and the available moment strength for a flat soffit, fully composite beam composed of a W14×61 spanning 24 ft and spaced 6 ft on center, supporting a 4-in. concrete slab. Use $f'_c = 3$ ksi and A992 steel. Determine (a) design strength by LRFD and (b) allowable strength by ASD.

7. Repeat Problem 1 with the shear stud capacity limited to $V'_q = 250$ kips. Determine (a) design strength by LRFD and (b) allowable strength by ASD.

8. Repeat Problem 2 with the shear stud capacity limited to $V'_q = 500$ kips. Determine (a) design strength by LRFD and (b) allowable strength by ASD.

9. Repeat Problem 3 with the shear stud capacity limited to $V'_q = 500$ kips. Determine (a) design strength by LRFD and (b) allowable strength by ASD.

10. Repeat Problem 4 with the shear stud capacity limited to $V'_q = 400$ kips. Determine (a) design strength by LRFD and (b) allowable strength by ASD.

11. Repeat Problem 5 with the shear stud capacity limited to $V'_q = 300$ kips. Determine (a) design strength by LRFD and (b) allowable strength by ASD.

12. Repeat Problem 6 with the shear stud capacity limited to $V'_q = 600$ kips. Determine (a) design strength by LRFD and (b) allowable strength by ASD.

13. A W12 composite beam spaced every 10 ft is used to support a uniform dead load of 1.0 k/ft and live load of 0.9 k/ft on a 20-ft span. Using a 4-in. flat soffit slab with $f'_c = 4$ ksi, $^3/_4$-in. shear studs, and A992 steel, determine the least-weight shape and the required number of shear connectors to support the load. Design by (a) LRFD and (b) ASD.

14. A W14 composite beam is to support a uniform dead load of 1.2 k/ft and live load of 1.2 k/ft. The beam spans 24 ft and is spaced 8 ft from adjacent beams. Using a 5-in. flat soffit slab and $^3/_4$-in. shear studs, determine the least-weight shape to support the load if $f'_c = 4$ ksi and A992 steel is used. Design by (a) LRFD and (b) ASD.

15. Compare the least-weight A992 W16 and W14 members required to support a uniform dead load of 2.4 k/ft and live load of 3.2 k/ft. The beams span 18 ft and are spaced 12 ft on center. They support a 6-in. concrete slab with $f'_c = 4$ ksi. Design by (a) LRFD and (b) ASD.

16. A series of W16×36 A992 composite beams are spaced at 10-ft intervals and span 24 ft. The beams support a $2^1/_2$-in. metal deck perpendicular to the beam with a slab whose total thickness is 5 in. Assuming full composite action, determine the available moment strength and the number of $^3/_4$-in. shear studs required. The deck has 6-in. wide ribs spaced at 12 in. Use $f'_c = 4$ ksi. Determine by (a) LRFD and (b) ASD.

17. Determine the available moment strength of a W18×35 A992 composite beam supporting a slab with a total thickness of 5 in. on a 3-in. metal deck perpendicular to the beam. The beam spans 28 ft and is spaced 12 ft from adjacent beams. Use $f'_c = 5$ ksi. Determine (a) design strength by LRFD and (b) allowable strength by ASD.

18. Determine the available moment strength for a W18×46 A992 member used as a partially composite beam to support 3 in. of concrete on a 3-in. metal deck for a total slab thickness of 6 in. The beam spans 30 ft and is spaced 11 ft from adjacent beams. Shear stud strength is $V'_q = 400$ kips, $f'_c = 5$ ksi. Determine (a) design strength by LRFD and (b) allowable strength by ASD.

19. A composite beam is to span 20 ft and support a 4-in. slab including a $1^1/_2$-in. metal deck. The deck span is 10 ft. The beam must accommodate a uniformly distributed dead load of 75 psf including the slab weight and live load of 100 psf. The deck has 2-in. ribs spaced 6 in. on center. Determine the required A992 W-shape and the number of $^3/_4$-in. shear studs. Use $f'_c = 3$ ksi. Design by (a) LRFD and (b) ASD.

20. Determine the required W-shape and $^3/_4$-in. shear studs for a composite girder that spans 30 ft and supports two concentrated dead loads of 12 kips and live loads of 20 kips at the third points. The $1^1/_2$-in. metal deck with 2-in. ribs spaced at 6 in. on center is parallel to the girder and supports a total slab of 5 in. Use $f'_c = 4$ ksi and A992 steel. Design by (a) LRFD and (b) ASD.

21. Determine the live load deflection for a W24×76 A992 composite beam with an 8 in. total thickness slab on a 3-in. metal deck. The beam spans 28 ft and is spaced at 10-ft intervals. The beam is to carry a live load of 3.4 k/ft. Assume Y2 = 6.0 in. and $\Sigma Q_n = 393$ kips.

22. Determine the live load deflection for a W16×26 A992 composite beam supporting a 6-in. slab on a $2^1/_2$-in. metal deck. The beam spans 24 ft and is spaced at 8 ft on center. The live load is 2.1 k/ft, Assume Y2 = 5.5 in. and $\Sigma Q_n = 384$ kips.

23. Determine the available compressive strength of a 20-ft effective length 18- × 18-in. composite column encasing an A992, W10×68 and eight #8, Gr. 60, reinforcing bars, $f'_c = 5$ ksi. Each face has three bars with their centers located 2.5 in. from the face of the concrete. Determine (a) design strength by LRFD and (b) allowable strength by ASD.

24. Determine the available compressive strength of a 22- × 22-in. composite column with an effective length of 16 ft. The concrete encases an A992, W12×120 and eight #9, Gr. 60, bars, $f'_c = 5$ ksi. Each face has three bars with their centers located 2.5 in. from the face of the concrete. Determine (a) design strength by LRFD and (b) allowable strength by ASD.

25. Determine the available compressive strength of a 20- × 22-in. composite column with an effective length of 12 ft. The concrete encases an A992 W12×136 and eight #10, Gr. 60, bars, $f'_c = 5$ ksi. Each face has three bars with their centers located 2.5 in. from the face of the concrete. Determine (a) design strength by LRFD and (b) allowable strength by ASD.

Chapter 10

University of Phoenix Stadium.
Photo courtesy Arizona Cardinals.

Connection Elements

10.1 INTRODUCTION

A steel building structure is essentially a collection of individual members attached to each other to form a stable and serviceable whole, called the frame. The assumed behavior of the connection between any two members determines how the structure is analyzed to resist gravity and lateral loads. This analysis, in turn, determines the moments, shears, and axial loads for which the beams, columns, and other members are designed. It is, therefore, essential that the designer understand the basic behavior of connections.

Members are attached to each other through a variety of connecting elements, such as plates, angles, and other shapes, using mechanical fasteners or welds. The characteristics of these connecting elements and fasteners must be understood to assess the response of the complete connection. With each connection, the load transfer mechanism must be understood so that the applicable limit states of the joint can be evaluated.

Table 10.1 lists the sections of the Specification and parts of the Manual discussed in this chapter.

10.2 BASIC CONNECTIONS

The wide variety of potential geometries and arrangements of members available for construction makes listing the corresponding potential connections quite complex. Every joint between members must be analyzed and designed according to the unique aspects of that joint.

Table 10.1 Sections of Specification and Parts of Manual Found in this Chapter

Specification	
B3	Design Basis
Chapter D	Design of Members for Tension
Chapter E	Design of Members for Compression
G2	Members with Unstiffened or Stiffened Webs
J2	Welds
J3	Bolts and Threaded Parts
J4	Affected Elements of Members and Connecting Elements
Manual	
Part 7	Design Considerations for Bolts
Part 8	Design Considerations for Welds
Part 9	Design of Connecting Elements

Figure 10.1 shows several examples of tension connections. The connections shown in Figures 10.1a, b, and c illustrate ways that a tension member can be spliced. In each case, the bolts are subjected to a shear force. The butt joint (Figure 10.1a) and the lap joint (Figure 10.1b) provide a connection between two members whereas the joint shown in Figure 10.1c shows the connection of a single member to a pair of members. This type of joint can also be considered as a portion of the butt joint shown in Figure 10.1a. The joint shown in Figure 10.1d represents a hanger connected to the lower flange of a beam; in this case the connection is accomplished with a WT-shape, and the bolts are subjected to a tensile load. The connection of a tension member to a gusset plate is shown in Figure 10.1e. Here again, the bolts are subjected to a shear force. All of these examples illustrate bolted connections. Similar connections can be accomplished with welds.

The connections illustrated in Figure 10.2 are bracket connections. The connection shown in Figure 10.2a shows a bracket attached to the flange of a column. In this case,

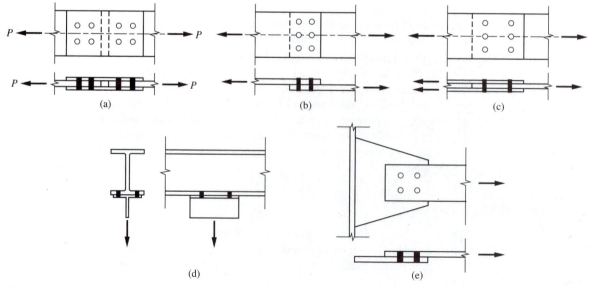

Figure 10.1 Tension Connections.

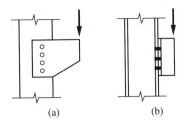

(a)   (b)   **Figure 10.2** Bracket Connections.

the bolts are subjected to shear and moment in the plane of the connection when loaded as shown. The bracket shown in Figure 10.2b, when loaded as shown, subjects the bolt group to shear in the plane of the connection and a moment out of the plane that results in a tensile force in the top bolts.

10.3 BEAM-TO-COLUMN CONNECTIONS

The connection of a beam to a column can also be accomplished in a variety of ways. Figure 10.3 illustrates several connections of W-shaped beams to W-shaped columns. The classification of these connections is a function of the forces being transferred between the members. The connections shown in Figure 10.3a through d are usually called *simple* or *shear connections* whereas those in Figure 10.3e through h are generally referred to as *fixed* or *moment connections*.

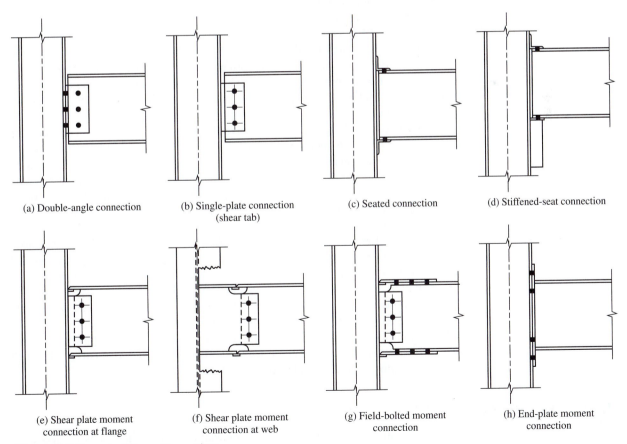

(a) Double-angle connection

(b) Single-plate connection (shear tab)

(c) Seated connection

(d) Stiffened-seat connection

(e) Shear plate moment connection at flange

(f) Shear plate moment connection at web

(g) Field-bolted moment connection

(h) End-plate moment connection

Figure 10.3 Beam-to-Column Connections.

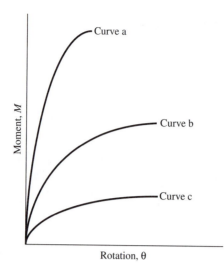

Figure 10.4 Beam-to-Column Moment-Rotation Curves.

Although normal practice tends to classify beam-to-column connections as simple or fixed, these connections actually exhibit a wide range of behaviors as discussed in Chapter 8. This behavior can be described through a plot of the moment-rotation characteristics of a particular connection. Typical moment-rotation relationships for three beam-to-column connections are presented in Figure 10.4. When a connection is very stiff, it deforms very little, even when subjected to large moments. This type of connection is represented by Curve a in Figure 10.4. At the other extreme, when a connection is quite flexible, it will rotate considerably but will not develop a significant moment, as shown by Curve c in Figure 10.4. Curve b in Figure 10.4 is representative of any connection whose moment rotation behavior occurs somewhere between Curves a and c. These connections have some appreciable stiffness but still exhibit a degree of flexibility; thus, significant rotation will occur along with significant moment resistance.

For the purposes of design, connections have usually been assumed to behave according to the simplified behaviors represented by the vertical axis of Figure 10.4 as a fixed connection and the horizontal axis of Figure 10.4 as a simple connection. Because connections do not actually behave in this way, those that follow Curves a and c and exhibit a behavior close to the idealized connection are called fixed connections and simple connections, respectively. Specification Section B3.6 divides connections into two categories: simple connections and moment connections. Within the moment connection category, it defines Fully Restrained Moment Connections (FR) and Partially Restrained Moment Connections (PR). FR connections transfer moment with a negligible rotation between the connected members, as shown in Curve a. PR connections transfer moment between the members but the rotation is not negligible, as demonstrated by Curve b.

It is the designers' responsibility to match connection behavior with the appropriate analysis model and to complete the connection design so that the actual connection behavior matches that used in the analysis. Often, this requires experience and judgment; the state-of-the-art is such that it is usually not possible to predict the actual M-θ curve with much accuracy for anything but the simplest of connections.

10.4 FULLY RESTRAINED CONNECTIONS

The basic assumption for frames with FR connections is that the beams and columns maintain their original relationship during the entire loading history. This is normally called a *rigid* or fixed *connection*. Figures 10.3e through h show examples of beam-to-column connections that are usually treated as FR connections. Although they may show some relative rotation between members, they have sufficient stiffness to justify ignoring this rotation.

Figure 10.3e shows a connection with a web plate shop welded to the column flange and field bolted to the beam web. The beam flanges have been beveled in the shop and are field welded to the column. Although the beam web is not continuously connected to the column, it has been repeatedly demonstrated that this connection can adequately transfer the full plastic moment of the beam to the column. Most of the moment strength is derived from the flange connections, which is equal to the flange force times beam depth. The small amount of moment in the web connection and local strain hardening in the flanges add to the connection's ability to reach the full plastic moment of the beam. This connection is generally known as the pre-Northridge connection because it was the standard connection for seismic applications prior to the 1994 Northridge, California earthquake. Because its performance under the seismic load of that event was not as favorable as expected, it is no longer used in seismic resisting frames. However, it is still used to resist moments due to gravity and wind loads.

Figure 10.3f is similar to Figure 10.3e except that the beam frames into the web of the column. To ensure that this connection has adequate ductility it is important to extend the flange connecting plates beyond the column flange and to design these plates a little thicker than the beam flange. Extending the connecting plate reduces the possibility of a tri-axial stress condition near the column flange tips. Thickening the plate reduces the average tension stress in the plate. It also facilitates welding to the beam flange.

The connection illustrated in Figure 10.3g is a flange plate connection. As with the pre-Northridge connection, the web is connected to transfer the beam shear force only. The flange force is first transferred to the top and bottom plates and then into the column flange. This connection is shown as a bolted connection but it is also possible to fabricate this as a welded connection. For fully welded connections, special care should be taken to address erection issues due to the requirement for field welding.

Figure 10.3h is an extended end plate connection. For this connection, a plate is shop welded to the end of the beam and then bolted to the column flange. Although very popular with some fabricators, others tend to avoid it. It must be fabricated with special care so that the end plates are parallel with each other. Also, it is not a very forgiving connection and can make erection difficult and expensive.

10.5   SIMPLE AND PARTIALLY RESTRAINED CONNECTIONS

A frame with PR connections must be analyzed accounting for the actual moment-rotation characteristics of the connection. These connections are now referred to as *partially restrained connections* but have historically been called semi-rigid connections. It is typically not possible to determine whether a connection should be classified as PR just by looking at it. Several connections that appear to be simple actually have the potential to resist significant moment. In the simple connection case, the analysis assumes that the connections are pinned, and free to rotate. The rotation capacity of the connection must be sufficient to accommodate the simple beam rotation of the beam to which it is connected.

There are basically two ways in which a simply connected frame can be designed to resist lateral loads and to provide stability for gravity loads. In one case, a positive bracing system is provided, such as diagonal steel bracing or a shear wall. In the second case, lateral stability is provided by the limited restraint offered by the connections and members themselves. This type of connection is called a flexible moment connection. Flexible moment connections are designed with a limited amount of moment resistance accompanied by a significant amount of rotation. The connections are flexible enough to rotate under gravity loads so that no gravity moments are transferred to the columns. At the same time they are assumed to have sufficient strength and stiffness to resist the lateral loads and to provide frame stability. This approach to frame design was addressed in Section 8.11. Connection design for these flexible moment connections will follow the same approach as other connections discussed later.

The design of PR connections requires that the frame be analyzed considering the true semi-rigid behavior of the connections. In this case, the actual M-θ curve of the connection must be known. The resulting analysis tends to be rather complex because of the nonlinear behavior of the connection. Although there are currently no commercially available computer programs for analysis of frames with PR connections, there are simplified approaches that will aid in the use of these connections.

Figure 10.3 shows examples of simple and PR connections. As mentioned earlier, it is not normally possible to tell by a visual inspection whether a connection should be treated as a PR connection. Figure 10.3a shows a double-angle connection also referred to as a *clip angle connection*. This connection has been used extensively over the years. In fact, it is usually considered the standard to which other simple connections are compared. Even though it is readily accepted as a simple connection, it has been shown that under certain circumstances it can be relied upon to resist some moment from lateral load.

Figure 10.3b shows a single plate framing connection that is often referred to as a *shear tab*. Care must be taken when designing these connections as simple connections to insure that the elements have sufficient flexibility to accommodate the simple beam rotation.

Figure 10.3c shows a seated connection and Figure 10.3d a stiffened seated connection. Either can be bolted or welded and they are usually used to frame a beam into the web of a W-shaped column section. Although they may appear to be stiffer than the standard double-angle connection, they are designed to rotate sufficiently without transferring a moment to the column so that they can be treated as simple connections.

10.6   MECHANICAL FASTENERS

The mechanical fasteners most commonly used today are bolts. The Specification provides for the use of common bolts and high-strength bolts. It also provides some direction for cases where bolts are to be used in conjunction with rivets in new work on historic structures. There are no provisions for rivets in new construction, however, because these connectors are no longer used in buildings.

10.6.1   Common Bolts

Common bolts are manufactured according to the ASTM A307 specification as discussed in Section 3.6.3. When used, they are usually found in simple connections for such elements as girts, purlins, light floor beams, bracing, and other applications where the loads are relatively small. Although permitted by the Specification, they are not recommended for normal steel-to-steel connections and should not be used where the loads are cyclic or vibratory, or where fatigue may be a factor.

Common bolts are also called *machine*, *unfinished* or *rough bolts*. They are identified by their square heads and nuts and should have the grade designation 307A or 307B on the heads. They are available in diameters from $1/4$ in. to 4 in.

These bolts are usually installed using a spud wrench. No specified pretension is required. Because no clamping force is assumed, it is only necessary to tighten the nut sufficiently to prevent it from backing off of the bolt. The design shear and tensile strength are given in Specification Section J3, Table J3.2.

10.6.2 High-Strength Bolts

Three types of high-strength bolts are currently permitted in steel structures according to Specification Section J3: ASTM A325, *High-Strength Bolts for Structural Joints*; ASTM A490, *Quenched and Tempered Alloy Steel Bolts for Structural Joints*; and F1852, *Twist Off Type Tension Control Structural Bolt/Nut/Washer Assemblies*. F1852 bolts have the strength characteristics of A325 bolts. Since publication of the Specification, ASTM has issued ASTM F2280, *Standard Specification for Twist Off Type Tension Control Structural Bolt/Nut/Washer Assemblies, Steel, Heat Treated, 150 ksi Minimum Tensile Strength*. These bolts have strength characteristics of A490 bolts. Details of the material and other properties of these bolts are described in Section 3.6.3. In most cases, the nominal strength of A490 bolts is 25% greater than that of A325 bolts because bolt strength is based on the tensile strength of the bolt material. All three types of bolts can be used for simple, FR, or PR connections and for both static and dynamic loading. Bolts have always been very popular for field installation. Their use in the shop has increased considerably with the introduction of automated equipment and the F1852 tension controll bolt.

A325 bolts are available in two types. Type 1, manufactured from a medium-carbon steel, is the most commonly used. It is available in sizes ranging from $1/2$ in. through $1\frac{1}{2}$ in. in diameter. Type 3 is a weathering steel bolt with corrosion characteristics similar to that of ASTM A242, A588, and A847 steels. Type 3 bolts are also available from $1/2$ through $1\frac{1}{2}$ in. in diameter. For $1/2$- to 1-in. bolts, $F_u = 120$ ksi, whereas for bolts larger than 1.0 in. $F_u = 105$ ksi.

A490 bolts are also available as Type 1 and Type 3 and in sizes ranging from $1/2$ in. to $1\frac{1}{2}$ in. in diameter. All A490 bolts have $F_u = 150$ ksi.

A325 Type 1 bolts are identified by the mark "A325" or by three radial lines 120 degrees apart on the bolt head. Type 3 bolts have the designation "A325" underlined. A490 bolts carry the symbol "A490" with the "A490" underlined for Type 3. Example bolt markings are shown in Figure 10.5. All bolts should also be marked with a symbol to designate the manufacturer as shown in the figure.

Figure 10.6a shows the principle parts and dimensions of a high-strength bolt: head, shank, bolt length, and thread length, whereas Figure 10.6b shows the principle parts of a tension control bolt.

Both A325 and A490 bolts can be installed with a spud wrench or, in cases where a clamping force is necessary, using an impact wrench. F1852 bolts are installed with a mechanical device that simultaneously holds the bolt shank and nut and rotates them relative to each other. The end of the bolt twists off when the prescribed tensile force is reached, insuring the required pretension.

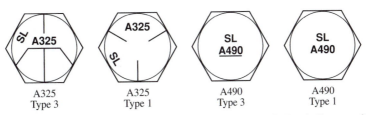

Figure 10.5 Example Bolt Identification Markings from St. Louis Screw and Bolt Company.

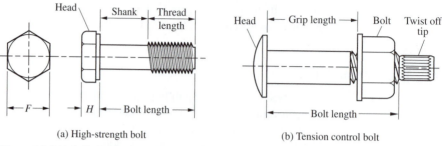

(a) High-strength bolt   (b) Tension control bolt

Figure 10.6 Bolt Definitions.

10.6.3   Bolt Holes

Because the hole into which a bolt is inserted will impact the strength of the bolts in place, it is important to address the hole requirements at this point. The Specification defines four types of bolt holes that are permitted in steel construction: standard, oversize, short slot, and long slot. Table 10.2 shows the nominal hole dimensions for each of these types and for bolts from $1/2$ in. diameter up. Figure 10.7 shows the four hole sizes for a $3/4$-in. bolt.

Standard holes or short-slotted holes transverse to the direction of load are the standard to be used unless one of the other types is permitted by the designer. This is because the other arrangements will adversely affect the final bolt strength. A standard hole has a diameter that is $1/16$ in. greater than the bolt diameter to accommodate placement of the bolt. Short-slotted holes have this same dimension in one direction but are elongated in the other direction to assist in fit-up of the connection parts. Any slot longer than a short slot should be classified as a long slot, even if it is not the full length of a long slot as shown in Table 10.2.

Oversize holes and long-slotted holes are specified when the increased tolerance is needed to accomplish the actual connection. If a design includes other than standard holes, the requirements of Specification Section J3.2 for washers come into play. For the examples in this book, only standard holes will be used.

In addition to prescribing the size of bolt holes, the Specification gives minimum and maximum hole spacing and edge distances. Figure 10.8 shows a plate with holes dimensioned with the standard variable names used in the Specification. The minimum hole spacing, s, for standard, oversized, or slotted holes must not be less than $2^2/_3$ times the bolt diameter. A spacing of 3 diameters, $3d$, is preferred. It will be shown later that even at a minimum spacing of $3d$, bolt strength may be less than what it could be if the spacing

Table 10.2   Nominal Hole Dimensions, in.

Bolt diameter	Standard (dia.)	Oversize (dia.)	Short slot (width × length)	Long slot (width × length)
			Hole dimensions	
$1/2$	$9/16$	$5/8$	$9/16 \times {}^{11}/_{16}$	$9/16 \times 1^1/_4$
$5/8$	$^{11}/_{16}$	$^{13}/_{16}$	$^{11}/_{16} \times 7/8$	$^{11}/_{16} \times 1^9/_{16}$
$3/4$	$^{13}/_{16}$	$^{15}/_{16}$	$^{13}/_{16} \times 1$	$^{13}/_{16} \times 1^7/_8$
$7/8$	$^{15}/_{16}$	$1^1/_{16}$	$^{15}/_{16} \times 1^1/_8$	$^{15}/_{16} \times 2^3/_{16}$
1	$1^1/_{16}$	$1^1/_4$	$1^1/_{16} \times 1^5/_{16}$	$1^1/_{16} \times 2^1/_2$
$\geq 1^1/_8$	$d + 1/16$	$d + 5/16$	$(d + 1/16) \times (d + 3/8)$	$(d + 1/16) \times (2.5 \times d)$

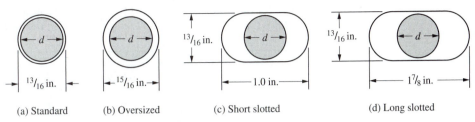

| (a) Standard | (b) Oversized | (c) Short slotted | (d) Long slotted |

Figure 10.7 Hole Sizes for a $^3/_4$ in. Diameter Bolt.

were just a little bit greater. The maximum spacing of bolts in a connection is 12 times the thickness of the connected part or 6 in. This maximum is not a strength requirement but rather one that is intended to keep a connection together and prevent any potential moisture build-up between the elements.

The minimum edge distances, L_e, specified are intended to facilitate construction and are not strength related. Table 10.3 shows the minimums from Specification Table J3.4. Because these dimensions will be shown to directly impact bolt strength, it is critical to provide edge distances that are compatible with the required strength of the connection. The table gives different edge distances for sheared and rolled edges. The use of these values will depend on the type of connecting element being used, such as a plate that might have been sheared or an angle that has a rolled toe. The maximum edge distance is the same as the maximum spacing and for the same reasons.

10.7   BOLT LIMIT STATES

Three basic limit states govern the response of bolts in bolted connections: shear through the shank or threads of the bolt, bearing on the material being connected, and tension in the bolt.

Cases where load reversals are expected or fatigue is a factor have an additional limit state to prevent slip in the connection. This limit state applies only to connections that are classified as slip-critical.

10.7.1   Bolt Shear

The most common application of bolts in connections is to resist shear. Shear through the shank of the bolt is the means whereby the load, P, in Figure 10.9a is transferred from one plate to the other. In this case, the bolt is sheared along one plane. Thus, it is said to be a *bolt in single shear*. The arrangement in Figure 10.9b shows two side plates connected to a central plate. In this case, the load, P, is transferred from the center plate to the side plates

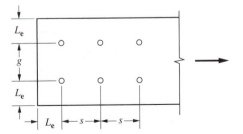

Figure 10.8 Hole Spacing, Guage, and Edge Distances.

Table 10.3   Minimum Edge Distance,[a] in., from Center of Standard Hole[b] to Edge of Connected Part Copyright © American Institute of Steel Construction, Inc. Reprinted with Permission. All rights reserved.

Bolt diameter (in.)	At sheared edges	At rolled edges of plates, shapes, or bars, or thermally cut edges[c]
$\frac{1}{2}$	$\frac{7}{8}$	$\frac{3}{4}$
$\frac{5}{8}$	$1\frac{1}{8}$	$\frac{7}{8}$
$\frac{3}{4}$	$1\frac{1}{4}$	1
$\frac{7}{8}$	$1\frac{1}{2}$[d]	$1\frac{1}{8}$
1	$1\frac{3}{4}$[d]	$1\frac{1}{4}$
$1\frac{1}{8}$	2	$1\frac{1}{2}$
$1\frac{1}{4}$	$2\frac{1}{4}$	$1\frac{5}{8}$
Over $1\frac{1}{4}$	$1\frac{3}{4} \times d$	$1\frac{1}{4} \times d$

[a]Lesser edge distances are permitted to be used provided provisions of Section J3.10, as appropriate, are satisfied.

[b]For oversized or slotted holes, see Table J3.5.

[c]All edge distances in this column are permitted to be reduced $\frac{1}{8}$ in. when the hole is at a point where required strength does not exceed 25% of the maximum strength in the element.

[d]These are permitted to be $1\frac{1}{4}$ in. at the ends of beam connection angles and shear end plates.

and the bolt is therefore loaded in double shear. A bolt in double shear has twice the shear strength as a bolt in single shear.

For the limit state of bolt shear, the nominal strength is based on the tensile strength of the bolt and the location of the shear plane with respect to the bolt threads. Section J3.6 provides that

$$R_n = F_n A_b$$

and

$$\phi = 0.75\,(\text{LRFD}) \qquad \Omega = 2.00\,(\text{ASD})$$

where

F_n = shear stress, F_{nv}, from Specification Table J3.2

A_b = area of the bolt shank

The information in Table 10.4 is taken from Specification Table J3.2. Each high-strength fastener has two descriptions: The first is for cases where the threads are not excluded

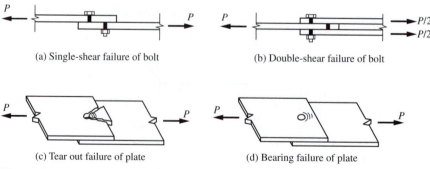

(a) Single-shear failure of bolt

(b) Double-shear failure of bolt

(c) Tear out failure of plate

(d) Bearing failure of plate

Figure 10.9 Bolt Failure Modes.

Table 10.4 Nominal Stress of Fasteners and Threaded Parts, ksi (MPa)

Description of fasteners	Nominal tensile stress, F_{nt}, ksi	Nominal shear stress in bearing-type connections, F_{nv}, ksi
A307 bolts	45	24
A325 or A325M bolts, when threads are not excluded from shear planes	90	48
A325 or A325M bolts, when threads are excluded from shear planes	90	60
A490 or A490M bolts, when threads are not excluded from shear planes	113	60
A490 or A490M bolts, when threads are excluded from shear planes	113	75
Threaded parts meeting the requirements of Section A3.4, when threads are not excluded from shear planes	$0.75F_u$	$0.40F_u$
Threaded parts meeting the requirements of Section A3.4, when threads are excluded from shear planes	$0.75F_u$	$0.50F_u$

from the shear plane and the second is for when the threads are excluded from the shear plane. Because in every case, the area of the bolt shank is used to determine the nominal strength, the reduced area when the shear plane passes through the threads is accounted for by reducing the nominal shear stress. When threads are excluded from the shear plane, the bolts are called either A325X or A490X bolts. In these cases, $F_{nv} = 0.5F_u$. When threads are not excluded from the shear plane, the bolts are referred to as either A325N or A490N bolts. In these cases, $F_{nv} = 0.4F_u$. Only one value is provided for A307 bolts and that value is based on the assumption that the threads occur in the shear plane.

Unless the designer can be sure that the final connection will result in the bolt threads being excluded from the shear plane, it is usually best to design the connection for the worst case of threads included in the shear plane.

10.7.2 Bolt Bearing

The available strength for the limit state of bearing is specified in Section J3.10. Because the material strength of a bolt is greater than that of the material it is bearing on, the only bearing check is for bearing on the material of the connected parts. The Specification provision considers two limit states for bearing strength at bolt holes: the limit state based on shear in the material being connected as shown in Figure 10.9c, and the limit state of material crushing as shown in Figure 10.9d.

When the clear distance from the edge of the hole to the edge of the part or next hole is less than 2 times the bolt diameter, the limit state of shear in the plate material, also referred to as *tear out*, will control. In this case, failure occurs by a piece of material tearing out of the end of the connection as shown in Figure 10.9c or by tearing between holes in the direction of force. The nominal strength for this failure mode, R_n, is provided by shear along the two planes. From statics

$$R_n = \text{(shear strength) (2 planes) (clear distance) (material thickness)}$$

$$R_n = 0.6F_u(2L_c)t = 1.2F_uL_ct$$

where

$0.6F_u =$ ultimate shear strength of the connected material, ksi

$t =$ thickness of the material, in.

$L_c =$ clear edge distance, measured from the edge of the hole to the edge of the material or the next hole

If the clear distance exceeds $2d$, bearing on the connected material will be the controlling limit state, as shown in Figure 10.9d. In this case, the limit state is that of hole distortion and the calculated bolt strength will be

$$R_n = 2.4dt\,F_u$$

where

$d =$ bolt diameter

$t =$ connected part thickness

$F_u =$ tensile strength of the connected part

These two expressions are provided in Specification Section J3.10 in a single expression as

$$R_n = 1.2L_c t F_u \leq 2.4dt\,F_u \tag{10.1}$$

If deformation at the bolt hole is not a design consideration at service loads, both of these limit states may be increased so that

$$R_n = 1.5L_c t F_u \leq 3.0dt\,F_u \tag{10.2}$$

When bolts are used in a connection with long slots and the force is perpendicular to the slot, bolt strength is reduced such that

$$R_n = 1.0L_c t F_u \leq 2.0dt\,F_u \tag{10.3}$$

As was the case for bolt shear, the resistance and safety factors for the limit state of bolt bearing are

$$\phi = 0.75\,(\text{LRFD}) \qquad \Omega = 2.00\,(\text{ASD})$$

10.7.3 Bolt Tension

For the limit state of bolt tension, strength is directly based on the tensile strength of the bolt material. Section J3.6 provides that

$$R_n = F_n A_b$$

and

$$\phi = 0.75\,(\text{LRFD}) \qquad \Omega = 2.00\,(\text{ASD})$$

where

$F_n =$ tensile stress, F_{nt}, from Specification Table J3.2

$A_b =$ area of the bolt shank

Table 10.4 shows the nominal tensile stress, F_{nt}, for bolts taken from Specification Table J3.2. Note that there is no distinction for the location of the shear plane, because the bolt is loaded axially and the limiting stresses occurs over the net tensile area. The area of the bolt shank is again used and the nominal tensile stress is given as $0.75F_u$.

EXAMPLE 10.1
Bolt Shear Strength

GOAL: Determine the available bolt shear strength.

GIVEN: (a) a single $3/4$-in. A325N bolt.
(b) a single $7/8$-in. A490X bolt.

SOLUTION

Part (a) a single $3/4$-in. A325N bolt

Step 1:   Determine the bolt shank area.

$$A_b = \frac{\pi d^2}{4} = \frac{\pi (0.75)^2}{4} = 0.442 \text{ in.}^2$$

Step 2:   Determine the nominal shear stress.
For an A325 bolt

$$F_u = 120 \text{ ksi}$$

and for the threads included (N)

$$F_{nv} = 0.4 F_u = 0.4(120) = 48 \text{ ksi}$$

Step 3:   For LRFD, $\phi = 0.75$ and the design strength is

$$\phi r_n = 0.75(48)(0.442) = 15.9 \text{ kips}$$

Step 3:   For ASD, $\Omega = 2.00$ and the allowable strength is

$$\frac{r_n}{\Omega} = \frac{(48)(0.442)}{2.00} = 10.6 \text{ kips}$$

Part (b) a single $7/8$-in. A490X bolt

Step 1:   Determine the bolt shank area.

$$A_b = \frac{\pi d^2}{4} = \frac{\pi (0.875)^2}{4} = 0.601 \text{ in.}^2$$

Step 2:   Determine the nominal shear stress.
For an A490 bolt

$$F_u = 150 \text{ ksi}$$

and for the threads excluded (X)

$$F_{nv} = 0.5 F_u = 0.5(150) = 75 \text{ ksi}$$

Step 3: For LRFD, $\phi = 0.75$ and the design strength is

$$\phi r_n = 0.75(75)(0.601) = 33.8 \text{ kips}$$

Step 3: For ASD, $\Omega = 2.00$ and the allowable strength is

$$\frac{r_n}{\Omega} = \frac{(75)(0.601)}{2.00} = 22.5 \text{ kips}$$

Manual Table 7-1 provides single-bolt strength values for a wide range of bolt sizes and strengths.

EXAMPLE 10.2
Lap Splice Connection Strength

GOAL: Determine the available shear and bearing strength for a four-bolt connection.

GIVEN: A lap joint using $1/2$-in. A36 plates is given in Figure 10.10. Use (a) $7/8$-in. A325X bolts and (b) $7/8$-in. A325N bolts.

SOLUTION

Part (a) $7/8$-in. A325X bolts

Step 1: Determine the nominal shear strength.

$$F_{nv} = 0.5(120) = 60 \text{ ksi}$$
$$A_b = 0.601 \text{ in.}^2$$
$$r_n = (60)(0.601) = 36.1 \text{ kips}$$

Step 2: For LRFD, the design shear strength for a single bolt is

$$\phi r_n = 0.75(36.1) = 27.1 \text{ kips}$$

Step 3: For the four bolts in shear

$$\phi R_n = 4(27.1) = 108 \text{ kips}$$

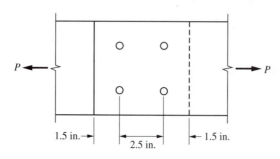

Figure 10.10 Lap Joint for Example 10.2.

Step 2: For ASD, the allowable shear strength for a single bolt is

$$\frac{r_n}{\Omega} = \frac{(36.1)}{2.00} = 18.1 \text{ kips}$$

Step 3: For the four bolts in shear

$$\frac{R_n}{\Omega} = 4(18.1) = 72.4 \text{ kips}$$

Step 4: Determine the nominal bearing strength.
The clear distance from the bolt hole to the end of the member

$$L_c = 1.5 - \left(\frac{1}{2}\right)(7/8 + 1/16) = 1.03 < 2d = 2(7/8) = 1.75 \text{ in.}$$

and between holes in the direction of force

$$L_c = 2.5 - (7/8 + 1/16) = 1.56 < 2d = 2(7/8) = 1.75 \text{ in.}$$

In both cases, because the clear distance is less than two bolt diameters, tear-out controls. Thus, for each end bolt

$$r_n = 1.2(58)(1.03)(0.5) = 35.8 \text{ kips}$$

and for each interior bolt

$$r_n = 1.2(58)(1.56)(0.5) = 54.3 \text{ kips}$$

Step 5: For LRFD, the design bearing strength for four bolts in bearing (tear out)

$$\phi R_n = 0.75(2(54.3) + 2(35.8)) = 135 \text{ kips}$$

Step 5: For ASD, the allowable bearing strength for four bolts in bearing (tear out)

$$\frac{R_n}{\Omega} = \frac{(2(54.3) + 2(35.8))}{2.00} = 90.1 \text{ kips}$$

Step 6: Determine the final connection strength.
The connection strength is the lowest value of strength for the limit states of bolt shear or bolt bearing; thus, the final connection bolt strength is

For LRFD

$$\phi R_n = 108 \text{ kips for the limit state of bolt shear}$$

to the end of the member

$$L_c = 1.25 - \left(\frac{1}{2}\right)(3/4 + 1/16) = 0.844 < 2d = 2(3/4) = 1.50 \text{ in.}$$

and between holes in the direction of force

$$L_c = 3.0 - (3/4 + 1/16) = 2.19 > 2d = 2(3/4) = 1.50 \text{ in.}$$

Thus, the clear distance is less than two bolt diameters for the end bolt and tear-out controls. But, between holes the clear distance is greater than two bolt diameters and bearing controls.

Thus, for each end bolt

$$r_n = 1.2(58)(0.844)(0.5) = 29.4 \text{ kips}$$

and for each interior bolt

$$r_n = 2.4(3/4)(0.5)(58) = 52.2 \text{ kips}$$

Thus, for the four bolts in bearing (tear-out), the nominal strength is

$$R_n = (2(29.4) + 2(52.2)) = 163 \text{ kips}$$

Step 4:   For LRFD, the design bearing strength is

$$\phi R_n = 0.75(163) = 122 \text{ kips}$$

Step 4:   For ASD, the allowable strength for bearing is

$$\frac{R_n}{\Omega} = \frac{163}{2.00} = 81.5 \text{ kips}$$

Step 5:   Determine the final connection strength.

The connection strength is determined by the lowest of bolt shear strength or bolt bearing strength; thus, the final connection strength based on the limit state of bolt bearing is

For LRFD

$$\phi R_n = 122 \text{ kips}$$

For ASD

$$\frac{R_n}{\Omega} = 81.5 \text{ kips}$$

10.7.4   Slip

The limit state of slip is associated with connections that are referred to as slip-critical. *Slip-critical connections* are permitted to be designed to prevent slip either as a serviceability limit state or at the required strength limit state. They should be used only when the connection is subjected to fatigue or the connection has oversized holes or slots parallel to the direction of load. In any case, the connection must also be checked for strength as a bearing type connection by the methods discussed in the previous sections. The nominal strength of a single bolt in a slip-critical connection is given in Specification Section J3.8 as

$$R_n = \mu D_u h_{sc} T_b N_s$$

where

μ = mean slip coefficient = 0.35 for Class A surfaces with other values found in Specification Section J3.8

D_u = 1.13

h_{sc} = 1.0 for standard-size holes with other values found in Specification Section J3.8

N_s = number of slip planes

T_b = minimum bolt tension specified

For connections in which slip prevention is a serviceability limit state

$$\phi = 1.00 \,(\text{LRFD}) \qquad \Omega = 1.50 \,(\text{ASD})$$

and for connections in which slip prevention is required at the strength level

$$\phi = 0.85 \,(\text{LRFD}) \qquad \Omega = 1.76 \,(\text{ASD})$$

Detailed use of the slip-critical connection is not addressed here. Examples of slip-critical connection design can be found in the AISC Manual Companion CD, Section II.

10.7.5   Combined Tension and Shear in Bearing-Type Connections

When bolts are subjected to simultaneous shear and tension, Specification Section J3.7 provides a nominal tensile stress modified to include the effects of shearing stress, F'_{nt}, to be used in determining the nominal bolt tensile strength such that

$$R_n = F'_{nt} A_b$$

and

$$\phi = 0.75 \,(\text{LRFD}) \qquad \Omega = 2.00 \,(\text{ASD})$$

When the required stress in either shear or tension is less than or equal to 20% of the corresponding available stress, the effects of the combined stresses can be ignored. If both required stresses exceed this 20% limit, the modified tensile stress is given as

$$F'_{nt} = 1.3 F_{nt} - \frac{F_{nt}}{F_{rv}} f_v \leq F_{nt}$$

where

F_{nt} = nominal tensile stress when only tension occurs

F_{rv} = available shear stress, ϕF_{nv} for LRFD or $\dfrac{F_{nv}}{\Omega}$ for ASD

F_{nv} = nominal shear stress when only shear occurs

f_v = required shear stress, either for LRFD or ASD

10.8 WELDS

Welding is a process of joining steel by melting additional metal into the joint between the two pieces to be joined. The ease with which various types of steel can be joined by welding, without exhibiting cracks and other flaws, is called *weldability*. Most structural steels used today accept welding without the occurrence of unwanted defects. The American Welding Society (AWS) defines weldability as "the capacity of a metal to be welded under fabrication conditions imposed, into a specific, suitably designed structure and to perform satisfactorily in the intended service."

Weldability depends primarily on the chemical composition of the steel and the thickness of the material. The impact on weldability of the various chemical elements in the composition of steel was discussed in Chapter 3.

10.8.1 Welding Processes

For structural steel, the four most popular welding processes and their abbreviations as designated by AWS are

Shielded Metal Arc (SMAW)

Submerged Arc (SAW)

Gas Shielded Metal Arc (GMAW)

Flux Cored Arc (FCAW)

Shielded Metal Arc Welding (SMAW)

Shielded Metal Arc Welding (*SMAW*) is one of the oldest welding processes. It is often called *manual* or *stick welding*. Figure 10.12a is a schematic representation of this welding process. A high voltage is induced between an electrode and the metal pieces that are to be joined. The electrode is the source of the metal introduced into the joint to make the weld. It is called the *consumable electrode*. When the welding operator strikes an arc between the electrode and the base metal, the resulting flow of current melts the electrode and the base metal adjacent to it. The electrode is coated with a special ceramic material called *flux*. This flux protects the molten metal from absorbing hydrogen and other impurities during the welding process. When the metal cools, a permanent bond exists between the electrode

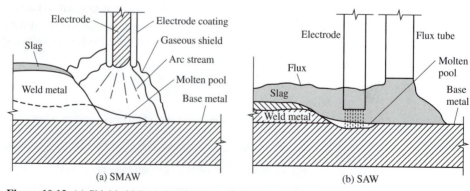

Figure 10.12 (a) Shielded Metal Arc Welding; (b) Submerged Arc Welding.

material and the base material. Because the flux cools at a different rate than the metal, it separates from the weld and is easily removed from the joint.

Submerged Arc Welding (SAW)

Submerged Arc Welding (SAW) is an automatic or semi-automatic process that is used primarily when long pieces of plate are to be joined. It is shown schematically in Figure 10.12b. SAW welds must be made in the near flat or horizontal position. The flux is a granular material introduced through a flexible tube on top of the electric arc. It is an economical process for applications in which repetitive and automated fabrication procedures lend efficiency to the work.

Gas Shielded Metal Arc Welding (GMAW)

Gas Shielded Metal Arc Welding (GMAW) is a process in which a continuous wire is fed into the joint to be welded. The molten metal is protected from the atmosphere by gas surrounding the wire. When used in the field, it is necessary to ensure that wind does not blow the gas away from the joint. This is the method often referred to as *MIG welding* for its use of inert gasses.

Flux Cored Arc Welding

Flux Cored Arc Welding (FCAW) is also a continuous wire process, except that the wire is essentially a thin hollow tube, filled with flux that protects the metal as the wire melts. It can be arranged as a semi-automatic process, and exceptionally high production rates can be attained.

10.8.2   Types of Welds

Four basic types of welds are used in steel construction, including fillet welds, groove welds, plug welds, and slot welds. Fillet and groove welds are shown in Figure 10.13. Plug and slot welds fill a hole or slot with weld material to attach one piece to another.

Figure 10.13a shows a fillet weld. The leg of the weld is measured along the interface between the weld metal and the base metal. The throat of the weld is the shortest dimension of the weld. Because most fillet welds are symmetrical, with a 45-degree surface, the throat

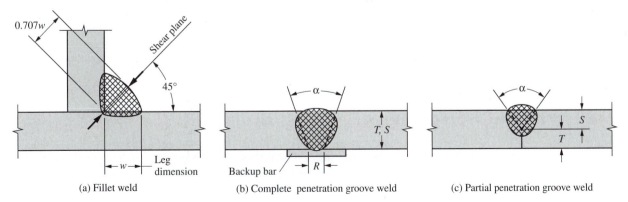

(a) Fillet weld   (b) Complete penetration groove weld   (c) Partial penetration groove weld

Figure 10.13 Fillet and Groove Welds in Section.

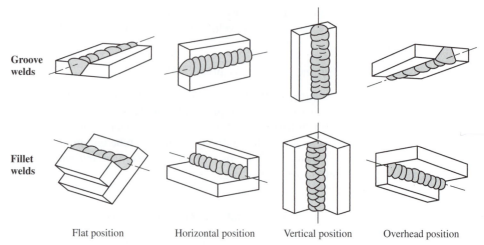

Groove welds

Fillet welds

Flat position   Horizontal position   Vertical position   Overhead position

Figure 10.14 Terminology for Fillet and Groove Weld Positions.

is 0.707 times the leg dimension as shown. The size of a fillet weld is given by its leg dimension, in increments of $1/16$ in.

A groove weld can be either a complete joint penetration groove weld (CJP), as shown in Figure 10.13b, or a partial joint penetration groove weld (PJP), as shown in Figure 10.13c. Both types of groove welds have been prequalified by AWS. This prequalification means that certain weld configurations, including the root opening, R; the angle of preparation, α; and the effective thickness, S, are deemed to be practical to build and will carry the intended load. AWS specifies provisions for prequalifying any weld configuration if circumstances indicate that it is practical. These prequalified complete and partial joint penetration groove welds are shown in detail in Manual Table 8-2. The configurations shown in Figure 10.13 are schematic representations.

Both fillet welds and groove welds can be laid down in a variety of different positions depending on the orientation of the pieces to be joined. The terminology for these positions is shown in Figure 10.14.

10.8.3   Weld Sizes

Specification Section J2 addresses effective areas and sizes for welds. The effective dimensions of groove welds are given in Specification Tables J2.1 and J2.2. The effective areas of fillet welds are given in Specification Section J2.2. The minimum sizes for fillet welds are based on the thinner of the parts being joined and are given in Specification Table J2.4. The maximum size of fillet welds for material less than $1/4$-in. thick is the thickness of the material, whereas for material $1/4$-in. thick or greater, the maximum size is the material thickness less $1/16$ in.

10.9   WELD LIMIT STATES

The only limit state to be considered for a weld is that of rupture. Yielding of the weld metal will occur but it occurs over such a short distance that it is not a factor in connection behavior. Strain hardening occurs and rupture takes place without excessive yielding deformation.

The ultimate tensile strength of an electrode may vary from 60 to 120 ksi, depending on the specified composition. AWS classifies electrodes according to the tensile strength of

Table 10.5   Matching Weld Electrodes for Commonly Used Steels

Group	Steel specification	Minimum yield strength	SAW matching electrode strength
I	A36	36	E60 or E70
	A53	35	
	A500 Gr A	33	
	A500 Gr B	42	
	A501	36	
	A529	42	
II	A572 Gr 42	42	E70
	A574 Gr 50	50	
	A588	50	
	A618	46–50	
	A913	50	
	A992	50	
IV	A852	70	E90

the weld metal and indicates electrode strength as F_{EXX}. In this notation, the E represents the electrode and the XX represents the tensile strength. Thus, a typical electrode used to weld A992 steel would have a strength of 70 ksi and be designated as an E70 electrode.

AWS and AISC specify that for a particular grade of structural steel, as indicated by yield strength, there is a matching electrode. Table 10.5 shows the matching electrodes for commonly used steels. Both organizations further specify that the steel can be joined by welding only with the matching electrode or one that is no more than one grade higher. This is to encourage yielding in the base metal before it occurs in the weld.

10.9.1   Fillet Weld Strength

For a fillet weld as shown in Figure 10.13a, load is transferred by shear through the throat of the weld and the weld rupture strength is a function of the properties of the electrode. Shear strength provisions for welds are found in Specification Section J2.4 and Table J2.5 where

$$R_n = F_w A_w$$

$$\phi = 0.75 \, (\text{LRFD}) \qquad \Omega = 2.00 \, (\text{ASD})$$

and

F_w = nominal strength of the weld metal per unit area = $0.6F_{EXX}$

A_w = effective area of the weld

F_{EXX} = weld electrode classification number, the weld strength

Because the limit state of all fillet welds is one of shear rupture through the throat, the effective area of the weld is the width of the weld at the throat, $0.707w$, times the length of the weld, L, so that

$$A_w = 0.707wL$$

The resulting nominal weld strength is

$$R_n = 0.6F_{EXX}(0.707wL)$$

For the most commonly used electrode, $F_{EXX} = 70\,\text{ksi}$, the design strength for LRFD can be determined as

$$\phi R_n = 0.75(0.6(70))(0.707wL) = 22.27wl$$

and the allowable strength for ASD can be determined as

$$\frac{R_n}{\Omega} = \frac{(0.6(70))(0.707wL)}{2.0} = 14.85wl$$

It is convenient in design to use the fillet weld strength for a fillet weld with a $1/16$-in. leg, which gives

Design strength for LRFD

$$\phi R_n = 22.27wl = 22.27\left(\frac{1}{16}\right)(1.0) = 1.392 \text{ kips per } 1/16 \text{ of weld per in. of length}$$

Allowable strength for ASD

$$\frac{R_n}{\Omega} = 14.85wl = 14.85\left(\frac{1}{16}\right)(1.0) = 0.928 \text{ kips per } 1/16 \text{ of weld per in. of length}$$

Therefore, a $1/4$-in. fillet weld has a design strength of 1.392×4 (sixteenths) $= 5.57$ kips per inch of length and an allowable strength of 0.928×4 (sixteenths) $= 3.71$ kips per inch of length.

When an in-plane load is applied to a fillet weld at an angle other than along the length of the weld, more strength is available than given by these calculations. The Specification provides an alternative for the weld nominal stress, based on the angle of the load to the longitudinal axis of the weld. Thus

$$F_w = 0.60F_{EXX}(1.0 + 0.5\sin^{1.5}\theta)$$

where

$$\theta = \text{angle of loading measured from the weld longitudinal axis}$$

This strength equation is intended to be used for welds or weld groups in which all elements are in line or parallel. When welds with different orientations are combined in the same joint, deformation of these different welds must be accounted for. Specification Section J2 provides two alternate approaches for combining welds that are not in line or parallel. The simplest case is for concentrically loaded fillet weld groups consisting of elements that are both longitudinal and transverse to the direction of the applied load. For this case, the nominal weld strength is taken as the larger of the simple sum of the welds without considering orientation, given by

$$R_n = R_{wl} + R_{wt}$$

or

$$R_n = 0.85R_{wl} + 1.5R_{wt}$$

where

R_{wl} = nominal strength of the longitudinally loaded weld without considering the angle of load

R_{wt} = nominal strength of the transversely loaded weld without considering the angle of load

---

EXAMPLE 10.4
Weld Strength and
Load Angle

SOLUTION

GOAL: Determine the available strength of the three welds given in Figure 10.15.

GIVEN: The welds are $^3/_4$-in. welds, 8.0 in. long, and loaded (a) along the length of the weld, (b) transversely to the weld, and (c) at a 45-degree angle to the weld. Use E70 electrodes.

Part (a) Weld loaded along its length

Step 1:   Determine the number of $^1/_{16}$ units for the given weld.
   A $^3/_4$-in. weld is twelve $^1/_{16}$-in. units across the leg.

Step 2:   Determine the strength of the weld when loaded along its length.
   The strength values already discussed can be used because the weld is loaded along its length.

For LRFD

$$\phi R_n = 8.0(12)(1.392) = 134 \text{ kips}$$

For ASD

$$\frac{R_n}{\Omega} = 8.0(12)(0.928) = 89.1 \text{ kips}$$

Part (b) Weld loaded at 90 degrees to the weld length

Step 1:   Determine the nominal weld strength using the alternate strength equation to account for the angle of load.

$$F_w = (0.60F_{EXX})(1.0 + 0.5 \sin^{1.5}\theta)$$
$$= (0.60F_{EXX})(1.0 + 0.5 \sin^{1.5}(90)) = (0.60F_{EXX})(1.5)$$

Therefore, the strength of the weld is increased by 1.5 over what it is when the load is along the weld. Thus

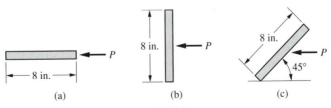

(a)   (b)   (c)

Figure 10.15 Welds for Example 10.4.

Step 2: For LRFD, the design strength is

$$\phi R_n = 1.5(134) = 201 \text{ kips}$$

Step 2: For ASD, the allowable strength is

$$\frac{R_n}{\Omega} = 1.5(89.1) = 134 \text{ kips}$$

Part (c) Weld loaded at 45 degrees to the weld

Step 1: Determine the nominal weld strength using the alternate strength equation to account for the angle of load.

$$F_w = (0.60F_{EXX})(1.0 + 0.5 \sin^{1.5}(45)) = (0.60F_{EXX})(1.30)$$

Therefore, the strength of the weld is increased by 1.30 over what it is when the load is along the weld. Thus

Step 2: For LRFD, the design strength is

$$\phi R_n = 1.30(134) = 174 \text{ kips}$$

Step 2: For ASD, the allowable strength is

$$\frac{R_n}{\Omega} = 1.30(89.1) = 116 \text{ kips}$$

EXAMPLE 10.5a
Weld Strength and Load Angle by LRFD

GOAL: Determine the design strength for C-shaped welds.

GIVEN: A C-shaped weld group is shown in Figure 10.16 to attach a tension plate to a gusset. Use E70 electrodes and a $7/8$-in. weld.

SOLUTION

Step 1: Determine the design strength for the two 2.0-in. welds parallel to the load.

$$\phi R_{wl} = 2(2.0)(14)(1.392) = 78.0 \text{ kips}$$

Step 2: Determine the design strength for the 6.0-in. weld transverse to the load.

$$\phi R_{wt} = 6.0(14)(1.392) = 117 \text{ kips}$$

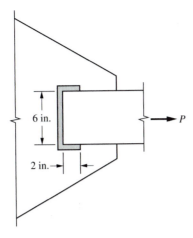

Figure 10.16 C-Shaped Weld for Example 10.5.

Step 3: Determine the connection design strength by adding the strength based on length of the welds.

$$\phi R_n = 78.0 + 117 = 195 \text{ kips}$$

Step 4: Determine the design strength considering the added contribution of the transverse welds while reducing the contribution of the longitudinal welds so that

$$\phi R_n = 0.85(78.0) + 1.5(117) = 242 \text{ kips}$$

Step 5: Determine the weld strength by selecting the largest from Steps 3 or 4.

$$\boxed{\phi R_n = 242 \text{ kips}}$$

EXAMPLE 10.5b
Weld Strength and Load Angle by ASD

GOAL: Determine the design strength for C-shaped welds.

GIVEN: A C-shaped weld group is shown in Figure 10.16 to attach a tension plate to a gusset. Use E70 electrodes and a $\frac{7}{8}$-in. weld.

SOLUTION

Step 1: Determine the allowable strength for the two 2.0-in. welds parallel to the load.

$$\frac{R_{wl}}{\Omega} = 2(2.0)(14)(0.928) = 52.0 \text{ kips}$$

Step 2: Determine the allowable strength for the 6.0-in. weld transverse to the load.

$$\frac{R_{wt}}{\Omega} = 6.0(14)(0.928) = 78.0 \text{ kips}$$

Step 3: Determine the connection allowable strength by adding the strength based on length of the welds.

$$\frac{R_n}{\Omega} = 52.0 + 78.0 = 130 \text{ kips}$$

> **Step 4:**   Determine the allowable strength considering the added contribution of the transverse welds while reducing the contribution of the longitudinal welds so that
>
> $$\frac{R_n}{\Omega} = 0.85(52.0) + 1.5(78.0) = 161 \text{ kips}$$
>
> **Step 5:**   Determine the weld strength by selecting the largest from Steps 3 or 4.
>
> $$\boxed{\frac{R_n}{\Omega} = 161 \text{ kips}}$$

10.9.2   Groove Weld Strength

A groove weld can be either a complete or partial joint penetration weld as shown in Figures 10.13b and c. The complete joint penetration grove weld (CJP) is not designed in the usual sense because the weld metal is always stronger than the base metal when properly matching electrodes are used. Therefore, the strength of the base metal controls the design.

In the case of a complete joint penetration groove weld, the nominal strength of the tension joint is the product of the yield strength of the base material and the cross-sectional area of the smallest piece joined. The nominal strength of a partial joint penetration groove weld in a tension joint is similar except that the full cross-sectional area of the joined pieces is not effective. In this case, AWS defines an effective throat dimension, S, which is a function of the configuration of the bevel as shown in Figure 10.13c and Manual Table 8-2.

10.10   CONNECTING ELEMENTS

The plates, angles, and other elements that go into making up a connection are called *connecting elements*. They, along with the region of the members actually involved in the connection, are treated in Specification Section J4. There are provisions for tension, compression, shear, and block shear. There are no special provisions for connecting elements in flexure.

10.10.1   Connecting Elements in Tension

Although the Specification addresses tension in connecting elements in Section J4.1, it does not alter the basic tension provisions found in Specification Chapter D. This means that two limit states are to be considered, the limit state of yielding and the limit state of rupture. Again, for the tension limit states, the resistance and safety factors are different for the two limit states so any comparison of strength must be made at the design or allowable strength level. The design strength is given by ϕR_n and the allowable strength by R_n/Ω, as has been the case throughout the Specification. For the limit state of yielding of connecting elements

$$R_n = F_y A_g$$

$$\phi = 0.90 \,(\text{LRFD}) \qquad \Omega = 1.67 \,(\text{ASD})$$

For the limit state of rupture of connecting elements

$$R_n = F_u A_e$$

$$\phi = 0.75 \,(\text{LRFD}) \qquad \Omega = 2.00 \,(\text{ASD})$$

The definition of terms are the same as for all other tension members previously considered except for the requirement that the effective net area, A_e, for bolted splice plates may not be taken greater than $0.85A_g$, regardless of the area deducted for holes.

10.10.2   Connecting Elements in Compression

Most connecting elements in compression are relatively short and have a fairly small slenderness ratio. In addition, determination of the appropriate effective length factor requires application of significant engineering judgment, usually amounting to making an educated guess for an appropriate factor. With this in mind, and in order to simplify connection design somewhat, the Specification provides, in Section J4.4, a simple relation for the compressive strength of connecting elements if the slenderness ratio is less than 25. For this case, $P_n = F_y A_g$ and the resistance and safety factors are the same as for other compression members as

$$\phi = 0.90 \,(\text{LRFD}) \qquad \Omega = 1.67 \,(\text{ASD})$$

If the slenderness ratio of the compression element is greater than 25, the element must be designed according to the column provisions of Specification Chapter E.

10.10.3   Connecting Elements in Shear

Member design for shear requires the consideration of the limit states of shear yielding and shear buckling. Connecting elements and the portion of members affected by the connection must be checked for the limit states of shear yielding and shear rupture. Shear yielding occurs on the gross area of the element whereas shear rupture occurs on a section containing holes. Thus, for shear yielding of the element

$$R_n = 0.6F_y A_g$$

$$\phi = 1.00 \,(\text{LRFD}) \qquad \Omega = 1.50 \,(\text{ASD})$$

The resistance and safety factors for this case are the same as those for the special case of rolled I-shaped members given in Specification Section G2.1.

For the limit state of shear rupture

$$R_n = 0.6F_u A_{nv}$$

$$\phi = 0.75 \,(\text{LRFD}) \qquad \Omega = 2.00 \,(\text{ASD})$$

where

A_{nv} = net area subjected to shear

As was the case for tension rupture, the net area is determined by removing the area of holes from the gross area.

10.10.4   Block Shear Strength

The limit state of block shear rupture can occur on the connecting elements or the affected members. It is a complex failure mode that combines shear and tension failures into a single mode of failure. Block shear was discussed in Section 4.7 as it pertained to tension members because it is a major factor in determining tension member strength. It can also be a factor in determining the strength of a beam end reaction, depending on the connection geometry. Thus, it is repeated here. The nominal strength for the limit state of block shear rupture is

$$R_n = 0.6F_u A_{nv} + U_{bs}F_u A_{nt} \le 0.6F_y A_{gv} + U_{bs}F_u A_{nt}$$

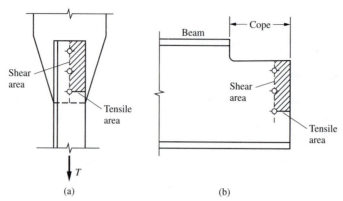

Figure 10.17 Example Block Shear Failure.

where

A_{gv} = gross shear area

A_{nt} = net tension area

A_{nv} = net shear area

U_{bs} = 1.0 for uniform tension stress distribution and U_{bs} = 0.5 for nonuniform tension stress

The resistance and safety factors for the limit state of block shear rupture are again

$$\phi = 0.75 \, (\text{LRFD}) \qquad \Omega = 2.00 \, (\text{ASD})$$

Figure 10.17 shows a single-angle tension member attached to a gusset plate and a coped beam end with the holes located in a single line. The tension area and shear area are identified for each and the area that would tear out is shaded.

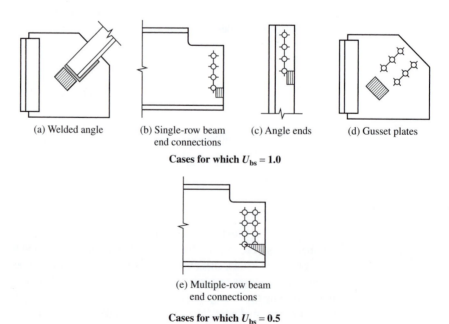

Figure 10.18 Block Shear Tensile Stress Distribution.

A review of the block shear equation given on page 336 shows that the expected failure mode will always include tension rupture whereas the shear failure mode will be the smaller of the shear rupture or shear yield. As noted on page 336, the tension stress distribution factor, U_{bs}, is a function of the variation of the tension stress over the tension area. Figure 10.18 shows several elements and the corresponding assumed tensile stress distribution. The only case identified by the Commentary where the tensile stress distribution is not uniform is that of a coped beam with two rows of bolts, as shown in Figure 10.18e.

EXAMPLE 10.6
Block Shear Strength

GOAL: Determine the block shear design strength and allowable strength for a coped beam.

GIVEN: A coped W16×40, A992 beam end is shown in Figure 10.19. Assume that the beam has standard holes for $\frac{5}{8}$-in. bolts.

SOLUTION

Step 1:   Determine the gross and net shear areas and net tension area for the beam.
Remember from the discussion of net area for tension members that for net area, an additional $\frac{1}{16}$ in. must be deducted to account for any hole damage from the punching operation.

$$A_{gv} = 11.0(0.305) = 3.36 \text{ in.}^2$$
$$A_{nv} = [11.0 - 3.5(5/8 + 1/8)](0.305) = 2.55 \text{ in.}^2$$
$$A_{nt} = [4.25 - 1.5(5/8 + 1/8)](0.305) = 0.953 \text{ in.}^2$$

Step 2:   Determine the shear yield and rupture strength and the tension rupture strength.
For this geometry, the tensile stress distribution is nonuniform; therefore, $U_{bs} = 0.5$.

$$\text{Shear Yield} = 0.6(50)(3.36) = 101 \text{ kips}$$
$$\text{Shear Rupture} = 0.6(65)(2.55) = 99.5 \text{ kips}$$
$$\text{Tension Rupture} = 0.5(65)(0.953) = 31.0 \text{ kips}$$

Step 3:   Determine the nominal block shear strength.
Because shear rupture is less than shear yield, combine the shear rupture with the tensile rupture. Thus

$$R_n = (99.5 + 31.0) = 131 \text{ kips}$$

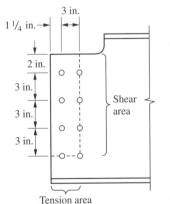

Tension area

Figure 10.19 Coped Beam End for Example 10.6.

Step 4:   For LRFD, the design strength is

$$\phi R_n = 0.75(131) = 98.3 \text{ kips}$$

Step 4:   For ASD, the allowable strength is

$$\frac{R_n}{\Omega} = \frac{(131)}{2.00} = 65.5 \text{ kips}$$

10.11   PROBLEMS

1. Develop a table showing the nominal shear strength for A325-N bolts for the following sizes: $5/8$, $3/4$, $7/8$, and 1 in.

2. Develop a table showing the nominal shear strength for A325-X bolts for the following sizes: $5/8$, $3/4$, $7/8$, and 1 in.

3. Develop a table showing the nominal shear strength for A490-N bolts for the following sizes: $5/8$, $3/4$, $7/8$, and 1 in.

4. Develop a table showing the nominal shear strength for A490-X bolts for the following sizes: $5/8$, $3/4$, $7/8$, and 1 in.

5. Develop a table showing the design shear strength for A325-N, A325-X, A490-N, and A490-X bolts for the following sizes: $5/8$, $3/4$, $7/8$, and 1 in.

6. Develop a table showing the allowable shear strength for A325-N, A325-X, A490-N, and A490-X bolts for the following sizes: $5/8$, $3/4$, $7/8$, and 1 in.

7. Determine the available strength of the $3/4$-in. A325-N bolts in the lap splice shown with two $1/2$-in. A36 plates. Determine (a) design strength by LRFD and (b) allowable strength by ASD.

8. Determine the available strength of the $3/4$-in. A325-X bolts in the lap splice shown for Problem 7, with two $1/2$-in. A36 plates. Determine (a) design strength by LRFD and (b) allowable strength by ASD.

9. Determine the available strength of the $7/8$-in. A325-N bolts in the lap splice shown for Problem 7, with two $1/2$-in. A36 plates. Determine (a) design strength by LRFD and (b) allowable strength by ASD.

10. Determine the available strength of the $7/8$-in. A490-N bolts in the lap splice shown with two $3/4$-in. A36 plates. Determine (a) design strength by LRFD and (b) allowable strength by ASD.

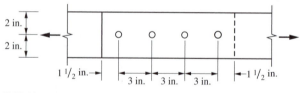

P10.10

11. Determine the available strength of the $7/8$-in. A490-X bolts in the lap splice shown for Problem 10 with two $3/4$-in. A36 plates. Determine (a) design strength by LRFD and (b) allowable strength by ASD.

12. Determine the available strength of the $3/4$-in. A325-N bolts in the butt splice shown with two $1/2$-in. side plates and a 1-in. main plate. Use A36 plates. Determine (a) design strength by LRFD and (b) allowable strength by ASD.

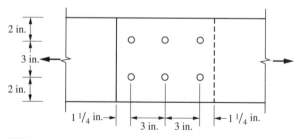

P10.7

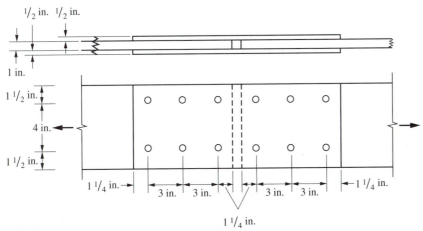

P10.12

13.   Determine the available strength of the four $^3/_4$-in. A325-N bolts in the single L3×3×$^1/_2$ A36 when the bolts are placed as shown. Determine (a) design strength by LRFD and (b) allowable strength by ASD.

15.   Determine the available strength of the six $^3/_4$-in. A325-N bolts in a 7-× $^3/_4$-in. A36 plate when the bolts are placed as shown. Determine (a) design strength by LRFD and (b) allowable strength by ASD.

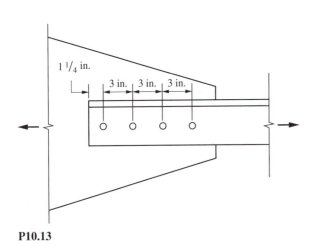

P10.13

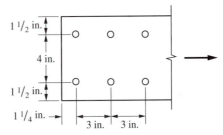

P10.15

16.   Determine the available strength of the eight $^3/_4$-in. A490-N bolts in a WT6×20, A992 steel plate when the bolts are placed as shown. Determine (a) design strength by LRFD and (b) allowable strength by ASD.

14.   Determine the available strength of the three $^3/_4$-in. A325-N bolts in the single L4×3×$^3/_8$ A36 when the bolts are placed as shown. Determine (a) design strength by LRFD and (b) allowable strength by ASD.

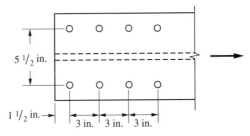

P10.16

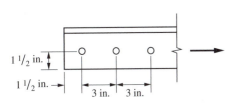

P10.14

17.   Determine the available strength of two $^3/_8$-in. welds that are loaded parallel to their length, are 10 in. long, and are made from E70 electrodes. Determine (a) design strength by LRFD and (b) allowable strength by ASD.

18. Determine the available strength of two $\frac{1}{4}$-in. welds that are loaded parallel to their length, are 8 in. long, and made from E70 electrodes. Determine (a) design strength by LRFD and (b) allowable strength by ASD.

19. If the welds of Problem 17 were loaded at their centroid and at 90 degrees to the weld length, determine (a) design strength by LRFD and (b) allowable strength by ASD.

20. If the welds of Problem 18 were loaded at their centroid and at 45 degrees to the weld length, determine (a) design strength by LRFD and (b) allowable strength by ASD.

21. Three $\frac{1}{4}$-in. welds are grouped to form a C and are loaded at their centroid. Determine the available weld strength if the single transverse weld is 9 in. and the two longitudinal welds are each 3 in. Use E70 electrodes. Determine (a) design strength by LRFD and (b) allowable strength by ASD.

22. Repeat Problem 21 with the transverse weld at 3 in. and the two longitudinal welds at 9 in. each. Determine (a) design strength by LRFD and (b) allowable strength by ASD.

23. Determine the available block shear strength for a coped W16×26, A992 beam with holes for $\frac{3}{4}$-in. bolts as shown. Determine (a) design strength by LRFD and (b) allowable strength by ASD.

24. Determine the available block shear strength for a coped W21×182, A992 beam with holes for $\frac{3}{4}$-in. bolts as shown. Determine (a) design strength by LRFD and (b) allowable strength by ASD.

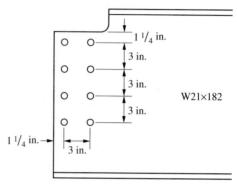

P10.24

25. Determine the available block shear strength for a coped W24×146, A992 beam with holes for $\frac{3}{4}$-in. bolts as shown. Determine (a) design strength by LRFD and (b) allowable strength by ASD.

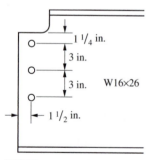

P10.23

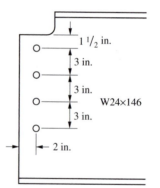

P10.25

Chapter 11

American Airlines Terminal, JFK International Airport
Photo courtesy Steven Rankel, PE.

Simple Connections

11.1 TYPES OF SIMPLE CONNECTIONS

This chapter addresses two types of connections, simple beam shear connections and simple bracing connections. The connecting elements and the connectors required for these connections have already been discussed in Chapter 10. The limit states that control the connection have also been discussed individually, although their link to connection design may not yet be completely clear. Connection design is a combination of element and connector selection with a checking of all appropriate limit states. The goal is to obtain a connection with sufficient strength and the appropriate stiffness to carry the load in a manner consistent with the model used in the structural analysis. In addition to these simple shear connections, beam bearing plates and column base plates will be discussed.

The limit states to be considered for a particular connection depend on the connection elements, the connection geometry, and the load path. They will be identified in the following sections as each connection type is considered. A summary of the potential limit states at this time, however, may prove useful. For bolts, the limit states of tensile rupture, shear rupture, bearing and tear-out, as well as slip, will be considered. For welds, the only limit state to be considered is shear rupture, although weld group geometry will add some complexity to that consideration. For connecting elements, the limit states are tension yielding and rupture, compression buckling, and shear yielding and rupture.

Table 11.1 lists the sections of the Specification and parts of the Manual discussed in this chapter.

Table 11.1 Sections of Specification and Parts of Manual Found in this Chapter

Specification	
B3.13	Gross and Net Area Determination
J2	Welds
J3	Bolts and Threaded Parts
J4	Affected Elements of Members and Connecting Elements
J10	Flanges and Webs with Concentrated Forces

Manual	
Part 7	Design Considerations for Bolts
Part 8	Design Considerations for Welds
Part 9	Design of Connecting Elements
Part 10	Design of Simple Shear Connections
Part 14	Design of Beam Bearing Plates, Column Base Plates, Anchor Rods, and Column Splices
Part 15	Design of Hanger Connections, Bracket Plates, and Crane-Rail Connections

11.2 SIMPLE SHEAR CONNECTIONS

A significant variety of potential connection geometries are associated with the various types of members to be connected. Five of the most commonly used simple shear connections are described in the following sections with design examples following. These connections are shown in Figures 11.1a through e as: double-angle, single-angle, single-plate commonly called a shear tab, unstiffened seated, and stiffened seated connections. Part 10 of the Manual includes many tables that can simplify connection design; however, the examples presented here show the required calculations when necessary to improve understanding. Once a calculation is sufficiently demonstrated, the Manual tables are used.

Several design considerations apply to all of the shear connections to be discussed and in some cases to other types of connections. It is helpful to address these before dealing with the specific connection. The first issue to consider is the location of the hinge within the connecting elements. It is critical that this hinge can actually occur in the real connection, because the analytical model of the connection assumes it behaves as a hinge or pin. The location of the hinge determines what forces and moments, if any, the individual elements must be designed for. In all cases, the hinge is located at the most flexible point within the connection. This may be at the face of the supporting member or at some other point within the connection. Several general design guidelines help insure that the connection behaves as desired. In most cases, this means that the hinge occurs at the face of the supporting member.

For double-angle connections, angle thickness should be limited to a maximum of $\frac{5}{8}$ in. The bolts in the outstanding legs, those connecting to the supporting member, should be spaced at as wide a gage as possible and for welded outstanding legs, the vertical welds should be spaced as far apart as possible. These characteristics insure that the connection behaves as a simple connection through bending of the outstanding legs.

For simple beam connections, the permitted tolerance for beam length must be considered. Although this tolerance is not normally a consideration for member design, it becomes important when the details of connecting members are considered. Beam length tolerance

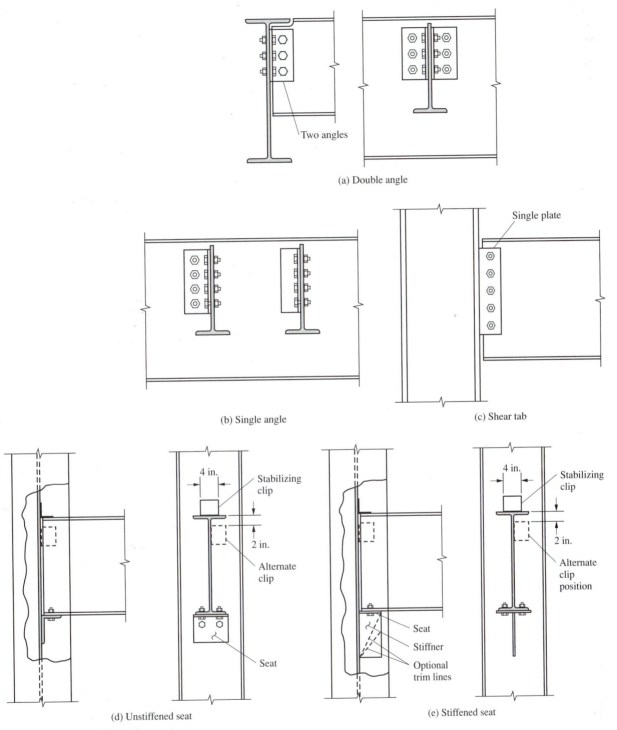

(a) Double angle

(b) Single angle

(c) Shear tab

(d) Unstiffened seat

(e) Stiffened seat

Figure 11.1 Simple Shear Connections.

is $\pm^1/_4$ in. To accommodate this, the beam is assumed to be held back $^1/_2$ in. from the face of the supporting member. In addition, when considering the edge distance from a bolt hole to the end of the member, the distance used in calculations should be taken as $^1/_4$ in. less than that actually detailed.

For welded connections, when a weld would end in the air, as would be the case for the welds on a shear tab, the effective length of the weld used in the calculations is the weld length less twice the weld size.

It is also helpful to remember the considerations for hole sizes. First, standard holes are sized $^1/_{16}$ in. larger than the bolt to be inserted. Then, when considering net sections for the limit states of tension rupture or shear rupture, Specification Section B3.13 requires that an additional $^1/_{16}$ in. be deducted to account for any material damage resulting from the hole-punching process. When a clear distance is calculated for the limit state of bearing, specifically the tear-out portion of the bearing check, the actual hole size is used.

These design considerations are used in the examples to follow.

11.3 DOUBLE-ANGLE CONNECTIONS: BOLTED-BOLTED

A double-angle shear connection, as shown in Figure 11.1a, is perhaps the most common simple shear connection used in steel construction. It is a fairly simple connection to fabricate and also a fairly easy connection for erection. When double-angle connections are to be installed back to back, there may be some problems, particularly when the supporting member is a column web. In the case of attachment to a column web, the safety requirements of OSHA call for special attention. One solution is to stagger the double angles. This connection can easily accommodate variation in beam length within acceptable tolerances.

The double-angle shear connection must be checked for the following limit states, grouped according to the elements that make up the connection: bolts, beam web, angles, and supporting member:

1. Bolts
   a. Shear rupture

2. Beam
   a. Bolt bearing on beam web
   b. Shear yielding of the web
   c. Block shear on coped beam web
   d. Coped beam flexural strength

3. Angles
   a. Bolt bearing on angles
   b. Shear rupture
   c. Shear yield
   d. Block shear

4. Supporting member
   a. Bolt bearing

Each of these limit states has been addressed previously in this book. In the examples that follow, these limit state checks are combined into a complete connection design.

EXAMPLE 11.1a
Bolted-Bolted
Double-Angle Shear
Connection by LRFD

GOAL: Design a bolted-bolted double-angle shear connection for an W18×50 beam.

GIVEN: The W18×50 beam must provide a required strength, $R_u = 83.0$ kips. The beam is A992 and the angles are A36. The beam flange is coped 2 in. Use $7/8$ in. A325-N bolts in standard holes in the legs on the beam web and short slots on the outstanding legs. The basic starting geometry is given in Figure 11.2.

SOLUTION

Step 1:   Determine the number of bolts required based on the shear rupture of the bolts.
From Manual Table 7-1, the design strength per bolt is

$$\phi r_n = 21.6 \text{ kips}$$

Because the bolts are in double shear, the total number of bolts required is

$$N = \frac{83.0}{2(21.6)} = 1.92$$

Therefore, try two bolts.

Step 2:   Check the bolt bearing on the web.
For the two-bolt connection, the top bolt is 1.25 in. from the beam cope and the second bolt is spaced 3.0 in. from the first. Determine the clear distances for each of these bolts.
For the top bolt

$$L_c = 1.25 - \frac{1}{2}(7/8 + 1/16) = 0.781 < 2(7/8) = 1.75$$

Thus, tear-out controls and the nominal bolt strength is

$$R_n = 1.2(0.781)(0.355)(65) = 21.7 \text{ kips}$$

For the second bolt

$$L_c = 3.0 - (7/8 + 1/16) = 2.06 > 2(7/8) = 1.75$$

Therefore, bearing controls, and the nominal bolt strength is

$$R_n = 2.4(7/8)(0.355)(65) = 48.5 \text{ kips}$$

Thus, for the two-bolt connection, the design strength is

$$\phi R_n = 0.75(21.7 + 48.5) = 52.7 < 83.0 \text{ kips}$$

Therefore, the two-bolt connection will not carry the load.

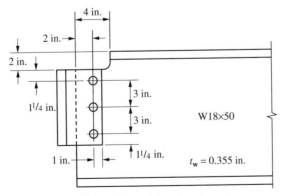

Figure 11.2 Connection Geometry for Example 11.1.

Step 3: Determine the number of bolts required considering bearing.

Adding a third bolt spaced at 3.0 in., as shown in Figure 11.2, gives a connection design strength for bolt bearing of

$$\phi R_n = 0.75(21.7 + 2(48.5)) = 89.0 > 83.0 \text{ kips}$$

Step 4: Consider the outstanding legs of the angles.

A similar calculation should be made for the bolts on the outstanding legs of the angles that connect to the supporting member. In this case, the bolts are in single shear but there are twice as many bolts so the load per bolt is half of the load in the bolts in the beam web. If the supporting member thickness is at least one-half of the beam web thickness and the strengths are the same, the bolts in the supporting member will be satisfactory. This is the assumed case for this example.

Step 5: Evaluate the minimum depth of the connection.

The beam web connection should be at least half the depth of the beam web measured as the distance between the fillets, T, given in Manual Table 1-1. This requirement is to prevent twisting of the simple supports. For this beam, $T = 15\frac{1}{2}$ in. so that the minimum angle depth should be $7\frac{3}{4}$ in. Thus, the $8\frac{1}{2}$-in. long angle will be an acceptable connection depth.

Step 6: Check shear yield of the beam web.

This is a check that should be carried out during the beam design process. At the point of connection design it is too late to be finding out that the beam will not be adequate. From Manual Table 3-2

$$\phi V_n = 192 \text{ kips} > 83.0 \text{ kips}$$

Step 7: Check the beam web for block shear.

The equations for block shear are found in Specification Section J4.3 and were presented in Section 10.10.4.

First calculate the required areas, remembering to account for the beam length tolerance in the tension area calculation,

$$A_{nt} = \left(1.75 - \frac{1}{2}\left(\frac{7}{8} + \frac{1}{8}\right)\right)(0.355) = 0.444 \text{ in.}^2$$

$$A_{gv} = 7.25(0.355) = 2.57 \text{ in.}^2$$

$$A_{nv} = \left(7.25 - 2.5\left(\frac{7}{8} + \frac{1}{8}\right)\right)(0.355) = 1.69 \text{ in.}^2$$

Consider shear yield and shear rupture and select the least nominal strength, thus

$$0.6F_y A_{gv} = 0.6(50)(2.57) = 77.1 \text{ kips}$$
$$0.6F_u A_{nv} = 0.6(65)(1.96) = 65.9 \text{ kips}$$

Selecting the shear rupture term and combining it with the tension rupture term gives a connection block shear design strength, recalling that $U_{bs} = 1.0$ for the case of uniform tensile stress distribution, we have

$$\phi R_n = 0.75(65.9 + 1.0(65)(0.444)) = 71.1 < 83.0 \text{ kips}$$

Thus, the given three-bolt connection is not adequate with block shear being the critical limit state to this point in our calculations.

Step 8: Revise the connection to meet the block shear strength requirements.

Consideration could be given to increasing the number of bolts and thereby increasing the length of the connection. However, because bolt shear required only two bolts, this would not be a particularly economical solution. If the connection were to be lowered on the beam end so that the distance from the center of the top bolt to the edge of the cope were 2.5 in., the connection would have more block shear strength.

Thus, the new shear areas become

$$A_{gv} = 8.5(0.355) = 3.02 \text{ in.}^2$$

$$A_{nv} = (8.5 - 2.5(7/8 + 1/8))(0.355) = 2.13 \text{ in.}^2$$

and the nominal shear yield and rupture strengths become

$$0.6F_y A_{gv} = 0.6(50)(3.02) = 90.6 \text{ kips}$$

$$0.6F_u A_{nv} = 0.6(65)(2.13) = 83.1 \text{ kips}$$

The resulting block shear design strength is

$$\phi R_n = 0.75(83.1 + 1.0(65)(0.444)) = 84.0 > 83.0 \text{ kips}$$

Step 9: Check the flexural strength of the coped beam.

It is a good idea to check this limit state during the initial design of the beam. It should be anticipated that a coped connection will be required during the design stage and it is at that stage that a change in beam section can most easily be accommodated.

Flexural strength of the coped beam is not addressed in the Specification directly but is covered in Part 9 of the Manual. The moment in the coped beam is taken as the shear force times the eccentricity from the face of the support to the edge of the cope, taken as 4.5 in. in this example.

$$M_u = 83.0(4.5) = 374 \text{ in.-kips}$$

To determine the flexural strength of the coped beam, the net section modulus is taken from Manual Table 9-2. With the depth of the cope,

$$d_c = 2.0 \text{ in.}, \; S_{net} = 23.4 \text{ in.}^3$$

For flexural rupture, $\phi = 0.75$ and

$$M_n = F_u S_{net} = 65(23.4) = 1520 \text{ in.-kips}$$

$$\phi M_n = 0.75(1520) = 1140 \text{ in.-kips} > 374 \text{ in.-kips}$$

For flexural local buckling, $\phi = 0.9$ and

$$M_n = F_{cr} S_{net}$$

The critical stress is given in Manual Part 9 as

$$F_{cr} = 26{,}210 \left(\frac{t_w}{h_o} \right)^2 fk$$

For this example

$$f = \frac{2c}{d} = \frac{2(4)}{18} = 0.444$$

$$k = 2.2 \left(\frac{h_o}{c} \right)^{1.65} = 2.2 \left(\frac{16.0}{4} \right)^{1.65} = 21.7$$

and

$$F_{cr} = 26{,}210 \left(\frac{0.355}{16.0} \right)^2 (0.444)(21.7) = 124 \text{ ksi} > F_y = 50 \text{ ksi}$$

Thus

$$M_n = 50(23.4) = 1170 \text{ in.-kips}$$

$$\phi M_n = 0.9(1170) = 1050 \text{ in.-kips} > 374 \text{ in.-kips}$$

So the coped beam has sufficient flexural strength.

Step 10:   Check bolt bearing on the A36 angle.

Assume a $5/16$-in. angle and maintain the 1.25-in. end distance as shown in Figure 11.2. The other bolts are spaced as originally shown at 3.0 in.

For the top bolt

$$L_c = 1.25 - \frac{1}{2}(7/8 + 1/16) = 0.781 < 2(7/8) = 1.75$$

Again, tear-out controls and the nominal bolt strength is

$$R_n = 1.2(0.781)\left(\frac{5}{16}\right)(58) = 17.0\,\text{kips}$$

For the second and third bolt

$$L_c = 3.0 - (7/8 + 1/16) = 2.06 > 2(7/8) = 1.75$$

and bearing controls, giving a nominal bolt strength of

$$R_n = 2.4(7/8)\left(\frac{5}{16}\right)(58) = 38.1\,\text{kips}$$

Thus, for the three-bolt connection, the design strength is

$$\phi R_n = 0.75(17.0 + 2(38.1)) = 69.9\,\text{kips} > \frac{83.0}{2} = 41.5\,\text{kips}$$

Therefore, the three-bolt connection in the angles is more than adequate.

Step 11:   Check the angles for shear rupture.

The net area of the angle on the vertical shear plane is

$$A_{nv} = (8.5 - 3(7/8 + 1/8))\left(\frac{5}{16}\right) = 1.72\,\text{in.}^2$$

and the design strength is

$$\phi V_n = (0.75)(0.6F_u A_{nv}) = (0.75)(0.6(58)(1.72)) = 44.9 > 41.5\,\text{kips}$$

So the angle is adequate for shear rupture.

Step 12:   Check the angles for shear yield.

The gross area of the angle on the vertical shear plane is

$$A_{gv} = (8.5)\left(\frac{5}{16}\right) = 2.66\,\text{in.}^2$$

and the design strength is

$$\phi V_n = (1.0)(0.6F_y A_g) = (1.0)(0.6(36)(2.66)) = 57.5 > 41.5\,\text{kips}$$

So the angle is also adequate for shear yield.

Step 13:   Check the angles for block shear.

The equations for block shear in the angle are the same as those for the web and as presented in Section 10.10.4.

First calculate the required areas,

$$A_{nt} = \left(1.0 - \frac{1}{2}(7/8 + 1/8)\right)\left(\frac{5}{16}\right) = 0.156\,\text{in.}^2$$
$$A_{gv} = 7.25\left(\frac{5}{16}\right) = 2.27\,\text{in.}^2$$
$$A_{nv} = (7.25 - 2.5(7/8 + 1/8))\left(\frac{5}{16}\right) = 1.48\,\text{in.}^2$$

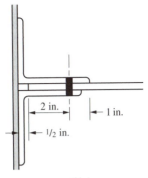

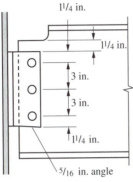

Figure 11.3 Final Connection Design for Example 11.1.

Consider shear yield and shear rupture and select the least nominal strength, thus

$$0.6 F_y A_{gv} = 0.6(36)(2.27) = 49.0 \text{ kips}$$

$$0.6 F_u A_{nv} = 0.6(58)(1.48) = 51.5 \text{ kips}$$

Selecting the shear yield term and combining it with the tension rupture term gives a connection block shear design strength, again $U_{bs} = 1.0$ for this case of uniform tensile stress distribution, of

$$\phi R_n = 0.75(49.0 + 1.0(58)(0.156)) = 43.5 > 41.5 \text{ kips}$$

Step 14:   Present the final connection design.

> The three-bolt connection, revised as shown in Figure 11.3, is adequate to carry the imposed load of 83.0 kips.

EXAMPLE 11.1b
Bolted-Bolted
Double-Angle Shear
Connection by ASD

GOAL: Design a bolted-bolted double-angle shear connection for an W18×50 beam.

GIVEN: The W18×50 beam must provide a required strength, $R_a = 55.0$ kips. The beam is A992 and the angles are A36. The beam flange is coped 2 in. Use $\frac{7}{8}$ in. A325-N bolts in standard holes in the legs on the beam web and short slots on the outstanding legs. The basic starting geometry is given in Figure 11.2.

shear force times the eccentricity from the face of the support to the edge of the cope, taken as 4.5 in. in this example.

$$M_a = 55.0(4.5) = 248 \text{ in.-kips}$$

To determine the flexural strength of the coped beam, the net section modulus is taken from Manual Table 9-2. With the depth of the cope,

$$d_c = 2.0 \text{ in., } S_{net} = 23.4 \text{ in.}^3$$

For flexural rupture, $\Omega = 2.00$ and

$$M_n = F_u S_{net} = 65(23.4) = 1520 \text{ in.-kips}$$

$$\frac{M_n}{\Omega} = \left(\frac{1520}{2.00}\right) = 760 \text{ in.-kips} > 248 \text{ in.-kips}$$

For flexural local buckling, $\Omega = 1.67$ and

$$M_n = F_{cr} S_{net}$$

The critical stress is given in Manual Part 9 as

$$F_{cr} = 26{,}210\left(\frac{t_w}{h_o}\right)^2 fk$$

For this example

$$f = \frac{2c}{d} = \frac{2(4)}{18} = 0.444$$

$$k = 2.2\left(\frac{h_o}{c}\right)^{1.65} = 2.2\left(\frac{16.0}{4}\right)^{1.65} = 21.7$$

and

$$F_{cr} = 26{,}210\left(\frac{0.355}{16.0}\right)^2 (0.444)(21.7) = 124 \text{ ksi} > F_y = 50 \text{ ksi}$$

Thus

$$M_n = 50(23.4) = 1170 \text{ in.-kips}$$

$$\frac{M_n}{\Omega} = \left(\frac{1170}{1.67}\right) = 701 \text{ in.-kips} > 248 \text{ in.-kips}$$

So the coped beam has sufficient flexural strength.

Step 10:   Check bolt bearing on the A36 angle.

Assume a 5/16-in. angle and maintain the 1.25-in. end distance as shown in Figure 11.2. The other bolts are spaced as originally shown at 3.0 in. For the top bolt

$$L_c = 1.25 - \frac{1}{2}(7/8 + 1/16) = 0.781 < 2(7/8) = 1.75$$

Again, tear-out controls and the nominal bolt strength is

$$R_n = 1.2(0.781)\left(\frac{5}{16}\right)(58) = 17.0 \text{ kips}$$

For the second and third bolt

$$L_c = 3.0 - (7/8 + 1/16) = 2.06 > 2(7/8) = 1.75$$

and bearing controls, giving a nominal bolt strength of

$$R_n = 2.4(7/8)\left(\frac{5}{16}\right)(58) = 38.1 \text{ kips}$$

Thus, for the three-bolt connection, the allowable strength is

$$\frac{R_n}{\Omega} = \frac{(17.0 + 2(38.1))}{2.00} = 46.6 \text{ kips} > \frac{55.0}{2} = 27.5 \text{ kips}$$

Therefore, the three-bolt connection in the angles is more than adequate.

Step 11:   Check the angles for shear rupture.
    The net area of the angle on the vertical shear plane is

$$A_{nv} = (8.5 - 3(7/8 + 1/8))\left(\frac{5}{16}\right) = 1.72 \text{ in.}^2$$

and the allowable strength is

$$\frac{V_n}{\Omega} = \frac{(0.6 F_u A_{nv})}{2.00} = \frac{(0.6(58)(1.72))}{2.00} = 29.9 > 27.5 \text{ kips}$$

So the angle is adequate for shear rupture.

Step 12:   Check the angles for shear yield.
    The gross area of the angle on the vertical shear plane is

$$A_{gv} = (8.5)\left(\frac{5}{16}\right) = 2.66 \text{ in.}^2$$

and the allowable strength is

$$\frac{V_n}{\Omega} = \frac{(0.6 F_y A_g)}{1.50} = \frac{(0.6(36)(2.66))}{1.50} = 38.3 > 27.5 \text{ kips}$$

So the angle is also adequate for shear yield.

Step 13:   Check the angles for block shear.
    The equations for block shear in the angle are the same as those for the web and as presented in Section 10.10.4.
    First calculate the required areas

$$A_{nt} = \left(1.0 - \frac{1}{2}(7/8 + 1/8)\right)\left(\frac{5}{16}\right) = 0.156 \text{ in.}^2$$

$$A_{gv} = 7.25\left(\frac{5}{16}\right) = 2.27 \text{ in.}^2$$

$$A_{nv} = (7.25 - 2.5(7/8 + 1/8))\left(\frac{5}{16}\right) = 1.48 \text{ in.}^2$$

Consider shear yield and shear rupture and select the least nominal strength, thus

$$0.6 F_y A_{gv} = 0.6(36)(2.27) = 49.0 \text{ kips}$$

$$0.6 F_u A_{nv} = 0.6(58)(1.48) = 51.5 \text{ kips}$$

Selecting the shear yield term and combining it with the tension rupture term gives a connection block shear allowable strength, again $U_{bs} = 1.0$ for this case of uniform tensile stress distribution, of

$$\frac{R_n}{\Omega} = \frac{(49.0 + 1.0(58)(0.156))}{2.00} = 29.0 > 27.5 \text{ kips}$$

Step 14:   Present the final connection design.

> The three-bolt connection, revised as shown in Figure 11.3, is adequate to carry the imposed load of 55.0 kips.

11.4   DOUBLE-ANGLE CONNECTIONS: WELDED-BOLTED

The double-angle shear connection can also be constructed by combining welding and bolting. In this case the angles are welded to the beam web, as shown in Figure 11.4. The limit states to be considered are:

1. Bolts
   a. Shear rupture
2. Weld
   a. Rupture
3. Beam
   a. Shear yielding of the web
   b. Block shear on coped beam web
   c. Coped beam flexural strength
   d. Web strength at the weld
4. Angles
   a. Bolt bearing on angles
   b. Shear rupture
   c. Shear yield
   d. Block shear

The limit states that were not considered for the bolted-bolted connection from Section 11.3 are those associated with the weld. These include block shear of the beam web as a result of the welded connection; weld rupture, which is influenced by the eccentricity of the force on the weld group; and the strength of the beam web at the weld.

Block shear for a welded connection differs only slightly from block shear for a bolted connection. The difference is in the lack of holes to be deducted when determining the net

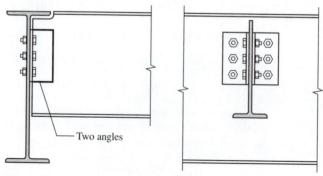

Two angles

Figure 11.4 Welded-Bolted Double-Angle Connection.

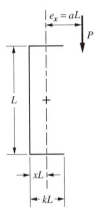

Figure 11.5 C-Shaped Weld Group.

area. Thus, the net shear area and gross shear area are the same. As a result, yielding is the controlling shear term in the block shear equation for this type of welded connection.

Weld rupture is a much more complex limit state to incorporate in this type of connection design. Chapter 10 discussed the strength of a weld loaded at its centroid and at any angle. The welds in the double-angle connection are loaded parallel to the length on one side of the angle and perpendicular to their lengths on the other two sides. Unfortunately, these welds are not loaded through their centroid so the simplified approach to combining them, previously shown in Chapter 10, cannot be used. The Manual uses the instantaneous center of rotation method to determine weld strength in cases like this. This approach accounts for the loading at an angle to the weld as well as the eccentricity of the load to the weld group. Figure 11.5 shows a C-shaped weld with the geometric variables labeled. In the typical connection design, the geometry can be set and Manual Table 8-8 can be used to determine the weld group strength. The application of this table is shown in Example 11.2.

The beam web strength at the weld is also a bit difficult to calculate. The usual approach is to determine the total strength of the weld and then proportion that force to the web based on a one-inch length of web and one-inch length of weld. This, too, is illustrated in Example 11.2.

EXAMPLE 11.2
Welded-Bolted
Double-Angle Shear
Connection

GOAL: Determine the available strength of the welded-bolted connection shown in Figure 11.6a.

GIVEN: Determine the design strength and allowable strength of the connection shown in Figure 11.6a for the three new limit states discussed for the welded-bolted double-angle connection.

SOLUTION

Step 1:   Determine the nominal strength for the limit state of block shear.
  For the tension area, the length is found by taking the 3-in. angle leg and subtracting the $\frac{1}{2}$-in. setback and the $\frac{1}{4}$-in. potential beam tolerance. Thus,

$$A_{nt} = 2.25(0.255) = 0.574 \text{ in.}^2$$

For the gross shear area, the angle is 8.5 in. long and set down from the cope $\frac{1}{2}$ in. Thus

$$A_{gv} = 9.0(0.255) = 2.30 \text{ in.}^2$$

Therefore, with $U_{bs} = 1.0$, the nominal block shear strength is

$$R_n = 0.6(50)(2.30) + (1.0)(65)(0.574) = 106 \text{ kips}$$

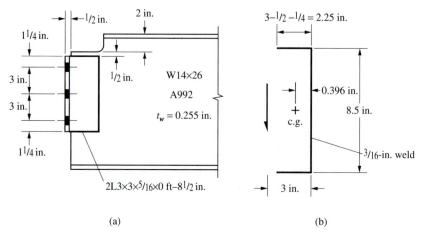

(a)

(b)

Figure 11.6 Connection for Example 11.2.

Step 2: For LRFD, the design strength is

$$\phi R_n = 0.75(106) = 79.5 \text{ kips}$$

Step 2: For ASD, the allowable strength is

$$\frac{R_n}{\Omega} = \frac{(106)}{2.00} = 53.0 \text{ kips}$$

Step 3: Determine the nominal strength for the limit state of weld rupture.

The geometry of the weld is given in Figure 11.6b. The angle is 8.5 in. long so the weld length is $L = 8.5$ in. The leg of the angle is 3.0 in. and the weld length is $kL = 3.0 - \frac{1}{2}$ in. setback $- \frac{1}{4}$ in. under run $= 2.25$ in. Thus

$$k = \frac{2.25}{8.5} = 0.265$$

From Manual Table 8-8, the location of the weld centroid can be determined. Enter the table with $k = 0.265$ and interpolate for x from the values at the bottom of the table, which yields $x = 0.0466$. With this, the weld centroid is determined as

$$xL = 0.0466(8.5) = 0.396 \text{ in.}$$

The eccentricity of the force is then determined as

$$a = \frac{e_x}{L} = \frac{(3.0 - 0.396)}{8.5} = 0.306$$

Using this value for a and the previously determined value for k, the value of C can be determined from the table as $C = 2.59$. As indicated in the table, the nominal strength of the weld group is then

$$R_n = CC_1DL$$

where C has been determined above. C_1 represents the electrode strength and is 1.0 for the E70XX electrodes used here. D is the number of sixteenths-of-an-inch in the fillet weld size and L is the defined length of the weld group. Thus, for this weld

$$R_n = (2.59)(1.0)(3)(8.5) = 66.0 \text{ kips for each angle}$$

Step 4:   For LRFD the design strength of a single angle is

$$\phi R_n = 0.75(66.0) = 49.5 \text{ kips for each angle}$$

Or, for the double angle connection

$$\boxed{\phi R_n = 2(49.5) = 99.0 \text{ kips}}$$

Step 4:   For ASD the allowable strength of a single angle is

$$\frac{R_n}{\Omega} = \frac{(66.0)}{2.00} = 33.0 \text{ kips for each angle}$$

Or, for the double angle connection

$$\boxed{\frac{R_n}{\Omega} = 2(33.0) = 66.0 \text{ kips}}$$

Step 5:   For LRFD determine the design rupture strength for the beam web at the weld.
The design rupture strength of the $3/16$-in. weld of unit length on both sides of the web, using the weld design strength determined in Chapter 10, is $2(3)(1.392) = 8.35$ kips. Using the strength determined above, the effective length of the weld is $99.0/8.35 = 11.9$ in.

The design rupture strength of a unit length of the beam web is

$$\phi(0.6F_u t_w) = 0.75(0.6(65)(0.255)) = 7.46 \text{ kips}$$

Therefore, the beam web design rupture strength at the weld is

$$\boxed{\phi R_n = (11.9)(7.46) = 88.8 \text{ kips}}$$

Step 5:   For ASD, determine the allowable rupture strength for the beam web at the weld.

The allowable rupture strength of the $^{3}/_{16}$-in. weld of unit length on both sides of the web, using the weld design strength determined in Chapter 10, is $2(3)(0.928) = 5.57$ kips. Using the strength determined above, the effective length of the weld is $66.0/5.57 = 11.9$ in.

The allowable rupture strength of a unit length of the beam web is

$$\frac{(0.6F_u t_w)}{\Omega} = \frac{(0.6(65)(0.255))}{2.00} = 4.97 \text{ kips}$$

Therefore, the beam web allowable rupture strength at the weld is

$$\phi R_n = (11.9)(4.97) = 59.1 \text{ kips}$$

Step 6:   Determine the controlling limit state for the three limit states considered in this example.

The connection is limited by the limit state of block shear.

Step 7:   For LRFD, the design strength is

$$\phi R_n = 79.5 \text{ kips}$$

Step 7:   For ASD, the allowable strength is

$$\frac{R_n}{\Omega} = 53.0 \text{ kips.}$$

11.5   DOUBLE-ANGLE CONNECTIONS: BOLTED-WELDED

This connection is shown in Figure 11.7 where the angles are bolted to the beam web and welded to the supporting member. The beam has a cope on the tension flange to permit the beam to be inserted in the space between the double angles like a knife being inserted into

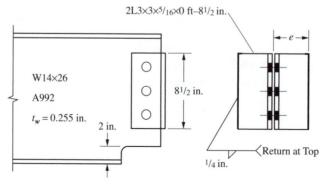

Figure 11.7 Double-Angle Bolted-Welded Connection for Example 11.3.

its sheath. This connection might be used on a beam-to-column flange connection but it would not be used as a beam-to-girder connection because of the interference of the girder flange.

For this connection, the new limit state of weld strength on the outstanding legs of the angles must be considered. In addition, consideration must be given to the flexural strength of the coped beam with the tension flange removed.

The strength of the weld on the outstanding leg of the angle is determined by an elastic method that assumes a uniform shear stress on the weld and a linearly varying tension stress over the lower 5/6th of the angle. The resulting weld stress is then determined by taking the square root of the sum of the squares of the tension and shear stresses. Part 10 of the Manual gives the resulting weld strength as

$$\phi R_n = 2 \left[\frac{1.392DL}{\sqrt{1 + \frac{12.96e^2}{L^2}}} \right] \text{ (LRFD) and } \frac{R_n}{\Omega} = 2 \left[\frac{0.928DL}{\sqrt{1 + \frac{12.96e^2}{L^2}}} \right] \text{ (ASD)}$$

where L is the length of the angle and e is the width of the outstanding leg.

EXAMPLE 11.3 **Bolted-Welded** **Double-Angle Shear** **Connection**	**GOAL:** Determine the available strength of the welds for a bolted-welded double angle connection.

GIVEN: The bolted-welded double angle connection is shown in Figure 11.7. Assume $1/4$-in. welds with a $1/2$-in. return on top as shown.

SOLUTION

Step 1: For LRFD, determine the design strength.
With the given information, $D = 4$, $L = 8.5$ in., and $e = 3.0$ in.; thus

$$\phi R_n = 2 \left[\frac{1.392DL}{\sqrt{1 + \frac{12.96e^2}{L^2}}} \right] = 2 \left[\frac{1.392(4)(8.5)}{\sqrt{1 + \frac{12.96(3)^2}{(8.5)^2}}} \right]$$

$$\boxed{\phi R_n = 58.5 \text{ kips}}$$

Step 1: For ASD, determine the allowable strength.
With the given information, $D = 4$, $L = 8.5$ in., and $e = 3.0$ in.; thus

$$\frac{R_n}{\Omega} = 2 \left[\frac{0.928DL}{\sqrt{1 + \frac{12.96e^2}{L^2}}} \right] = 2 \left[\frac{0.928(4)(8.5)}{\sqrt{1 + \frac{12.96(3)^2}{(8.5)^2}}} \right]$$

$$\boxed{\frac{R_n}{\Omega} = 39.0 \text{ kips}}$$

11.6   DOUBLE ANGLE CONNECTIONS: WELDED-WELDED

The double-angle welded-welded connection is not a particularly common connection because it requires field welding. If it is desirable for a particular situation, however, the limit states are those that have already been discussed. The procedures are the same and all potential limit states must be checked.

11.7   SINGLE-ANGLE CONNECTIONS

The single-angle connections shown in Figure 11.1b illustrate a bolted-bolted connection and a bolted-welded connection. In both cases, the connection is shown on a beam-to-girder connection. This is a particularly efficient connection because it eliminates erection problems when transverse beams frame into a girder at the same point on opposite sides of the girder web. It is also efficient because it has fewer parts than the double-angle connection. This connection is growing in popularity with both fabricators and erectors. It aids in erection efficiency because the beam can be installed from one side with the angle pre-attached to the girder. The disadvantages of this connection are that the components, such as angles, bolts, and welds, are larger than for the double-angle connection. For this connection, the bolts in the beam web are in single shear and, if the controlling limit state were bolt shear, it would require twice as many bolts than if it were a double-angle connection. Because all of the beam force must pass through only one angle, the angle likely needs to be larger. Greater weld size and weld length is also required. However, the single angle is still the best choice in many situations, particularly when limit states like block shear might control the strength of the connection.

The single-angle connection easily behaves as a simple shear connection, as it is modeled in analysis, because it is more flexible than the previously considered double-angle connections. Because of this increased flexibility, however, this connection is not recommended for laterally unsupported beams that rely on their end connections for lateral stability.

The limit states to be checked for the single-angle connection are the same as those for the double-angle connection with some modifications and additions. The major modifications have to do with the eccentricities induced in the connecting elements. For the supported beam, as long as there is only one row of bolts, no eccentricities are considered and this portion of the connection is treated as for the double-angle connections. For the outstanding leg, the bolts or welds must be designed to account for the connection eccentricity. This eccentricity also adds the limit states of flexural yielding and flexural rupture for the outstanding leg of the angle.

Figures 11.8a and b show how the eccentricity is measured for a bolted and a welded outstanding leg. Note that the eccentricity is measured, in both cases, from the center line of the supported beam web. Figures 11.8c and d illustrate, with a bold line, the location and cross section for the moment in the angle for the limit states of flexural yielding (Figure 11.8c) and flexural rupture (Figure 11.8d).

The limit state of flexural yielding is calculated based on the plastic moment of the element. Thus, the plastic section modulus for the angle leg without holes, $Z = t_p L^2/4$, is used and the nominal moment strength is given as

$$M_n = R_n e_a = F_y Z.$$

For the flexural yielding limit state, $\phi = 0.9$ and $\Omega = 1.67$.

For the limit state of flexural rupture, the plastic section modulus of the net section is needed. This can be either determined by calculation or obtained from Manual Table 15-2.

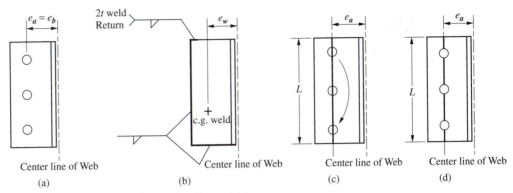

Figure 11.8 Single-Angle Connection Eccentricities.

The resulting nominal moment strength is then

$$M_n = R_n e_a = F_u Z_{net}$$

For the flexural rupture limit state, $\phi = 0.75$ and $\Omega = 2.00$.

To account for the eccentricity of the load on the bolt group, an equivalent number of bolts must be determined. This can be accomplished using Manual Table 7-7 and is illustrated in Example 11.4. To account for the eccentricity on the weld group, Manual Table 8-11 can be used. This table is for an L-shaped weld group with the welds on only two sides. The top of the leg is kept free to insure sufficient rotation capacity. Application of this table is also demonstrated in Example 11.4.

EXAMPLE 11.4a
Single-Angle Shear
Connection by LRFD

GOAL: Determine the design strength of a bolted-bolted and a bolted-welded single-angle connection.

GIVEN: A single-angle connection is shown in Figure 11.9 for the bolted outstanding leg case (Figure 11.9b) and the welded outstanding leg case (Figure 11.9c). The angle is A36 steel, $3\frac{1}{2} \times 3\frac{1}{2} \times \frac{3}{8} \times 12$ in. The bolts are $\frac{3}{4}$-in. A325N, the weld is $\frac{3}{16}$ in. E70XX, and the beam is a W16×31, A992 steel.

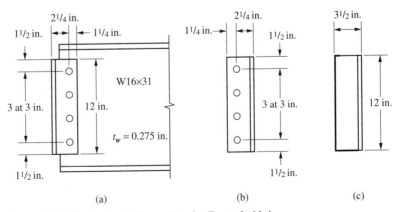

Figure 11.9 Single-Angle Connection for Example 11.4.

SOLUTION

Part (a)   Consider first the connection to the supported beam.

Step 1:   Determine the bolt shear rupture strength.

From Manual Table 7-1, $\phi r_n = 15.9$ kips. Therefore, for the four bolts, the design shear strength is

$$\phi R_n = 4(15.9) = 63.6 \text{ kips}$$

Step 2:   Determine the bolt bearing strength on the angle.

The bottom bolt is 1.5 in. from the bottom of the angle with the remaining bolts spaced at 3.0 in. Determine the clear distances for each of the bolts.

For the bottom bolt

$$L_c = 1.5 - \frac{1}{2}(3/4 + 1/16) = 1.09 < 2(3/4) = 1.5$$

Thus, tear out controls and the nominal bolt strength is

$$R_n = 1.2(1.09)(0.375)(58) = 28.4 \text{ kips}$$

For the other bolts

$$L_c = 3.0 - (3/4 + 1/16) = 2.19 > 2(3/4) = 1.5$$

Therefore, bearing controls, and the nominal bolt strength is

$$R_n = 2.4(3/4)(0.375)(58) = 39.2 \text{ kips}$$

Thus, for the four-bolt connection, the design strength is

$$\phi R_n = 0.75(28.4 + 3(39.2)) = 110 \text{ kips}$$

Step 3:   Determine the bolt-bearing strength on the beam web.

Because the clear distance is greater than two times the bolt diameter

$$R_n = 2.4(3/4)(0.275)(65) = 32.2 \text{ kips}$$

and

$$\phi R_n = 0.75(4(32.2)) = 96.6 \text{ kips}$$

Step 4:   Determine the shear yield strength of the angle.

$$A_g = 12.0(0.375) = 4.50 \text{ in.}^2$$

$$\phi V_n = 1.0(0.6(36))(4.5) = 97.2 \text{ kips}$$

Step 5:   Determine the shear rupture strength of the angle.

$$A_{nv} = (12.0 - 4(3/4 + 1/8))(0.375) = 3.19 \text{ in.}^2$$

$$\phi V_n = 0.75(0.6(58))(3.19) = 83.3 \text{ kips}$$

Step 6:   Determine the block shear strength of the angle.

First calculate the required areas

$$A_{nt} = \left(1.25 - \frac{1}{2}(3/4 + 1/8)\right)(0.375) = 0.305 \text{ in.}^2$$

$$A_{gv} = 10.5(0.375) = 3.94 \text{ in.}^2$$

$$A_{nv} = (10.5 - 3.5(3/4 + 1/8))(0.375) = 2.79 \text{ in.}^2$$

Consider shear yield and shear rupture and select the least strength, thus

$$0.6F_y A_{gv} = 0.6(36)(3.94) = 85.1 \text{ kips}$$

$$0.6F_u A_{nv} = 0.6(58)(2.79) = 97.1 \text{ kips}$$

Selecting the shear rupture term and combining it with the tension rupture term, with $U_{bs} = 1.0$ for this case of uniform tensile stress distribution, gives

$$\phi R_n = 0.75(85.1 + 1.0(58)(0.305)) = 77.1 \text{ kips}$$

Step 7:   Determine the design strength of the leg attached to the supported member.
The design strength is controlled by bolt shear where

$$\boxed{\phi R_n = 63.6 \text{ kips}}$$

Part (b)   Consider the bolted outstanding leg.

Step 8:   Check the eccentric shear using Manual Table 7-7 to account for the eccentricity on the bolt group.
The eccentricity of the load on the line of bolts is the bolt distance from the angle heel plus one half of the beam web, thus

$$e_x = 2.25 + \frac{0.275}{2} = 2.39 \text{ in.}$$

For the four-bolt connection with bolt spacing of 3.0 in., Manual Table 7-7 gives the effective number of bolts as $C = 3.12$. Therefore

$$\phi R_n = 3.12(15.9) = 49.6 \text{ kips}$$

The strength of the bolts in the outstanding leg will be less than that in the leg on the supported beam because the outstanding leg must accommodate an eccentricity that is not present in the leg on the beam. Therefore, there really was no reason to have checked the bolt shear on the beam, except that it will be needed when the welded connection is checked.

Step 9:   Determine the flexural yielding strength of the outstanding leg.
The plastic section modulus is determined for the rectangle formed by the length and thickness of the angle, and the nominal moment strength is determined by multiplying the plastic section modulus by the yield stress, thus

$$Z = \frac{(0.375)(12^2)}{4} = 13.5 \text{ in.}^3$$

and

$$M_n = 36(13.5) = 486 \text{ in.-kips}$$

Because the moment is the shear force times the eccentricity

$$\phi R_n = \frac{\phi M_n}{e} = \frac{0.9(486)}{2.39} = 183 \text{ kips}$$

Step 10:   Determine the flexural rupture strength of the outstanding leg.
The net plastic section modulus is determined for the rectangle less the holes. Although this can readily be calculated, it can also be obtained from Manual Table 15-2 where $Z_{net} = 9.56 \text{ in.}^3$. Thus

$$M_n = 58(9.56) = 554 \text{ in.-kips}$$

and

$$\phi R_n = \frac{\phi M_n}{e} = \frac{0.75(554)}{2.39} = 174 \text{ kips}$$

Step 11:   Determine the controlling limit state strength for the bolted outstanding legs.

For the bolted outstanding leg, the strength is controlled by the eccentric shear of the bolts where

$$\phi R_n = 49.6 \text{ kips}$$

Because this is less than the value for the leg attached to the beam, this is the strength of the bolted-bolted single-angle connection.

Part (c)   Consider the welded outstanding leg.

Step 12:   Determine the eccentric weld rupture strength.

Manual Table 8-10 will be used to determine the eccentric weld rupture strength. The weld for the single-angle connection is applied to the bottom edge of the angle, not the top. This insures that the angle is sufficiently flexible to behave as a simple connection as modeled. Manual Table 8-10 shows this weld on the top but this does not impact the use of the table because the geometry is the same whether the horizontal weld is at the top or bottom of the connection.

Based on the dimensions given in Figure 11.9c, $L = 12.0$, $kL = 3.5$, thus $k = 0.292$. The weld is a $\frac{3}{16}$-in. weld with E70 electrodes. From the table, interpolating between $k = 0.2$ and 0.3, yields $x = 0.0336$. Therefore, the eccentricity is

$$e_x = kl + \frac{t_w}{2} - xl$$
$$e_x = 3.5 + \frac{0.275}{2} - 0.0336(12.0) = 3.23 \text{ in.}$$

and

$$a = \frac{e_x}{L} = \frac{3.23}{12.0} = 0.269$$

With a double interpolation between $k = 0.2$ and 0.3 and $a = 0.25$ and 0.30, the coefficient C is determined as

$$C = 2.17$$

Therefore, the nominal weld strength is

$$R_n = CC_1DL = 2.17(1.0)(3)(12.0) = 78.1 \text{ kips}$$

And the design strength is

$$\phi R_n = 0.75(78.1) = 58.6 \text{ kips}$$

Step 13:   Determine the design strength for the limit state of flexural yielding.

For the limit state of flexural yielding of the angle the strength is determined as was shown for the bolted outstanding leg, thus

$$\phi R_n = 183 \text{ kips}$$

Step 14:   Determine the controlling limit state's strength for the welded outstanding legs.

For the welded outstanding leg, the design strength is controlled by eccentric shear on the weld where

$$\phi R_n = 58.6 \text{ kips}$$

Because this is less than the value for the leg attached to the beam, this is the design strength of the bolted-welded single-angle connection.

EXAMPLE 11.4b
Single-Angle Shear
Connection by ASD

GOAL: Determine the allowable strength of a bolted-bolted and a bolted-welded single-angle connection.

GIVEN: A single-angle connection is shown in Figure 11.9 for the bolted outstanding leg case (Figure 11.9b) and the welded outstanding leg case (Figure 11.9c). The angle is A36 steel, $3\frac{1}{2} \times 3\frac{1}{2} \times \frac{3}{8} \times 12$ in. The bolts are $\frac{3}{4}$-in. A325N, the weld is $\frac{3}{16}$ in. E70XX, and the beam is a W16×31, A992 steel.

SOLUTION

Part (a) Consider first the connection to the supported beam.

Step 1: Determine the bolt shear rupture strength.
From Manual Table 7-1, $\frac{r_n}{\Omega} = 10.6$ kips. Therefore for the four bolts, the allowable shear strength is

$$\frac{R_n}{\Omega} = 4(10.6) = 42.4 \text{ kips}$$

Step 2: Determine the bolt bearing strength on the angle.
The bottom bolt is 1.5 in. from the bottom of the angle with the remaining bolts spaced at 3.0 in. Determine the clear distances for each of these bolts.
For the bottom bolt

$$L_c = 1.5 - \frac{1}{2}(3/4 + 1/16) = 1.09 < 2(3/4) = 1.5$$

Thus, tear out controls and the nominal bolt strength is

$$R_n = 1.2(1.09)(0.375)(58) = 28.4 \text{ kips}$$

For the other bolts

$$L_c = 3.0 - (3/4 + 1/16) = 2.19 > 2(3/4) = 1.5$$

Therefore, bearing controls, and the nominal bolt strength is

$$R_n = 2.4(3/4)(0.375)(58) = 39.2 \text{ kips}$$

Thus, for the four-bolt connection, the allowable strength is

$$\frac{R_n}{\Omega} = \frac{(28.4 + 3(39.2))}{2.00} = 73.0 \text{ kips}$$

Step 3: Determine the bolt-bearing strength on the beam web.
Because the clear distance is greater than two times the bolt diameter

$$R_n = 2.4(3/4)(0.275)(65) = 32.2 \text{ kips}$$

and

$$\frac{R_n}{\Omega} = \frac{(4(32.2))}{2.00} = 64.4 \text{ kips}$$

Step 4: Determine the shear yield strength of the angle.

$$A_g = 12.0(0.375) = 4.50 \text{ in.}^2$$

$$\frac{V_n}{\Omega} = \frac{(0.6(36))(4.5)}{1.5} = 64.8 \text{ kips}$$

Step 5:   Determine the shear rupture strength of the angle.

$$A_{nv} = (12.0 - 4(3/4 + 1/8))(0.375) = 3.19 \text{ in.}^2$$

$$\frac{V_n}{\Omega} = \frac{(0.6(58))(3.19)}{2.00} = 55.5 \text{ kips}$$

Step 6:   Determine the block shear strength of the angle.
First calculate the required areas

$$A_{nt} = \left(1.25 - \frac{1}{2}(3/4 + 1/8)\right)(0.375) = 0.305 \text{ in.}^2$$

$$A_{gv} = 10.5(0.375) = 3.94 \text{ in.}^2$$

$$A_{nv} = (10.5 - 3.5(3/4 + 1/8))(0.375) = 2.79 \text{ in.}^2$$

Consider shear yield and shear rupture and select the least strength, thus

$$0.6F_y A_{gv} = 0.6(36)(3.94) = 85.1 \text{ kips}$$

$$0.6F_u A_{nv} = 0.6(58)(2.79) = 97.1 \text{ kips}$$

Selecting the shear rupture term and combining it with the tension rupture term, with $U_{bs} = 1.0$ for this case of uniform tensile stress distribution, gives

$$\frac{R_n}{\Omega} = \frac{(85.1 + 1.0(58)(0.305))}{2.00} = 51.4 \text{ kips}$$

Step 7:   Determine the allowable strength of the leg attached to the supported member.
The allowable strength is controlled by bolt shear where

$$\boxed{\frac{R_n}{\Omega} = 42.4 \text{ kips}}$$

Part (b)   Consider the bolted outstanding leg.

Step 8:   Check the eccentric shear using Manual Table 7-7 to account for the eccentricity on the bolt group.
The eccentricity of the load on the line of bolts is the bolt distance from the angle heel plus one half of the beam web, thus

$$e_x = 2.25 + \frac{0.275}{2} = 2.39 \text{ in.}$$

For the four-bolt connection with bolt spacing of 3.0 in., Manual Table 7-7 gives the effective number of bolts as $C = 3.12$. Therefore

$$\frac{R_n}{\Omega} = 3.12(10.6) = 33.1 \text{ kips}$$

The strength of the bolts in the outstanding leg will be less than that in the leg on the supported beam because the outstanding leg must accommodate an eccentricity that is not present in the leg on the beam. Therefore, there really was no reason to have checked the bolt shear on the beam, except that it will be needed when the welded connection is checked.

Step 9:   Determine the flexural yielding strength of the outstanding leg.
The plastic section modulus is determined for the rectangle formed by the length and thickness of the angle, and the nominal moment strength is determined by multiplying the plastic section modulus by the yield stress, thus

$$Z = \frac{(0.375)(12^2)}{4} = 13.5 \text{ in.}^3$$

and

$$M_n = 36(13.5) = 486 \text{ in.-kips}$$

Because the moment is the shear force times the eccentricity

$$\frac{R_n}{\Omega} = \frac{M_n}{\Omega e} = \frac{(486)}{1.67(2.39)} = 122 \text{ kips}$$

Step 10: Determine the flexural rupture strength of the outstanding leg.

The net plastic section modulus is determined for the rectangle less the holes. Although this can readily be calculated, it can also be obtained from Manual Table 15-2 where $Z_{net} = 9.56 \text{ in.}^3$. Thus

$$M_n = 58(9.56) = 554 \text{ in.-kips}$$

and

$$\frac{R_n}{\Omega} = \frac{M_n}{\Omega e} = \frac{(554)}{2.00(2.39)} = 116 \text{ kips}$$

Step 11: Determine the controlling limit state strength for the bolted outstanding legs.

For the bolted outstanding leg, the strength is controlled by the eccentric shear of the bolts where

$$\boxed{\frac{R_n}{\Omega} = 33.1 \text{ kips}}$$

Because this is less than the value for the leg attached to the beam, this is the strength of the bolted-bolted single-angle connection.

Part (c)   Consider the welded outstanding leg.

Step 12: Determine the eccentric weld rupture strength.

Manual Table 8-10 will be used to determine the eccentric weld rupture strength. The weld for the single-angle connection is applied to the bottom edge of the angle, not the top. This insures that the angle is sufficiently flexible to behave as a simple connection as modeled. Manual Table 8-10 shows this weld on the top but that does not impact the use of the table because the geometry is the same whether the horizontal weld is at the top or bottom of the connection.

Based on the dimensions given in Figure 11.9c, $L = 12.0$, $kL = 3.5$, thus $k = 0.292$. The weld is a $^3/_{16}$-in. weld with E70 electrodes. From the table, interpolating between $k = 0.2$ and 0.3, yields $x = 0.0336$. Therefore, the eccentricity is

$$e_x = kl + \frac{t_w}{2} - xl$$

$$e_x = 3.5 + \frac{0.275}{2} - 0.0336(12.0) = 3.23 \text{ in.}$$

and

$$a = \frac{e_x}{L} = \frac{3.23}{12.0} = 0.269$$

With a double interpolation between $k = 0.2$ and 0.3 and $a = 0.25$ and 0.30, the coefficient C is determined as

$$C = 2.17$$

Therefore, the nominal weld strength is

$$R_n = CC_1DL = 2.17(1.0)(3)(12.0) = 78.1 \text{ kips}$$

And the allowable strength is

$$\frac{R_n}{\Omega} = \frac{(78.1)}{2.00} = 39.1 \text{ kips}$$

Step 13:   Determine the allowable strength for the limit state of flexural yielding.
For the limit state of flexural yielding of the angle the strength is determined as was shown for the bolted outstanding leg, thus

$$\frac{R_n}{\Omega} = 122 \text{ kips}$$

Step 14:   Determine the controlling limit state's strength for the welded outstanding legs.
For the welded outstanding leg, the allowable strength is controlled by eccentric shear on the weld where

$$\frac{R_n}{\Omega} = 39.1 \text{ kips}$$

Because this is less than the value for the leg attached to the beam, this is the allowable strength of the bolted-welded single-angle connection.

11.8   SINGLE-PLATE SHEAR CONNECTIONS

The single-plate shear connection, also called a *shear tab connection*, is shown in Figure 11.1c. It consists of a plate, shop welded to the support, and field bolted to the beam and is similar to the single-angle connection when it comes to erection. The shear tab consists of only a single-plate, which is about as simple as can be expected. It is welded to the supporting member and must be bolted to the supported beam in order to accommodate the required rotation. Even when bolted to the beam, this connection is stiffer than the single- or double-angle connections and requires careful detailing to insure sufficient flexibility.

The behavior of this connection is similar to that of a double-angle connection except that it achieves its rotation capacity through the bending of the tab and deformation of the plate or beam web in bearing at the bolt holes. Because of the complexity of assessing some of the limit states for this connection, AISC has developed two design approaches including a somewhat prescriptive approach for what is called the *conventional configuration* and a detailed limit states checking procedure for all others, which is referred to as the *extended configuration*.

The limit states that must be checked are the same for either configuration; the difference is that in the conventional configuration, physical limitations have been set so that most of those limit states do not govern. The potential limit states are:

1. Bolts
   a. Shear rupture
2. Beam
   a. Bearing on the web
   b. Shear yielding of web
3. Plate
   a. Bearing on the plate
   b. Elastic yield moment
   c. Shear yield

   d. Shear rupture
   e. Block shear rupture
   f. Buckling
   g. Plastic flexural yielding with shear interaction
4. Weld
   a. Weld rupture with eccentricity

Of these 11 limit states, those associated with flexure and buckling of the plate are new to the discussion of simple connection design, and the weld rupture limit state is treated a little bit differently than those weld limit states already discussed.

The conventional configuration of the shear tab results in a connection that is very simple to design. This is the type of connection that is treated here. For other configurations, the detailed procedures are given in Part 10 of the Manual. The dimensional limitations of the conventional shear tab require:

1. Only a single vertical row of bolts limited to 2 to 12 bolts
2. The distance from the bolt line to the weld line cannot exceed $3\frac{1}{2}$ in.
3. Only standard or short slotted holes can be used
4. The horizontal edge distance, L_{eh}, must be at least $2d_b$ for both the plate and beam web where d_b is the bolt diameter
5. The vertical edge distance must satisfy the Specification minimum from Table J3.4
6. Either the plate or beam web must have $t \leq (d_b/2 + \frac{1}{16})$

If the connection is additionally limited to a maximum of 9 bolts, or Manual Table 10-9 is used, eccentricity can be ignored. Once these limitations are satisfied, the connection need be checked only for

   a. Bolt shear rupture
   b. Bolt bearing
   c. Block shear rupture
   d. Plate shear yielding
   e. Plate shear rupture

EXAMPLE 11.5a
Shear Tab Conventional Configuration by LRFD

GOAL: Determine the design strength of a conventional configuration shear tab connection.

GIVEN: The shear tab connection is given in Figure 11.10. The beam is a W16×50, A992 framing into the flange of a W14×90, A992 column with an A36 $\frac{1}{4}×4\frac{1}{2}×12$ plate. Use four $\frac{3}{4}$-in. A325N bolts in standard holes.

SOLUTION

Step 1:   Determine whether the given shear tab meets the limitations for the conventional configuration.
   Limitations for the conventional configuration:
   1. 4 bolts—is between 2 and 12
   2. a = 3.0 in.—does not exceed $3\frac{1}{2}$ in.
   3. Standard holes—standard or short-slotted are permitted
   4. L_{eh} = 1.5 in.—at least $2d_b$ = 1.5 in.
   5. L_v = 1.5 in. > 1.25 in. from Table J3.4
   6. $t_{plate} = \frac{1}{4}$ − less than $\left(\dfrac{d_b}{2} + \frac{1}{16}\right) = \left(\dfrac{\frac{3}{4}}{2} + \frac{1}{16}\right) = \frac{7}{16}$

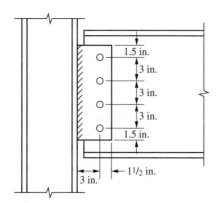

Figure 11.10 Shear Tab Connection for Example 11.5.

Step 2:   Determine the bolt design shear strength.
From Manual Table 7-1

$$\phi R_n = 4(15.9) = 63.6 \text{ kips}$$

Step 3:   Determine the bolt design bearing strength on the plate.
For the top bolt

$$L_c = 1.5 - \frac{1}{2}(3/4 + 1/16) = 1.09 < 2(3/4) = 1.5$$

Thus, tear-out controls and the nominal bolt strength is

$$R_n = 1.2(1.09)(0.250)(58) = 19.0 \text{ kips}$$

For the other bolts

$$L_c = 3.0 - (3/4 + 1/16) = 2.19 > 2(3/4) = 1.5$$

Therefore, bearing controls, and the nominal bolt strength is

$$R_n = 2.4(3/4)(0.250)(58) = 26.1 \text{ kips}$$

Thus, for the four bolts, the design strength is

$$\phi R_n = 0.75(19.0 + 3(26.1)) = 73.0 \text{ kips}$$

Step 4:   Determine the bolt design bearing strength on the web.
For the beam web, the material is A992 and the web thickness is 0.380 in. Because both the strength and thickness are greater than the comparable values for the plate, the beam web does not control.

Step 5:   Determine the design block shear strength of the plate.
Calculating the required areas

$$A_{nt} = \left(1.5 - \frac{1}{2}(3/4 + 1/8)\right)(0.250) = 0.266 \text{ in.}^2$$

$$A_{gv} = 10.5(0.250) = 2.63 \text{ in.}^2$$

$$A_{nv} = (10.5 - 3.5(3/4 + 1/8))(0.250) = 1.86 \text{ in.}^2$$

Consider shear yield and shear rupture and select the least strength, thus

$$0.6F_y A_{gv} = 0.6(36)(2.63) = 56.8 \text{ kips}$$

$$0.6F_u A_{nv} = 0.6(58)(1.86) = 64.7 \text{ kips}$$

Selecting the shear yield term and combining it with the tension rupture term gives a connection design block shear strength, with $U_{bs} = 1.0$, of

$$\phi R_n = 0.75(56.8 + 1.0(58)(0.266)) = 54.2 \text{ kips}$$

Step 6:   Determine the design shear yield strength of the plate.

$$\phi R_n = 1.0(0.6(36))(12.0)(0.250) = 64.8 \text{ kips}$$

Step 7:   Determine the design shear rupture strength of the plate.

$$\phi R_n = 0.75(12.0 - 4(3/4 + 1/8))(0.250)(58) = 92.4 \text{ kips}$$

Step 8:   Determine the design weld rupture strength.

The conventional configuration requires that the plate be welded to the supporting member through a pair of fillet welds on each side of the plate with the weld leg width, $w = \frac{5}{8}t_p$. This develops the strength of either an A36 or an A992 plate and therefore does not require any further limit states check.

Step 9:   Determine the controlling limit state and design strength of the connection.

The design strength is controlled by the limit state of block shear rupture of the plate where

$$\boxed{\phi R_n = 54.2 \text{ kips}}$$

A check of Manual Table 10-9 shows that this is quite close to the tabulated value for the 11.5-in. plate given there, as would be expected.

EXAMPLE 11.5b
Shear Tab Conventional
Configuration by ASD

GOAL: Determine the allowable strength of a conventional configuration shear tab connection.

GIVEN: The shear tab connection is given in Figure 11.10. The beam is a W16×50, A992 framing into the flange of a W14×90, A992, column with an A36 $\frac{1}{4} \times 4\frac{1}{2} \times 12$ plate. Use four $\frac{3}{4}$-in. A325N bolts in standard holes.

SOLUTION

Step 1:   Determine whether the given shear tab meets the limitations for the conventional configuration.

Limitations for the conventional configuration:
1. 4 bolts—is between 2 and 12
2. a = 3.0 in.—does not exceed $3\frac{1}{2}$ in.
3. Standard holes—standard or short-slotted are permitted
4. $L_{eh} = 1.5$ in.—at least $2d_b = 1.5$ in.
5. $L_v = 1.5$ in. > 1.25 in. from Table J3.4
6. $t_{plate} = 1/4 -$ less than $\left(\dfrac{d_b}{2} + 1/16\right) = \left(\dfrac{3/4}{2} + 1/16\right) = 7/16$

Step 2:   Determine the allowable bolt shear strength.

From Manual Table 7-1

$$\frac{R_n}{\Omega} = 4(10.6) = 42.4 \text{ kips}$$

Step 3:   Determine the allowable bolt bearing strength on the plate.

For the top bolt

$$L_c = 1.5 - \frac{1}{2}(3/4 + 1/16) = 1.09 < 2(3/4) = 1.5$$

Thus, tear-out controls and the nominal bolt strength is

$$R_n = 1.2(1.09)(0.250)(58) = 19.0 \text{ kips}$$

For the other bolts

$$L_c = 3.0 - (3/4 + 1/16) = 2.19 > 2(3/4) = 1.5$$

Therefore bearing controls, and the nominal bolt strength is

$$R_n = 2.4(3/4)(0.250)(58) = 26.1 \text{ kips}$$

Thus, for the four bolts, the allowable strength is

$$\frac{R_n}{\Omega} = \frac{(19.0 + 3(26.1))}{2.00} = 48.7 \text{ kips}$$

Step 4: Determine the allowable bolt bearing strength on the web.

For the beam web, the material is A992 and the web thickness is 0.380 in. Because both the strength and thickness are greater than the comparable values for the plate, the beam web does not control.

Step 5: Determine the allowable block shear strength of the plate.

Calculating the required areas

$$A_{nt} = \left(1.5 - \frac{1}{2}(3/4 + 1/8)\right)(0.250) = 0.266 \text{ in.}^2$$

$$A_{gv} = 10.5(0.250) = 2.63 \text{ in.}^2$$

$$A_{nv} = (10.5 - 3.5(3/4 + 1/8))(0.250) = 1.86 \text{ in.}^2$$

Consider shear yield and shear rupture and select the least strength, thus

$$0.6F_y A_{gv} = 0.6(36)(2.63) = 56.8 \text{ kips}$$

$$0.6F_u A_{nv} = 0.6(58)(1.86) = 64.7 \text{ kips}$$

Selecting the shear yield term and combining it with the tension rupture term gives a connection allowable block shear strength, with $U_{bs} = 1.0$, of

$$\frac{R_n}{\Omega} = \frac{(56.8 + 1.0(58)(0.266))}{2.00} = 36.1 \text{ kips}$$

Step 6: Determine the allowable shear yield strength of the plate.

$$\frac{R_n}{\Omega} = \frac{(0.6(36))(12.0)(0.250)}{1.5} = 43.2 \text{ kips}$$

Step 7: Determine the allowable shear rupture strength of the plate.

$$\frac{R_n}{\Omega} = \frac{(12.0 - 4(3/4 + 1/8))(0.250)(58)}{2.00} = 61.6 \text{ kips}$$

Step 8: Determine the allowable weld rupture strength.

The conventional configuration requires that the plate be welded to the supporting member through a pair of fillet welds on each side of the plate with the weld leg width, $w = \frac{5}{8}t_p$. This develops the strength of either an A36 or an A992 plate and therefore does not require any further limit states check.

Step 9: Determine the controlling limit state and the allowable strength of the connection.

The allowable strength is controlled by the limit state of block shear of the plate with

$$\boxed{\frac{R_n}{\Omega} = 36.1 \text{ kips}}$$

A check of Manual Table 10-9 shows that this is quite close to the tabulated value for the 11.5-in. plate given there, as would be expected.

11.9   SEATED CONNECTIONS

An unstiffened seated connection is shown in Figure 11.1d and a stiffened seated connection is shown in Figure 11.1e. These connections are typically used to attach a beam to the web of a column. They can also be used to add capacity to other types of existing connections in a retrofit situation. Because of their simplicity, they are fairly easy connections to erect. They have very few parts, a seat angle, a connection to the supporting member through welds or bolts, a limited connection to the supported member, and a top connection to insure stability of the beam. All of the force is transferred through bearing of the beam on the seat and then through the connection of the seat to the supporting member. When the seat lacks sufficient strength in bending of the top leg, it can be stiffened to produce the stiffened seated connection.

The seat can be welded or bolted to the supporting member and is usually bolted to the supported member. The connection to the supported member is not designed for a specific strength when only a vertical force is being transferred. The seated connection performs excellently as a simple connection. It can rotate sufficiently about the bottom of the beam without imposing any significant moment to the supporting member.

The simplicity of this connection results in relatively few limit states to be checked. Because the transfer of force between the beam and seat is through the bearing of the beam on the seat, the limit states of beam web yielding and beam web crippling must be checked. These limit states were introduced in Section 6.14. The outstanding leg of the seat angle must be checked for the limit states of flexural yielding and shear yielding, and the connection to the supporting member, bolts, or welds must be checked for their appropriate limit states. In summary, the potential limit states are:

1. Beam
   a. Web yielding
   b. Web crippling
2. Seat angle
   a. Flexural yielding
   b. Shear yielding
3. Connector
   a. Bolt or weld shear

The nominal strength for the limit state of web yielding was discussed in Sections 6.14 and 7.4. If the strength equation for a force applied at the end of a member is rewritten to solve for the minimum required bearing length, it can be used for design. Thus, from Specification Section J10.2, for the limit state of web local yielding

$$N_{\min} = \frac{R_n}{F_y t_w} - 2.5k$$

For the limit state of web yielding, $\phi = 1.00$, and $\Omega = 1.50$.

Similarly, from Specification Section J10.3, for the limit state of web crippling, an equation for the minimum required bearing length can be determined. However, there are three different strength equations, depending on the relationship N/d. Because it is most likely that N will be less than $d/2$ and very likely it will be less than $0.2d$, only Specification Equation J10-5a is treated here. The other situations can both be handled in the same way if need be. Thus, rearranging Equation J10-5a to solve for the minimum required bearing

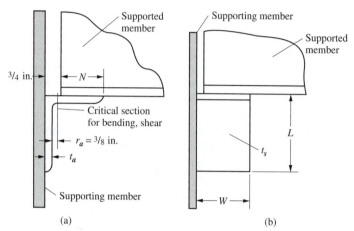

Figure 11.11 Seated Beam Connection.

length for the limit state of web crippling yields, for $N/d \leq 0.2$

$$N_{\min} = \frac{d}{3}\left[\frac{R_n}{0.40t_w^2}\sqrt{\frac{t_w}{EF_{yt_f}}} - 1\right]\left(\frac{t_f}{t_w}\right)^{1.5}$$

For the limit state of web crippling, $\phi = 0.75$, and $\Omega = 2.00$.

Because the unstiffened angle is a very flexible connection, the load levels usually considered are quite low. This tends to result in very small minimum required bearing lengths and, in some calculations, a negative minimum required bearing length. To offset this potential problem, the minimum bearing length for seated connections is taken as k_{det}. A review of Manual Table 1-1 shows two values for k: k_{des} is a dimension used in design calculations and is the smaller of the two values, k_{det} is a dimension normally used in detailing and is used here because it is the larger of these two values. These two ks are the result of differences in production by different mills. They represent the extremes of the values actually found and are selected within a calculation to give a conservative answer.

The outstanding leg of the angle must be capable of supporting the beam reaction applied at an eccentricity from the critical section of the angle. This requires checking the limit state of flexural yielding of the leg. The angle shown in Figure 11.11a is an unstiffened seat angle. The critical section for both flexure and shear is taken as $^3/_8$ in. out from the face of the vertical leg, the radius of the fillet. The eccentricity is measured from this line to the mid-point of the minimum required bearing length of the beam on the angle, N. Thus, the eccentricity is

$$e = \frac{N}{2} + 3/4 - (t_a + 3/8) = \frac{N}{2} + 3/8 - t_a$$

The nominal beam reaction strength, for the limit state of flexural yielding, is based on the plastic section modulus of the seat leg of the angle. Thus

$$Z = \frac{Lt_a^2}{4}$$

and

$$R_n = \frac{M_n}{e} = \frac{F_y Z}{e} = \frac{F_y\left(\dfrac{Lt_a^2}{4}\right)}{e}$$

Because this is based on flexural yielding, $\phi = 0.9$, and $\Omega = 1.67$.

For the limit state of shear yielding, the gross area of the angle leg at this same location is used. Because this is beyond the angle fillet, the angle thickness is used to determine the gross area and

$$R_n = 0.6F_yLt_a$$

For the limit state of shear yielding, $\phi = 1.0$ and $\Omega = 1.5$.

EXAMPLE 11.6a
Seated Connection Design by LRFD

GOAL: Design an unstiffened welded seated connection.

GIVEN: An unstiffened welded seated connection is shown in Figure 11.12. A W16×36 beam is framing into a W14×90 column. The beam has an LRFD required strength of $V_u = 35$ kips. The beam and column are A992 and the angle is A36. From Manual Table 1-1, $k_{det} = 1\frac{1}{8}$ in.

SOLUTION

Step 1: Determine the minimum required bearing length for web yielding.

$$N_{min} = \frac{R_n}{F_y t_w} - 2.5k = \frac{35/(1.0)}{(50)(0.295)} - 2.5(0.832) = 0.293 \text{ in.} < k_{det} = 1\tfrac{1}{8} \text{ in.}$$

Step 2: Determine the minimum required bearing length for web crippling, assuming that $N/d < 0.2$

$$N_{min} = \frac{15.9}{3}\left[\frac{35/0.75}{(0.40)(0.295)^2}\sqrt{\frac{0.295}{29{,}000(50)(0.430)}} - 1\right]\left(\frac{0.430}{0.295}\right)^{1.5} = -0.726 < k_{det}$$

Thus, $N = k_{det} = 1.125$ in. and $N/d = 1.125/15.3 = 0.0735 < 0.2$ so the correct equation has been used and the bearing length is taken as the minimum required length of 1.125 in.

Step 3: Determine the eccentricity to be used in calculating angle thickness.
Assume an angle thickness of $\frac{1}{2}$ in.

$$e = \frac{N}{2} + 3/8 - t_a = \frac{1.125}{2} + \frac{3}{8} - \frac{1}{2} = 0.438 \text{ in.}$$

Step 4: Determine the minimum required angle thickness based on the limit state of flexural yielding.

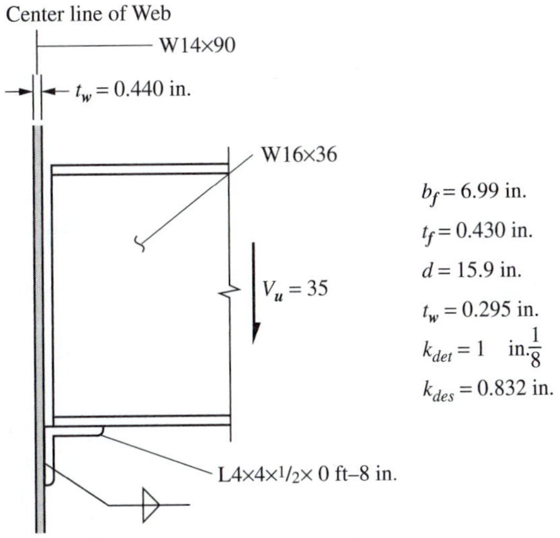

Center line of Web
W14×90
$t_w = 0.440$ in.
W16×36
$V_u = 35$
$b_f = 6.99$ in.
$t_f = 0.430$ in.
$d = 15.9$ in.
$t_w = 0.295$ in.
$k_{det} = 1\frac{1}{8}$ in.
$k_{des} = 0.832$ in.
L4×4×1/2× 0 ft–8 in.

Figure 11.12 Unstiffened Seated Beam Connection for Example 11.6.

Assume an 8.0-in. long angle so that it extends beyond the beam flange on both sides and

$$t_{req} = \sqrt{\frac{4V_n e}{F_y L}} = \sqrt{\frac{4(35/0.9)(0.438)}{(36)(8.0)}} = 0.486 \text{ in.}$$

Step 5: Determine the minimum angle thickness for the limit state of shear yielding

$$t_{min} = \frac{V_n}{0.6 F_y L} = \frac{(35/1.0)}{0.6(36)(8.0)} = 0.203 \text{ in.}$$

Step 6: Check the selected angle thickness.

The $\frac{1}{2}$-in. angle provides a thickness greater than each of the minimums determined in Steps 4 and 5. Thus, the $\frac{1}{2}$-in. angle is adequate.

Step 7: Determine the required weld size.

Use Manual Table 8-4 to account for the eccentricity.

The eccentricity for the weld is taken from the center of bearing to the face of the supporting column which yields

$$e = \frac{3}{4} + \frac{1.125}{2} = 1.31 \text{ in.}$$

Assuming an angle with a 4.0-in. outstanding leg,

$$a = \frac{e}{L} = \frac{1.31}{4.0} = 0.328$$

From Manual Table 8-4, the coefficient is determined for $k = 0$ through interpolation as $C = 2.97$, so the minimum weld is

$$D_{min} = \frac{R_u}{\phi C C_1 L} = \frac{35}{0.75(1.0(2.97)(4.0))} = 3.93$$

Therefore, the calculated minimum weld is $\frac{1}{4}$ in. and the angle is a $4 \times 4 \times \frac{1}{2}$, as shown in Figure 11.12.

Note: In addition to the seat angle, a top clip angle is needed to provide lateral stability. This angle is not normally designed to support any load and is usually a $\frac{1}{4}$-in. angle attached with two bolts to the beam and the supporting member.

EXAMPLE 11.6b
Seated Connection
Design by ASD

GOAL: Design an unstiffened welded seated connection.

GIVEN: An unstiffened welded seated connection is shown in Figure 11.12. A W16×36 beam is framing into a W14×90 column. The beam has an ASD required strength of $V_a = 23$ kips. The beam and column are A992 and the angle is A36. From Manual Table 1-1, $k_{det} = 1\frac{1}{8}$ in.

SOLUTION

Step 1: Determine the minimum required bearing length for web yielding.

$$N_{min} = \frac{R_n}{F_y t_w} - 2.5k = \frac{1.5(23)}{(50)(0.295)} - 2.5(0.832) = 0.259 \text{ in.} < k_{det} = 1\frac{1}{8} \text{ in.}$$

Step 2: Determine the minimum required bearing length for web crippling, assuming that $N/d < 0.2$

$$N_{min} = \frac{15.9}{3}\left[\frac{23(2.00)}{(0.40)(0.295)^2}\sqrt{\frac{0.295}{29,000(50)(0.430)}} - 1\right]\left(\frac{0.430}{0.295}\right)^{1.5} = -0.849 < k_{det}$$

Thus, $N = k_{det} = 1.125$ in. and $N/d = 1.125/15.3 = 0.0735 < 0.2$ so the correct equation has been used and the bearing length is taken as the minimum required length of 1.125 in.

Step 3:   Determine the eccentricity to be used in calculating angle thickness.
Assume an angle thickness of $^1/_2$ in.

$$e = \frac{N}{2} + 3/8 - t_a = \frac{1.125}{2} + \frac{3}{8} - \frac{1}{2} = 0.438 \text{ in.}$$

Step 4:   Determine the minimum required angle thickness based on the limit state of flexural yielding.
Assume an 8.0-in. long angle so that it extends beyond the beam flange on both sides and

$$t_{req} = \sqrt{\frac{4V_n e}{F_y L}} = \sqrt{\frac{4(1.67(23))(0.438)}{(36)(8.0)}} = 0.483 \text{ in.}$$

Step 5:   Determine the minimum angle thickness based on the limit state of shear yielding.

$$t_{min} = \frac{V_n}{0.6 F_y L} = \frac{1.50(23)}{0.6(36)(8.0)} = 0.200 \text{ in.}$$

Step 6:   Check the selected angle thickness.
The $^1/_2$-in. angle provides a thickness greater than each of the minimums determined in Steps 4 and 5. Thus, the $^1/_2$-in. angle is adequate.

Step 7:   Determine the required weld size.
Use Manual Table 8-4 to account for the eccentricity.
The eccentricity for the weld is taken from the center of bearing to the face of the supporting column which yields

$$e = \frac{3}{4} + \frac{1.125}{2} = 1.31 \text{ in.}$$

Assuming an angle with a 4.0-in. outstanding leg

$$a = \frac{e}{L} = \frac{1.31}{4.0} = 0.328$$

From Manual Table 8-4, the coefficient is determined for $k = 0$ through interpolation as $C = 2.97$, so the minimum weld is

$$D_{min} = \frac{\Omega R_u}{CC_1 L} = \frac{2.00(23)}{(1.0(2.97)(4.0))} = 3.87$$

Therefore, the calculated minimum weld is $^1/_4$ in. and the angle is a $4 \times 4 \times ^1/_2$, as shown in Figure 11.12.
Note: In addition to the seat angle, a top clip angle is needed to provide lateral stability. This angle is not normally designed to support any load and is usually a $^1/_4$-in. angle attached with two bolts to the beam and the supporting member.

A stiffened seated connection is shown in Figure 11.1e, and a detail of the stiffened seated connection is shown in Figure 11.11b. This type of connection is used when the loads are too large to be supported by an unstiffened seat. The stiffener can be a single plate, the stem of a Tee, or the back-to-back legs of a pair of angles. A plate on top of the stiffener provides for the bearing surface and the location of the bolts required to attach the beam flange to the seat. The limit states for this connection are the same as for the unstiffened connection already discussed but also include the additional limit state of punching shear in the supporting member.

Punching shear on a column web will not be critical if the following parameters are met:

1. The simplified approach is applicable to columns in the following depths with weights per foot no less than: W14×43, W12×40, W10×33, W8×24, W6×20, and W5×16.
2. The width of the stiffener W is no greater than 7.0 in.
3. The beam is bolted, not welded, to the bearing plate at a point no greater than $W/2$ or $2\frac{5}{8}$ in. from the column face.
4. The top angle must have a minimum thickness of $\frac{1}{4}$ in.

The eccentricity of the beam reaction is taken as $0.8W$ when determining the strength of the weld or bolt group connecting the seat to the supporting member. Part 10 of the Manual provides tables for the design of stiffened seated connections.

11.10 LIGHT BRACING CONNECTIONS

Bracing connections have as many potential variations as do the simple shear connections discussed above. Figure 11.13a shows a bolted-welded tension brace connection and Figure 11.13b shows a welded-bolted tension brace connection. It is also permissible to have a welded-welded connection or a bolted-bolted connection although these are not illustrated here.

The limit states for these connections have already been addressed. For the bolted-welded connection, shown in Figure 11.13a, they are:

1. Angles
   a. Tension yielding
   b. Tension rupture
   c. Bolt bearing and tear-out
   d. Block shear rupture
2. Bolts
   a. Shear rupture

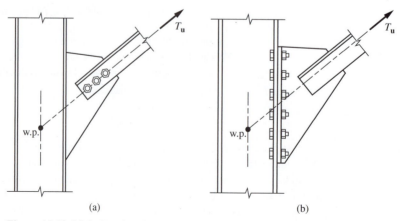

(a)　　　(b)

Figure 11.13 Light Bracing Connections.

3. Gusset plate
  a. Tension yielding
  b. Tension rupture
  c. Bolt bearing and tear-out

4. Welds
  a. Weld rupture for combined tension and shear

For the welded-bolted connection, Figure 11.13b, the limit states are:

1. Angles
  a. Tension yielding
  b. Tension rupture

2. Welds
  a. Weld rupture

3. Tee stem
  a. Tension yielding
  b. Tension rupture
  c. Block shear
  d. Shear yielding

4. Tee flange
  a. Flange bending
  b. Shear yielding
  c. Shear rupture
  d. Bolt bearing and tear-out
  e. Block shear

5. Bolts
  a. Combined shear and tension

6. Column flange
  a. Flange bending
  b. Bolt bearing and tear-out

7. Column web
  a. Web yielding

Although they appear to be simple connections, light bracing connections require checking for quite a number of different limit states. Three of these limit states have not previously been addressed (1) tension rupture and tension yield on the Whitmore Section, (2) limit state of bolt rupture due to combined shear and tension, and (3) high-strength bolts in tension with prying action.

Figure 11.14 shows a single-angle brace attached to a gusset plate with welds along the sides of the angle. Research has shown that the distribution of stresses from the brace through the welds into the gusset is such that the entire width of the gusset is not effective if it exceeds the width defined by a 30-degree angle from the beginning of the connection to the end of the connection. This width is defined as the Whitmore Section. When the gusset is wider than the Whitmore Section, only the Whitmore Section can be considered to resist the force and when the Whitmore Section is wider than the available plate dimension, only the width of the plate at the connection end can be considered to resist the force. If the connection is bolted rather than welded as shown, the Whitmore Section distribution starts at the first bolt and proceeds to the last bolt in the connection. This is illustrated in Example 11.7.

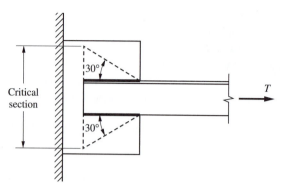

Figure 11.14 Whitmore Section in a Gusset Plate.

EXAMPLE 11.7
Tapered Gusset Plate Strength

GOAL: Determine the available strength of a gusset plate connected to a double-angle brace.

GIVEN: The gusset plate portion of the connection is shown in Figure 11.15. The brace is a double angle bolted to a tapered gusset with the dimensions as shown in Figure 11.15b. The plate is A36 steel.

SOLUTION

Step 1:   Determine the width of the plate at the location of the last bolt.
   This is also the location of the Whitmore Section. By proportions, as seen in Figure 11.15c

$$\frac{x}{1.0} = \frac{7.25}{10.25}$$

Therefore, $x = 0.707$ in. and the width of the plate at this location is $W = 5.0 + 2(0.707) = 6.41$ in.

Step 2:   Determine the width of the Whitmore Section using the geometry shown in Figure 11.15d.

$$W = 2(6\tan(30°)) = 6.93 \text{ in.}$$

Step 3:   Determine the width to be used to determine the plate strength.
   Because the actual plate width at this location is less than the Whitmore Section, the actual plate width is used to determine the strength of the plate.

Step 4:   Determine the nominal strength for the limit state of plate yielding.
   The gross area at the critical location is

$$A_g = 6.41(0.5) = 3.21 \text{ in.}^2$$

and the nominal tensile strength is

$$T_n = (36)(3.21) = 116 \text{ kips}$$

Step 5:   For LRFD, the design tensile strength is

$$\phi T_n = 0.9(116) = 104 \text{ kips}$$

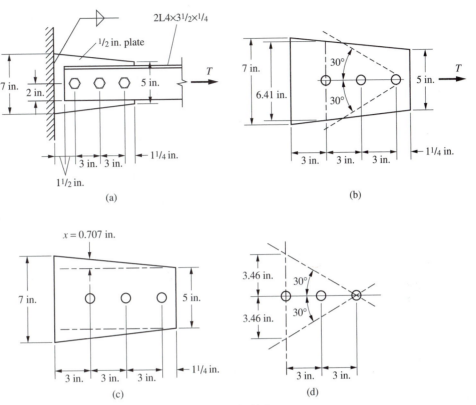

Figure 11.15 Tapered Gusset Plate for Example 11.7.

Step 5: For ASD, the allowable tensile strength is

$$T_n = \frac{(116)}{1.67} = 69.5 \text{ kips}$$

Step 6: Determine the nominal strength for the limit state of plate rupture.
The net area at the same location is

$$A_n = (6.41 - (3/4 + 1/8))(0.5) = 2.77 \text{ in.}^2 > 0.85A_g = 0.85(3.21) = 2.73 \text{ in.}^2$$

Therefore, using the maximum permitted net area for a connecting element, from Specification Section J4.1, the nominal tensile strength is

$$T_n = (58)(2.73) = 158 \text{ kips}$$

Step 7: For LRFD, the design tensile strength is

$$\phi T_n = 0.75(158) = 119 \text{ kips}$$

Step 7: For ASD, the allowable tensile strength is

$$\frac{T_n}{\Omega} = \frac{(158)}{2.00} = 79.0 \text{ kips}$$

Step 8: Determine the strength of the gusset plate based on the controlling limit state.
   The design strength of the gusset is $\phi T_n = 104$ kips, based on the limit state of yielding. The allowable tensile strength of the gusset is $T_n/\Omega = 69.5$ kips, based on the limit state of yielding.

EXAMPLE 11.8
Uniform Width Gusset
Plate Strength

GOAL: Determine the available strength of a gusset plate of uniform width.

GIVEN: Determine the available strength of a gusset plate for the same situation as in Example 11.7, except that the plate is a uniform width of 8.0 in. The plate is again A36 steel.

SOLUTION

Step 1: Determine the controlling width at the critical section.
   The width of the plate at the location of the last bolt is given as 8.0 in.
   The width of the Whitmore Section is determined using the same geometry as shown in Figure 11.15d, which again yields

$$W = 2(6 \tan(30°)) = 6.93 \text{ in.}$$

   In this case, the actual plate width is greater than the Whitmore Section so the Whitmore Section width is used to determine the strength of the plate.

Step 2: Determine the nominal tensile strength for the limit state of plate yielding.
   The gross area at the critical location is

$$A_g = 6.93(0.5) = 3.47 \text{ in.}^2$$

   The nominal tensile strength is

$$T_n = (36)(3.47) = 125 \text{ kips}$$

Step 3: For LRFD, the design tensile strength is

$$\phi T_n = 0.9(125) = 113 \text{ kips}$$

Step 3:   For ASD, the allowable tensile strength is

$$\frac{T_n}{\Omega} = \frac{(125)}{1.67} = 74.9 \text{ kips}$$

Step 4:   Determine the nominal strength for the limit state of plate rupture.
The net area at the same location is

$$A_n = (6.93 - (3/4 + 1/8))(0.5) = 3.03 \text{ in.}^2 > 0.85A_g = 0.85(3.47) = 2.95 \text{ in.}^2$$

Therefore, using the maximum permitted net area for a connecting element

$$T_n = (58)(2.95) = 171 \text{ kips}$$

Step 5:   For LRFD, the design strength is

$$\phi T_n = 0.75(171) = 128 \text{ kips}$$

Step 5:   For ASD, the allowable strength is

$$\frac{T_n}{\Omega} = \frac{(171)}{2.00} = 85.5 \text{ kips}$$

Step 6:   Determine the strength of the gusset plate based on the controlling limit state.
The design strength of the uniform-width gusset is limited to $\phi T_n = 113$ kips, again based on the limit state of yielding. The allowable strength is $T_n/\Omega = 74.9$ kips based on the limit state of yielding. Note that there is no advantage to using a plate wider than 6.93 in.

The next limit state to address is high-strength bolts in combined shear and tension. A bolt loaded in combined shear and tension has a reduced capacity to resist shear in a bearing-type connection due to the presence of tension. In a slip-critical connection, the tension reduces the contact force and, thus, lowers the shear required to cause the connection to slip. These reductions must be accounted for in the design of connections where these combined limit states occur.

Tests have shown that the interaction of shear and tension in a bearing-type connection can be fairly well predicted through an elliptical interaction curve. However, for simplicity, the Specification has adopted three straight lines to approximate the ellipse. Both the ellipse and straight line are shown in Figure 11.16. Section J3.7 gives two equations for this interaction, one for ASD and one for LRFD, where the nominal tensile stress including the effects of shear-tension interaction is

$$F'_{nt} = 1.3 F_{nt} - \frac{F_{nt}}{\phi F_{nv}} f_v \leq F_{nt} \text{ (LRFD)}$$

$$F'_{nt} = 1.3 F_{nt} - \frac{\Omega F_{nt}}{F_{nv}} f_v \leq F_{nt} \text{ (ASD)}$$

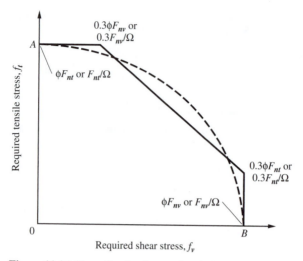

Figure 11.16 Shear-Tension Interaction for Bolts.

where

F_{nt} = nominal tensile stress for tension alone from Specification Table J3.2
F_{nv} = nominal shear stress for shear alone from Specification Table J3.2
f_v = required shear stress

These two equations can be combined and written in terms of nominal strength if the required shear stress, f_v, is combined with ϕ and Ω to give the nominal shear stress including the effects of shear-tension interaction. Thus

$$F'_{nv} = \frac{f_v}{\phi} \text{ (LRFD) or } = \Omega f_v \text{ (ASD)}$$

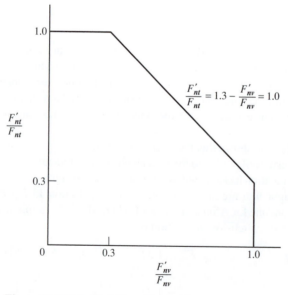

Figure 11.17 Modified Shear-Tension Interaction for Bolts.

so that

$$F'_{nt} = 1.3F_{nt} - \frac{F_{nt}}{F_{nv}} F'_{nv} \le F_{nt}$$

If this equation is then divided by F'_{nt}, another form of the interaction equation results as

$$\frac{F'_{nt}}{F_{nt}} = 1.3 - \frac{F'_{nv}}{F_{nv}} \le 1.0$$

This relationship is shown in Figure 11.17.

In a slip-critical connection, the shear-tension interaction equation serves a different purpose. In this case, shear is assumed to be transferred by friction between the plies. The strength of the connection, therefore, is a linear function of the force compressing the plies. This force is the initial pretension, T_b, minus the applied load, T. The specified slip-critical shear value is, therefore, reduced by the factor, $(1 - T/T_b)$. The actual reduction factor is provided in Specification equation J3-5, again with one for ASD and one for LRFD.

EXAMPLE 11.9a
Bolts in Combined Shear and Tension by LRFD

GOAL: Determine the strength of a connection using bolts in combined shear and tension and compare to the applied load.

GIVEN: An inclined hanger that supports a dead load of 10 kips and a live load of 50 kips is shown in Figure 11.18. The connection uses four 1.0-in. A325-N bolts.

SOLUTION

Step 1: Determine the required strength for the appropriate load combination.

$$R_u = 1.2(10.0) + 1.6(50.0) = 92.0 \text{ kips}$$

Step 2: Determine the force assigned to each bolt in tension and shear.

$$\text{Bolt Tension} = \sin(30°)\left(\frac{92.0}{4}\right) = 11.5 \text{ kips}$$

$$\text{Bolt Shear} = \cos(30°)\left(\frac{92.0}{4}\right) = 19.9 \text{ kips}$$

$$\text{Bolt Tensile Stress} = f_t = \frac{11.5}{0.785} = 14.6 \text{ ksi}$$

$$\text{Bolt Shear Stress} = f_v = \frac{19.9}{0.785} = 25.4 \text{ ksi}$$

Step 3: Determine the reduced nominal tensile stress.
The nominal shear and tensile stress from Specification Table J3.2

$$F_{nv} = 48 \text{ ksi}$$

$$F_{nt} = 90 \text{ ksi}$$

and the nominal shear stress including the effects of tension-shear interaction is

$$F'_{nv} = \frac{f_v}{\phi} = \frac{25.4}{0.75} = 33.9 \text{ ksi}$$

Thus

$$F'_{nt} = 1.3F_{nt} - \frac{F_{nt}}{F_{nv}} F'_{nv} = 1.3(90) - \left(\frac{90}{48}\right)(33.9) = 53.4 \le 90$$

Step 4:   Check the design tensile stress vs. the required tensile stress.

$$\phi F'_{nt} = 0.75(53.4) = 40.1 \text{ ksi} > 14.6 \text{ ksi}$$

Thus, by LRFD, the bolts are adequate.

EXAMPLE 11.9b
Bolts in Combined Shear and Tension by ASD

GOAL: Determine the strength of a connection using bolts in combined shear and tension and compare to the applied load.

GIVEN: An inclined hanger that supports a dead load of 10 kips and a live load of 50 kips is shown in Figure 11.18. The connection uses four 1.0-in. A325-N bolts.

SOLUTION

Step 1:   Determine the required strength for the appropriate load combination.

$$R_a = 10.0 + 50.0 = 60.0 \text{ kips}$$

Step 2:   Determine the force assigned to each bolt in tension and shear.

$$\text{Bolt Tension} = \sin(30°)\left(\frac{60.0}{4}\right) = 7.50 \text{ kips}$$

$$\text{Bolt Shear} = \cos(30°)\left(\frac{60.0}{4}\right) = 13.0 \text{ kips}$$

$$\text{Bolt Tensile Stress} = f_t = \frac{7.50}{0.785} = 9.55 \text{ ksi}$$

$$\text{Bolt Shear Stress} = f_v = \frac{13.0}{0.785} = 16.6 \text{ ksi}$$

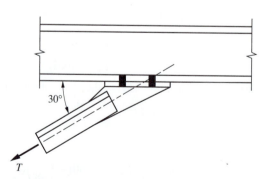

Figure 11.18 Connection for Example 11.9.

Step 3: Determine the reduced nominal tensile stress.
The nominal shear and tensile stress from Specification Table J3.2

$$F_{nv} = 48 \, \text{ksi}$$

$$F_{nt} = 90 \, \text{ksi}$$

and the nominal shear stress including the effects of tension-shear interaction is

$$F'_{nv} = \Omega f_v = 2.00(16.6) = 33.2 \, \text{ksi}$$

Thus

$$F'_{nt} = 1.3 F_{nt} - \frac{F_{nt}}{F_{nv}} F'_{nv} = 1.3(90) - \left(\frac{90}{48}\right)(33.2) = 54.8 \leq 90$$

Step 4: Check the allowable tensile stress vs. the required tensile stress.

$$\boxed{\frac{F'_{nt}}{\Omega} = \frac{54.8}{2.00} = 27.4 \, \text{ksi} > 9.55 \, \text{ksi}}$$

Thus, by ASD, the bolts are adequate.

When high-strength bolts are installed with an initial pretension, they act as a clamp, holding the two connected elements together. Figure 11.19 shows a typical tension hanger where the bolts are expected to carry the applied tension load. Any pretension from the bolt actually causes a compressive force to develop between the connected parts. Application of the applied load reduces the contact force but has little effect on the bolt tension, as long as contact is maintained between the plates. Once the plates are separated, the initial conditions have no influence and the bolt force must equal the applied load.

If the attached element, in this case the flange of the Tee, is permitted to deform, as shown in Figure 11.20, additional forces develop at the tips of the flange. These additional forces, q, are the result of prying action and are called the prying forces. There is a relationship between the thickness of the flange and the prying force. When t is large, the plate does not bend and no prying action takes place. When t is small, bending of the plate may be extensive and the prying force may be large. Prying action may be completely eliminated in a design by selecting a sufficiently thick plate, although this may not be a practical solution. It may also be avoided if washers are used to keep the flange from coming in contact with the support; however, this, too, is normally not desirable.

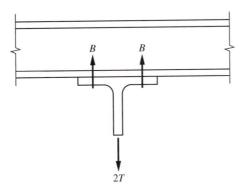

Figure 11.19 Hanger Connection with Bolts in Tension.

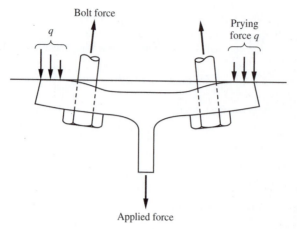

Figure 11.20 Tee Deformation with Prying Action.

The details for design of this type of connection including prying action are given in Part 9 of the Manual. It suggestes that the minimum plate thickness to eliminate prying action be determined. If this is a reasonable thickness, no further action is required. If this thickness is not reasonable for the details of the design, a design that takes into account prying action should be undertaken with a goal of having a reasonable combination of strength and stiffness that results in an economical connection.

Figure 11.21 shows a WT section used as a hanger attached to the supporting member with bolts. The dimensions given are used to determine a relationship between the flexural strength of the flange and the applied load. The applied load is $2T$ so that the load per bolt is T. It is not a simple matter to determine the actual moment in the flange but the design approach given assumes that b' will be a good representation of the moment arm so that the moment is $M_r = Tb'$. It has also been found that the strength should be calculated in terms of F_u rather than F_y. So, using a tributary width of plate associated with each bolt, p, the nominal moment

$$M_n = F_u \frac{pt^2_{min}}{4}$$

$$\phi = 0.9 \,(\text{LRFD}) \qquad \Omega = 1.67 \,(\text{ASD})$$

Setting the required strength equal to the available strength yields the following equations

$$t_{min} = \sqrt{\frac{4.44 T_u b'}{pF_u}} \,(\text{LRFD}) \qquad t_{min} = \sqrt{\frac{6.66 T_a b'}{pF_u}} \,(\text{ASD})$$

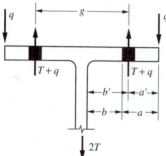

Figure 11.21 Force Equilibrium Considering Prying Action.

EXAMPLE 11.10a
Hanger Connection by LRFD

GOAL: Determine whether the WT hanger connection is adequate without considering prying action.

GIVEN: A WT9×48.5 section, A992 steel, is used as shown in Figure 11.21 to carry a dead load of 20 kips and a live load of 60 kips. Four $^7/_8$-in. diameter A325 bolts are used in a 9-in. long fitting.

SOLUTION

Step 1: Determine the moment arm, b', based on the properties of the section.

$$t_f = 0.870 \text{ in.}, \ t_w = 0.535 \text{ in.}, \ b_f = 11.1 \text{ in.}, \ \text{gage} = 4 \text{ in.}, \ p = 9/2 = 4.5 \text{ in.}$$

$$b = \frac{\text{gage} - t_w}{2} = \frac{(4.0 - 0.535)}{2} = 1.73 \text{ in.}$$

$$b' = b - \frac{d_b}{2} = 1.73 - \frac{^7/_8}{2} = 1.29 \text{ in.}$$

Step 2: Determine the force per bolt.

$$T_u = 1.2(20.0) + 1.6(60.0) = 120 \text{ kips}$$

$$T_u = \frac{120}{4} = 30.0 \text{ kips/bolt}$$

Step 3: Determine the minimum flange thickness to ignore prying action.

$$t_{\min} = \sqrt{\frac{4.44 T_u b'}{p F_u}} = \sqrt{\frac{4.44(30.0)(1.29)}{4.5(65)}} = 0.766 \text{ in.}$$

Step 4: Compare the available thickness with the required thickness.

$$\boxed{t_{\min} = 0.766 \text{ in.} < t_f = 0.870 \text{ in.}}$$

Because the actual flange thickness is greater than the minimum, the WT9×48.5 is adequate without considering prying action.

EXAMPLE 11.10b
Hanger Connection by ASD

GOAL: Determine whether the WT hanger connection is adequate without considering prying action.

GIVEN: A WT9×48.5 section, A992 steel, is used as shown in Figure 11.21 to carry a dead load of 20 kips and a live load of 60 kips. Four $^7/_8$-in. diameter A325 bolts are used in a 9-in. long fitting.

SOLUTION

Step 1: Determine the moment arm, b', based on the properties of the section.

$$t_f = 0.870 \text{ in.}, \ t_w = 0.535 \text{ in.}, \ b_f = 11.1 \text{ in.}, \ \text{gage} = 4 \text{ in.}, \ p = 9/2 = 4.5 \text{ in.}$$

$$b = \frac{\text{gage} - t_w}{2} = \frac{(4.0 - 0.535)}{2} = 1.73 \text{ in.}$$

$$b' = b - \frac{d_b}{2} = 1.73 - \frac{^7/_8}{2} = 1.29 \text{ in.}$$

Step 2: Determine the force per bolt.

$$T_a = 20.0 + 60.0 = 80.0 \text{ kips}$$

$$T_a = \frac{80.0}{4} = 20.0 \text{ kips/bolt}$$

Step 3: Determine the minimum flange thickness to ignore prying action.

$$t_{\min} = \sqrt{\frac{6.66 T_a b'}{p F_u}} = \sqrt{\frac{6.66(20.0)(1.29)}{4.5(65)}} = 0.766 \text{ in.}$$

Step 4: Compare the available thickness with the required thickness.

$$\boxed{t_{\min} = 0.766 \text{ in.} < t_f = 0.870 \text{ in.}}$$

Because the actual flange thickness is greater than the minimum, the WT9×48.5 is adequate without considering prying action.

11.11 BEAM-BEARING PLATES AND COLUMN BASE PLATES

The connections discussed throughout this chapter transfer force through a series of connecting elements to a supporting member. Two other types of simple connections deserve mention here, the beam bearing plate and the column base plate. These plates transfer a force through direct bearing from one member to another member or directly to a support. Although these plates are used in two very different applications, the actual behavior of each is quite similar.

For design of the plate, three properties must be determined: the width and breadth which results in an appropriate area, and the thickness. The area of the plate is determined by assessing the limit states of the supporting member or material and those of the member applying the force to the plate. The thickness of the plate is determined through the limit state of flexural yielding of the plate.

To determine the required plate thickness for either type of plate, two primary assumptions are made: (1) the plate exerts a uniform pressure on the supporting material, and (2) the plate is treated as a cantilevered strip that is 1.0 in. wide. For a bending cross section 1.0 in. wide with a thickness, t_p, the nominal flexural strength for the limit state of yielding is

$$M_n = F_y Z = F_y \left(\frac{1.0 t_p^2}{4} \right)$$

For a uniform contact pressure between the plate and the supporting material, f_p, and a cantilever length, l, the required moment strength for the cantilever is

$$M_r = \frac{f_p l^2}{2}$$

For LRFD, the required plate thickness can be obtained by setting the design moment equal to the required moment where the required moment is obtained using f_u, thus

$$\phi M_n = M_r$$

$$\frac{\phi F_y t_p^2}{4} = \frac{f_u l^2}{2}$$

which yields

$$t_p = 1.49l\sqrt{\frac{f_u}{F_y}}$$

Similarly, for ASD, using f_a

$$\frac{M_n}{\Omega} = M_r$$

$$\frac{F_y t_p^2}{4\Omega} = \frac{f_a l^2}{2}$$

which yields

$$t_p = 1.83l\sqrt{\frac{f_a}{F_y}}$$

The determination of the cantilever distance, l, to be used in the case of a beam-bearing plate or a column-base plate is addressed in Manual Part 14.

11.12   PROBLEMS

For Problems 1 through 6, use $\frac{5}{16}$-in. A36 angles, $\frac{3}{4}$-in. A325-N bolts in standard holes, and uncoped beams.

1. Design an all-bolted double-angle connection for a W18×50, A992 beam to carry a dead load reaction of 15 kips and a live load reaction of 45 kips. The beam is connected to the flange of a W14×109. Design by (a) LRFD and (b) ASD.

2. Design an all-bolted double-angle connection for a W27×102, A992 beam to carry a dead load reaction of 30 kips and a live load reaction of 90 kips. The beam is connected to the web of a W36×135. Design by (a) LRFD and (b) ASD.

3. Design an all-bolted double-angle connection for a W24×146, A992 beam to carry a dead load reaction of 25 kips and a live load reaction of 75 kips. The beam is connected to the flange of a W14×132. Design by (a) LRFD and (b) ASD.

4. Design an all-bolted double-angle connection for a W16×67, A992 beam to carry a dead load reaction of 20 kips and a live load reaction of 60 kips. The supporting member is not critical. Design by (a) LRFD and (b) ASD.

5. Design an all-bolted double-angle connection for a W18×143, A992 beam to carry a dead load reaction of 25 kips and a live load reaction of 75 kips. The supporting member is not critical. Design by (a) LRFD and (b) ASD.

6. Design an all-bolted double-angle connection for a W8×40, A992 beam to carry a dead load reaction of 8 kips and a live load reaction of 24 kips. The supporting member is not critical. Design by (a) LRFD and (b) ASD.

For Problems 7 through 12, use $\frac{5}{16}$-in. A36 angles, $\frac{3}{4}$-in. A325-N bolts in standard holes, and assume that the beams are coped so that the edge distance is $1\frac{1}{4}$ in. Assume that the supporting member is not critical.

7. Design an all-bolted double-angle connection for a W30×191, A992 beam spanning 40 ft and carrying a total uniformly distributed dead load of 60 kips and live load of 180 kips. Design by (a) LRFD and (b) ASD.

8. Design an all-bolted double-angle connection for a W18×76, A992 beam to support a dead load reaction of 16 kips and a live load reaction of 48 kips. Design by (a) LRFD and (b) ASD.

9. Design an all-bolted double-angle connection for a W21×68, A992 beam spanning 20 ft and carrying a total uniformly distributed dead load of 28 kips and live load of 84 kips. Design by (a) LRFD and (b) ASD.

10. Design an all-bolted double-angle connection for a W24×84, A992 beam to support a dead load reaction of 25 kips and a live load reaction of 75 kips. Design by (a) LRFD and (b) ASD.

11. Design an all-bolted double-angle connection for a W12×87, A992 beam to support a dead load reaction of 14 kips and a live load reaction of 42 kips. Design by (a) LRFD and (b) ASD.

12. Design an all-bolted double-angle connection for a W16×67, A992 beam spanning 20 ft and carrying a total uniformly distributed dead load of 23 kips and live load of 69 kips. Design by (a) LRFD and (b) ASD.

For Problems 13 through 18, use 70-ksi welding electrodes and a connection welded to the beam web being supported and bolted to the supporting member. Use $\frac{5}{16}$-in. A36 angles and $\frac{3}{4}$-in. A325-N bolts in standard holes.

13. Design a welded-bolted double-angle connection for an uncoped W18×50, A992 beam to carry a dead load reaction of 15 kips and a live load reaction of 45 kips. The beam is connected to the flange of a W14×109. Design by (a) LRFD and (b) ASD.

14. Design a welded-bolted double-angle connection for an uncoped W27×102, A992 beam to carry a dead load reaction of 30 kips and a live load reaction of 90 kips. The beam is connected to the web of a W36×135. Design by (a) LRFD and (b) ASD.

15. Design a welded-bolted double-angle connection for an uncoped W24×146, A992 beam to carry a dead load reaction of 25 kips and a live load reaction of 75 kips. The beam is connected to the flange of a W14×132. Design by (a) LRFD and (b) ASD.

16. Design a welded-bolted double-angle connection for a coped W18×76, A992 beam to support a dead load reaction of 16 kips and a live load reaction of 48 kips. Assume the beam is coped so that the edge distance is $1\frac{1}{4}$ in. Design by (a) LRFD and (b) ASD.

17. Design a welded-bolted double-angle connection for a coped W12×87, A992 beam to support a dead load reaction of 14 kips and a live load reaction of 42 kips. Assume the beam is coped so that the edge distance is $1\frac{1}{4}$ in. Design by (a) LRFD and (b) ASD.

18. Design a welded-bolted double-angle connection for a coped W16×67, A992 beam spanning 20 ft and carrying a total uniformly distributed dead load of 23 kips and live load of 69 kips. Assume the beam is coped so that the edge distance is $1\frac{1}{4}$ in. Design by (a) LRFD and (b) ASD.

For Problems 19 through 21, use $\frac{3}{8}$-in. A36 angles and $\frac{3}{4}$-in. A325-N bolts in standard holes.

19. Design a bolted-bolted single-angle connection for an uncoped W18×50, A992 beam to carry a dead load reaction of 8 kips and a live load reaction of 24 kips. The beam is connected to the web of a W360×150. Design by (a) LRFD and (b) ASD.

20. Design a welded-bolted single-angle connection for a coped W12×87, A992 beam to support a dead load reaction of 7 kips and a live load reaction of 21 kips. Assume the beam

is coped so that the edge distance is $1\frac{1}{4}$ in. Design by (a) LRFD and (b) ASD.

21. Design a welded-welded single-angle connection for a coped W16×67, A992 beam spanning 20 ft and carrying a total uniformly distributed dead load of 12 kips and live load of 36 kips. Assume the beam is coped so that the edge distance is $1\frac{1}{4}$ in. Design by (a) LRFD and (b) ASD.

For Problems 22 through 25, use a $\frac{3}{8}$-in. thick, A36, shear tab and $\frac{3}{4}$-in. A325-N bolts. Assume that the supporting member is not critical.

22. Design a shear tab connection for an uncoped W18×50, A992 beam to carry a dead load reaction of 10 kips and live load reaction of 30 kips.

23. Design a shear tab connection for an uncoped W27×102, A992 beam to carry a dead load reaction of 15 kips and a live load reaction of 45 kips.

24. Design a shear tab connection for a coped W21×68, A992 beam spanning 20 ft and carrying a total uniformly distributed dead load of 23 kips and live load of 70 kips. Assume that the edge distance at the cope is $1\frac{1}{4}$ in.

25. Design a shear tab connection for a coped W18×76, A992 beam to carry a dead load reaction of 10 kips and a live load reaction of 30 kips. Assume an edge distance of $1\frac{1}{4}$ in.

26. Design a welded seated connection for a W16×26, A992 beam framing into the web of a W14×99 column. The seat must carry a dead load reaction of 6 kips and a live load reaction of 18 kips. Use an equal leg A36 angle and E70 electrode.

27. Design a welded seated connection for a W18×40, A992 beam framing into the web of a W14×109 column. The seat must carry a dead load reaction of 10 kips and a live load reaction of 30 kips. Use an equal leg A36 angle and E70 electrode.

28. Design the connection for an A36 double-angle tension member connected to a uniform-width A36 gusset plate. The angles are 4 × 4 × $\frac{1}{2}$ and carry a dead load of 10 kips and a live load of 30 kips. The angles are connected to the gusset plate by a single line of $\frac{3}{4}$-in. A325-N bolts. The gusset plate is welded perpendicular to the axis of the member with welds from E70 electrodes. Design by (a) LRFD and (b) ASD.

29. An inclined WT hanger is used to support a tension member carrying a dead load of 8 kips and a live load of 24 kips. The force is applied at an angle of 45 degrees from the horizontal and is transferred by four- A325-N bolts. Determine whether the bolts have sufficient strength to carry the applied load by (a) LRFD and (b) ASD.

30. A WT7×24, A992 steel is used as a tension hanger with four $\frac{3}{4}$-in. A325-N bolts in the flanges similar to that shown in Figure 11.21. The hanger must resist a dead load of 11 kips and a live load of 33 kips. Determine whether prying action must be included to determine the connection strength by (a) LRFD and (b) ASD.

Chapter 12

Orange County Convention Center
Photo courtesy Walter P Moore.

Moment Connections

12.1 TYPES OF MOMENT CONNECTIONS

Although they are called moment connections, these connections are expected to transfer both shear and moment between the connected members. The moment connections defined by the Specification are either Type FR (fully restrained) or Type PR (partially restrained). The impact of these connection types on the behavior of a steel frame was discussed in Chapter 8. Figure 8.18, shown here as Figure 12.1, shows three moment-rotation curves for connections with distinctly different behavior. For fully restrained connections, the moment is transferred while the relative rotation of the members remains zero. For partially restrained connections, the moment is transferred while some predictable relative rotation is permitted. For the simple shear connection as discussed in Chapter 11, no moment is expected to be transferred and the connection is assumed to rotate freely. The rigid and simple connection behavior shown in Figure 12.1 illustrates that real connection behavior does not exactly follow the ideal behavior demonstrated by the vertical axis for a rigid connection and the horizontal axis for the simple connection.

Five moment connections that are common for connecting beams framing into the strong axis of columns are illustrated in Figure 12.2. For the first four examples, which

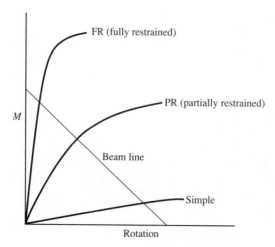

Figure 12.1 Connection Behavior.

include the direct welded flange (Figure 12.2a), the welded flange plate (Figure 12.2b), the bolted flange plate (Figure 12.2c), and the bolted Tee (Figure 12.2d), shear is transferred through a web connection similar to those discussed in Chapter 11 whereas the moment is transferred through the various flange connections. In the extended end plate connection, Figure 12.2e, shear and moment are combined and transferred through the connecting plate and bolts.

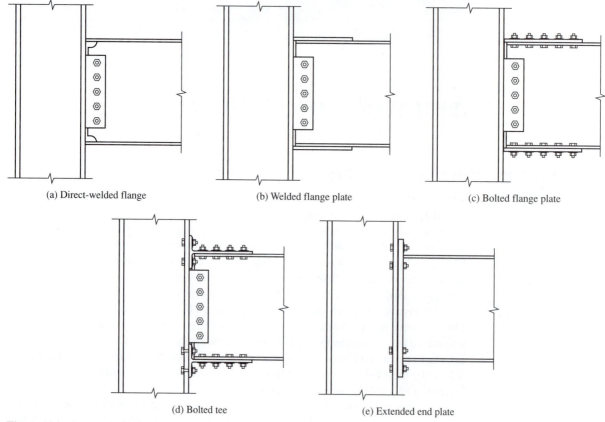

Figure 12.2 Moment Connections.

Because shear is resisted by the web of a wide flange beam, it is logical that the shear force is transferred through the web connection to the supporting member. Similarly, because the moment is resisted primarily through the flange of a wide flange beam, the flange connections primarily transfer the moment to the supporting member. Because of the moment resistance provided by the flanges, there is no need to consider eccentricity in the design of the web shear connection. Thus, the web plate or angles are sized to resist only shear, simplifying the connection design. The flange connection is designed to resist the full moment, even though the flanges do not actually carry this full moment. Through strain hardening, and in combination with some moment strength of the web connection, the flange connections are capable of developing the full moment.

The welded flange connection, Figure 12.2a, is the most direct moment connection and requires the fewest number of parts. The flanges are field welded to the supporting member with complete joint penetration groove welds. The web connection is usually a single plate welded to the column and bolted to the beam. In this arrangement, the flange force, P_f, is determined by dividing the moment by the distance between flange midpoints. Thus

$$P_f = \frac{M_r}{(d - t_f)}$$

The flange plated connections, Figure 12.2b and c, transfer the flange forces to the corresponding plates through either bolt shear or weld shear. The plate force is then transferred to the supporting member through welds. The flange plate connectors, bolts or welds, are sized to resist the force developed at the plate-flange interface. Thus

$$P_f = \frac{M_r}{d}$$

The bolted Tee connection, Figure 12.2d, is needed when the connection to the supporting member must be bolted. Although this connection is not as clean and simple as the flange-plated connections, it provides a solution for when there is a compelling reason to require an all-bolted connection. The connection to the beam flange is treated as with the flange plate connections and the connection to the support is treated similar to the tension connection discussed in Chapter 11.

The extended end plate connection shown in Figure 12.2e represents a connection that may take a variety of forms. The end plate is fully welded to the end of the beam and then bolted to the support. The end plate must extend beyond the beam flange on the tension side so that a minimum of four bolts can be symmetrically spaced with the flange located at the bolt centroid. If an extended end plate connection is called upon to resist a moment that is always in the same direction, it may be extended on only one side. However, if the moment is expected to reverse, the plate must be extended beyond both the top and bottom flanges.

Table 12.1 lists the sections of the Specification and parts of the Manual discussed in this chapter.

12.2   LIMIT STATES

The limit states that control the strength of these connections are the same as those that have already been considered for the shear connections. Their specific application depends on the complete connection geometry and the forces that the elements are expected to carry. These limit states include

1. Bolts
   a. Shear rupture
   b. Tension

Table 12.1 Sections of Specification and Parts of Manual Found in this Chapter

Specification	
B4	Classification of Sections for Local Buckling
D3	Area Determination
F13	Proportions of Beams and Girders
J2	Welds
J4	Affected Elements of Members and Connecting Elements
J10	Flanges and Webs with Concentrated Forces
Manual	
Part 7	Design Considerations for Bolts
Part 8	Design Considerations for Welds
Part 9	Design of Connecting Elements
Part 10	Design of Simple Shear Connections
Part 15	Design of Hanger Connections, Bracket Plates, and Crane-Rail Connections

    c. Shear-tension interaction

    d. Bearing/tear out

  2. Welds

    a. Tension rupture

    b. Shear rupture

  3. Plates

    a. Compression buckling

    b. Tension yielding

    c. Tension rupture

    d. Shear yielding

    e. Shear rupture

    f. Block shear

  4. Beam

    a. Flexure of reduced section

    b. Shear yield

    c. Shear rupture

In addition to these limit states, which are all associated with the beam side of the connection, the designer must consider the impact of the connection on the column to which it is attached. These limit states include

  5. Column

    a. Flange local bending

    b. Web local yielding

    c. Web local crippling

    d. Web compression buckling

    e. Web panel zone shear

12.3 MOMENT CONNECTION DESIGN

Design of moment connections is presented in two parts. First, examples are given for a direct welded beam-to-column connection, a welded flange plate connection, and a bolted flange plate connection. These examples treat the beam side of the connection without considering the column to which the connection is attached.

This is followed with a discussion of the limit states associated with the column and examples are given to illustrate that design process.

12.3.1   Direct Welded Flange Connection

The direct welded flange moment connection provides an FR connection with very few connecting elements. As the name implies, the flanges are directly welded to the supporting member, usually the flange of a column. These welds are either complete joint penetration groove welds or a pair of fillet welds on each side of the beam flanges. The groove weld provides a weld that can be made in the downward position for both flanges whereas the fillet welds require overhead welding on the bottom of each flange. The only limit state to consider for the flange connection is tension or shear rupture of the weld.

The web connection is usually made with a single plate, the same as the shear tab simple connection discussed in Chapter 11. However, unlike the shear tab connection, the web plate in this FR connection does not need to account for any eccentricity because the flanges are designed to carry all of the moment. The limit states for the web connection are those previously discussed for the shear tab.

EXAMPLE 12.1a
Direct Welded Moment Connection by LRFD

GOAL: Design a direct welded beam-to-column moment connection.

GIVEN: A direct welded beam-to-column moment connection is shown in Figure 12.2a. The beam is a W24×76 and the column is a W14×109. Bolts are $^3/_4$-in, A325-N and the electrodes are E70. The shapes are A992 steel and the plate is A36. The required strength is $M_u = 500$ ft-kips and $V_u = 60.0$ kips.

SOLUTION

Step 1:   Obtain the beam and column properties from Manual Table 1-1.

$$\text{Beam} - \text{W24} \times 76 \quad d = 23.9 \text{ in.} \quad b_f = 8.99 \text{ in.}$$
$$t_w = 0.440 \text{ in.} \quad t_f = 0.680 \text{ in.}$$
$$Z = 200 \text{ in.}^3$$

$$\text{Column} - \text{W14} \times 109 \quad d = 14.3 \text{ in.} \quad b_f = 14.6 \text{ in.}$$
$$t_w = 0.525 \text{ in.} \quad t_f = 0.860 \text{ in.}$$

Step 2:   Check the flexural strength of the beam.
This check should have been made during design of the beam. Because the beam section is not reduced because of bolt holes in the flange, M_p can be determined using the gross section plastic section modulus as

$$\phi M_n = \phi M_p = \frac{0.9(50)(200)}{12} = 750 \text{ ft-kips} > 500 \text{ ft-kips}$$

Thus, the flexural strength is adequate.

Step 3:   Design the flange-to-column weld.
The flange-to-column weld can be either a complete joint penetration groove weld (CJP) or fillet welds. CJP welds are used in this example. Because they will develop the full strength of the beam flanges, no further calculations are needed.

Step 4:   Design the web plate.
First consider the shear rupture of the bolts to determine the minimum number of bolts required.
For a $^3/_4$-in. A325-N bolt, $\phi r_n = 15.9$ kips, therefore

$$\text{Required number of bolts} = \frac{60}{15.9} = 3.77$$

Thus, try a four-bolt connection with bolt spacing of 3.0 in. and end distances of 1.5 in. Thus, $L = 12.0$ in., which is greater than $T/2 = 10.4$ in. Assume that the plate has $t = \frac{3}{8}$ in.

Step 5:   Determine the bolt bearing strength.
    For the last bolt, determine the clear distance.

$$L_c = 1.5 - \frac{1}{2}(3/4 + 1/16) = 1.09 < 2(3/4) = 1.5$$

Thus, tear-out controls and the bolt nominal strength is

$$R_n = 1.2(1.09)(0.375)(58) = 28.4 \text{ kips}$$

For the other bolts

$$L_c = 3.0 - (3/4 + 1/16) = 2.19 > 2(3/4) = 1.5$$

Therefore, bearing will control, and the bolt nominal strength is

$$R_n = 2.4(3/4)(0.375)(58) = 39.2 \text{ kips}$$

Thus, for the four-bolt connection, the design strength is

$$\phi R_n = 0.75(28.4 + 3(39.2)) = 110 > 60.0 \text{ kips}$$

Step 6:   Check the plate for shear yield.

$$A_{gv} = (0.375)(12.0) = 4.50 \text{ in.}^2$$
$$\phi V_n = 1.0(0.6(36))(4.50) = 97.2 > 60.0 \text{ kips}$$

Step 7:   Check the plate for shear rupture.

$$A_{nv} = (12.0 - 4(3/4 + 1/8))(0.375) = 3.19 \text{ in.}^2$$
$$\phi V_n = 0.75(0.6(58))(3.19) = 83.3 > 60.0 \text{ kips}$$

Step 8:   Check the block shear of the plate.
    First calculate the required areas.

$$A_{nt} = \left(1.5 - \frac{1}{2}(3/4 + 1/8)\right)(0.375) = 0.398 \text{ in.}^2$$
$$A_{gv} = 10.5(0.375) = 3.94 \text{ in.}^2$$
$$A_{nv} = (10.5 - 3.5(3/4 + 1/8))(0.375) = 2.79 \text{ in.}^2$$

Consider shear yield and shear rupture and select the least strength, thus

$$0.6 F_y A_{gv} = 0.6(36)(3.94) = 85.1 \text{ kips}$$
$$0.6 F_u A_{nv} = 0.6(58)(2.79) = 97.1 \text{ kips}$$

Selecting the shear yield term and combining it with the tension rupture term gives a connection block shear strength, with $U_{bs} = 1.0$, of

$$\phi R_n = 0.75(85.1 + 1.0(58)(0.398)) = 81.1 > 60.0 \text{ kips}$$

Step 9:   Check the beam web for bolt bearing.
    Because the beam is not coped, there is no need to check the clear distance for the top bolt. Thus, for each bolt $L_c = 2.19 > 2(\frac{3}{4}) = 1.5$ in. so the bolt nominal strength based on bearing is

$$R_n = 2.4(3/4)(0.440)(65) = 51.5 \text{ kips}$$

and for the four-bolt connection the design strength is

$$\phi R_n = 0.75(4(51.5)) = 155 > 60.0 \text{ kips}$$

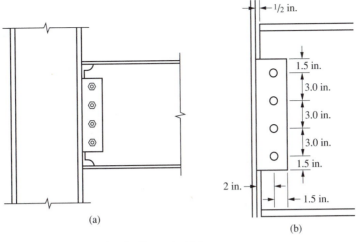

Figure 12.3 Connection for Example 12.1.

Step 10: Select the plate-to-column weld.

Based on a fillet weld on each side of the plate, with a weld design strength of 1.392 kips/in./sixteenth

$$D = \frac{60.0}{(2(1.392)(12.0))} = 1.80 \text{ sixteenths}$$

Therefore, use a $^3/_{16}$-in. weld, the minimum weld for the $^3/_8$-in. plate, as given in Specification Table J2.4.

Step 11: Final design.

Figure 12.3 shows the final design using

four $^3/_4$-in. A325N bolts in a $3^1/_2$- × 12.0- × $^3/_8$-in. plate

---

EXAMPLE 12.1b
Direct Welded Moment Connection by ASD

GOAL: Design a direct welded beam-to-column moment connection.

GIVEN: A direct welded beam-to-column moment connection is shown in Figure 12.2a. The beam is a W24×76 and the column is a W14×109. Bolts are $^3/_4$-in., A325-N and the electrodes are E70. The shapes are A992 steel and the plate is A36. The required strength is $M_a = 333$ ft-kips and $V_a = 40.0$ kips.

SOLUTION

Step 1: Obtain the beam and column properties from Manual Table 1-1.

$$\text{Beam} - \text{W24×76} \quad d = 23.9 \text{ in.} \quad b_f = 8.99 \text{ in.}$$
$$t_w = 0.440 \text{ in.} \quad t_f = 0.680 \text{ in.}$$
$$Z = 200.0 \text{ in.}^3$$

$$\text{Column} - \text{W14×109} \quad d = 14.3 \text{ in.} \quad b_f = 14.6 \text{ in.}$$
$$t_w = 0.525 \text{ in.} \quad t_f = 0.860 \text{ in.}$$

Step 2:   Check the flexural strength of the beam.

This check should have been made during design of the beam. Because the beam section is not reduced because of bolt holes in the flange, M_p can be determined using the gross section plastic section modulus as

$$\frac{M_n}{\Omega} = \frac{M_p}{\Omega} = \frac{(50)(200)}{1.67}\left(\frac{1}{12}\right) = 499 \text{ ft-kips} > 333 \text{ ft-kips}$$

Thus, the flexural strength is adequate.

Step 3:   Design the flange-to-column weld.

The flange-to-column weld can be either a complete joint penetration groove weld (CJP) or fillet welds. CJP welds are used in this example. Because they will develop the full strength of the beam flanges, no further calculations are needed.

Step 4:   Design the web plate.

First consider the shear rupture of the bolts to determine the minimum number of bolts required.

For a $^3/_4$-in., A325-N bolt, $\frac{r_n}{\Omega} = 10.6$ kips, therefore

$$\text{Required number of bolts} = \frac{40.0}{10.6} = 3.77$$

Thus, try a four-bolt connection with bolt spacing of 3.0 in. and end distances of 1.5 in. Thus, $L = 12.0$ in., which is greater than $T/2 = 10.4$ in. Assume that the plate has $t = ^3/_8$ in.

Step 5:   Determine the bolt bearing strength.

For the last bolt, determine the clear distance.

$$L_c = 1.5 - \frac{1}{2}(3/4 + 1/16) = 1.09 < 2(3/4) = 1.5$$

Thus tear-out controls and the bolt nominal strength is

$$R_n = 1.2(1.09)(0.375)(58) = 28.4 \text{ kips}$$

For the other bolts

$$L_c = 3.0 - (3/4 + 1/16) = 2.19 > 2(3/4) = 1.5$$

Therefore, bearing will control, and the bolt nominal strength is

$$R_n = 2.4(3/4)(0.375)(58) = 39.2 \text{ kips}$$

Thus, for the four-bolt connection, the allowable strength is

$$\frac{R_n}{\Omega} = \frac{(28.4 + 3(39.2))}{2.00} = 73.0 > 40.0 \text{ kips}$$

Step 6:   Check the plate for shear yield.

$$A_{gv} = (0.375)(12.0) = 4.50 \text{ in.}^2$$

$$V_n = \frac{(0.6(36))(4.50)}{1.5} = 64.8 > 40.0 \text{ kips}$$

Step 7:   Check the plate for shear rupture.

$$A_{nv} = (12.0 - 4(3/4 + 1/8))(0.375) = 3.19 \text{ in.}^2$$

$$\frac{V_n}{\Omega} = \frac{(0.6(58))(3.19)}{2.00} = 55.5 > 40.0 \text{ kips}$$

Step 8:   Check the block shear of the plate.
First calculate the required areas.

$$A_{nt} = \left(1.5 - \frac{1}{2}(3/4 + 1/8)\right)(0.375) = 0.398 \text{ in.}^2$$

$$A_{gv} = 10.5(0.375) = 3.94 \text{ in.}^2$$

$$A_{nv} = (10.5 - 3.5(3/4 + 1/8))(0.375) = 2.79 \text{ in.}^2$$

Consider shear yield and shear rupture and select the least strength, thus

$$0.6F_y A_{gv} = 0.6(36)(3.94) = 85.1 \text{ kips}$$

$$0.6F_u A_{nv} = 0.6(58)(2.79) = 97.1 \text{ kips}$$

Selecting the shear yield term and combining it with the tension rupture term gives a connection block shear allowable strength, with $U_{bs} = 1.0$, of

$$\frac{R_n}{\Omega} = \frac{(85.1 + 1.0(58)(0.398))}{2.00} = 54.1 > 40.0 \text{ kips}$$

Step 9:   Check the beam web for bolt bearing.
Because the beam is not coped, there is no need to check the clear distance for the top bolt. Thus, for each bolt $L_c = 2.19 > 2(\frac{3}{4}) = 1.5$ in. so the bolt nominal strength based on bearing is

$$R_n = 2.4(3/4)(0.440)(65) = 51.5 \text{ kips}$$

and for the four-bolt connection, the allowable strength is

$$\frac{R_n}{\Omega} = \frac{(4(51.5))}{2.00} = 103 > 40.0 \text{ kips}$$

Step 10:   Select the plate-to-column weld.
Based on a fillet weld on each side of the plate, with a weld allowable strength of 0.928 kips/in./sixteenth

$$D = \frac{40.0}{(2(0.928)(12.0))} = 1.80 \text{ sixteenths}$$

Therefore, use a $\frac{3}{16}$-in. weld, the minimum weld for the $\frac{3}{8}$-in. plate, as given in Specification Table J2.4.

Step 11:   Final design.
Figure 12.3 shows the final design using

> four $\frac{3}{4}$-in., A325N bolts in a $3\frac{1}{2}$- $\times$ 12.0- $\times$ $\frac{3}{8}$-in. plate

12.3.2   Welded Flange Plate Connection

The welded flange plate connection replaces the directly welded flanges with plates that are welded to the supported beam flange and to the supporting column. The web connection is usually the typical single plate shear connection. To accommodate the plate-to-beam flange weld, the top flange plate must be kept to a width at least 1.0 in. less than the beam flange width and the bottom flange plate is at least 1.0 in. greater in width than the beam flange width. Manual Figure 8-11 provides minimum shelf dimensions for specific fillet weld sizes.

The limit states associated with the tension flange plate are yielding, rupture, and block shear whereas those associated with the compression flange plate are yielding, local plate buckling, and compression buckling. These limit states are evaluated in Example 12.2.

EXAMPLE 12.2a
Welded Flange Plate
Moment Connection
by LRFD

GOAL: Design a welded flange plate beam-to-column moment connection.

GIVEN: A welded flange plate beam-to-column moment connection is shown in Figure 12.2b. The beam is a W18×50 and the column is a W14×90. Bolts are $\frac{7}{8}$-in. A325-N and the electrodes are E70. The shapes are A992 steel and the plates are A36. The LRFD required strength is $M_u = 250$ ft-kips and $V_u = 45$ kips. Assume the moment will cause the top flange to be in tension and the bottom flange to be in compression.

SOLUTION

Step 1:   Obtain the beam and column properties from Manual Table 1-1.

$$\text{Beam} - \text{W18} \times 50 \quad d = 18.0 \text{ in.} \quad b_f = 7.50 \text{ in.}$$
$$t_w = 0.355 \text{ in.} \quad t_f = 0.570 \text{ in.}$$
$$\text{Column} - \text{W14} \times 90 \quad d = 14.0 \text{ in.} \quad b_f = 14.5 \text{ in.}$$
$$t_w = 0.440 \text{ in.} \quad t_f = 0.710 \text{ in.}$$

Step 2:   Determine the force to be carried in each flange plate.
Conservatively assume the moment arm is the depth of the beam. Thus

$$P_u = \frac{M_u}{d} = \frac{250(12)}{18.0} = 167 \text{ kips}$$

Step 3:   Determine the minimum plate area based on the limit state of yielding.

$$A_p = \frac{P_u}{\phi F_y} = \frac{167}{0.9(36)} = 5.15 \text{ in.}^2$$

The top flange plate should be narrower than the beam flange to facilitate welding in the down position. Therefore, try a $\frac{7}{8}$- $\times$ 6.5-in. plate. $A_p = 5.69$ in.^{2}

Step 4:   Check the plate for tension rupture.
The shear lag factor, U, for a welded joint is given in Specification Table D3.1 Case 4. Here it is noted that the lowest value for U is 0.75. This value is used as a conservative approach at this time. Thus

$$P_n = UF_u A_n = 0.75(58)(5.69) = 248 \text{ kips}$$

and

$$\phi P_n = 0.75(248) = 186 > 167 \text{ kips}$$

Thus, the tension rupture limit state does not control regardless of the final weld length.

Step 5:   Select the fillet weld size based on weld rupture.
The minimum size weld for a $\frac{7}{8}$-in. plate attached to a 0.570-in. beam flange, based on Specification Table J2.4, is $\frac{1}{4}$ in. Therefore, determine the required length of a pair of $\frac{1}{4}$-in. fillet welds on the sides of the flange plate.

$$L = \frac{167}{2(1.392(4))} = 15.0 \text{ in.}$$

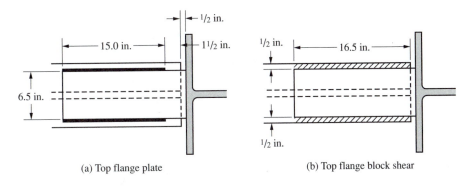

(a) Top flange plate

(b) Top flange block shear

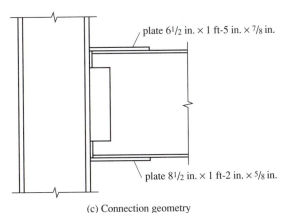

plate 6½ in. × 1 ft-5 in. × ⅞ in.

plate 8½ in. × 1 ft-2 in. × ⅝ in.

(c) Connection geometry

Figure 12.4 Connection for Example 12.2.

This length appears to be reasonable for this connection. Thus, the top plate is 6.5- × 17.0- × ⅞-in., as shown in Figure 12.4.

Step 6:   Consider the block shear rupture of the top flange of the beam.

Because the plate is welded to the flange, the critical shear limit state is shear yielding. For the two blocks on each side of the web, as shown in Figure 12.4b, the required areas are

$$A_{gv} = 16.5(0.570) = 9.41 \text{ in.}^2$$
$$A_{nt} = \tfrac{1}{2}(0.570) = 0.285 \text{ in.}^2$$

and the design strength is

$$\phi R_n = 2[0.75(0.6(50)(9.41) + 1.0(65))(0.285)] = 451 > 167 \text{ kips}$$

Step 7:   Determine the required compression flange plate.

This plate must be checked for local buckling and overall buckling. As a starting point, assume that yielding will be the controlling limit state. Thus, the same area will be required as for the tension plate. However, this plate should be wider than the beam flange so that the welds can again be placed in the downward position. Assume a plate width of 8.5 in. Thus

$$t_p = \frac{5.15}{8.5} = 0.606 \text{ in.}$$

Select a ⅝-in. plate for further consideration.

Step 8:   Check the compression plate for local buckling.

Local buckling of the compression plate is checked with the width/thickness limits from Specification Table B4.1. The width of plate between welds is treated as a stiffened plate and the width that projects beyond the weld is treated as an unstiffened plate.

For the stiffened plate, Case 14

$$\frac{b}{t} = \frac{7.5}{0.625} = 12.0 < 1.49\sqrt{\frac{E}{F_y}} = 42.3$$

For the unstiffened plate, Case 3

$$\frac{b}{t} = \frac{0.5}{0.625} = 0.80 < 0.56\sqrt{\frac{E}{F_y}} = 15.9$$

So the plate strength is not limited by local buckling.

Step 9:   Determine the compressive strength of the plate.

The plate is assumed to have a length for compression buckling of 2.0 in. from the column flange to the end of the weld, as shown in Figure 12.4a. The effective length factor is taken as 0.65, the value recommended in the Commentary for a fixed-fixed column. Determine the slenderness ratio for this plate.

$$r = \sqrt{\frac{I}{A}} = \sqrt{\frac{h^2}{12}} = \sqrt{\frac{(0.625)^2}{12}} = 0.180$$

$$\frac{KL}{r} = \frac{0.65(2.0)}{0.180} = 7.22$$

For compression elements that are part of connections, Specification Section J4.4 indicates that, when the slenderness ratio is less than 25, $F_{cr} = F_y$. Thus, the selection of this plate for yielding, as was originally done, is correct and the 8.5- × $^5/_8$-in. plate is acceptable for the compression limit states.

Step 10:   Determine the welds required to connect the flange plates to the column flange.

The force to be transferred is the same for both plates. A comparison of the plate width with the column flange width shows that they are compatible because each plate is narrower than the column flange width, $b_f = 14.5$ in. In addition, the force is perpendicular to the weld so the weld strength can be increased by 1.5. Thus, for fillet welds on both the top and bottom of the plate

$$D = \frac{167}{1.5(1.392)(2b_p)} = \frac{40.0}{b_p}$$

For the top flange plate

$$D = \frac{40.0}{6.5} = 6.15 \text{ sixteenths, therefore use a pair of } ^7/_{16}\text{-in. welds}$$

For the bottom flange plate

$$D = \frac{40.0}{8.5} = 4.71 \text{ sixteenths, therefore use a pair of } ^5/_{16}\text{-in. welds}$$

Step 11:   Final design.

The web connection design that was demonstrated in Example 12.1a must also be carried out here. The final geometry for the welded flange plate connection is shown in Figure 12.4.

EXAMPLE 12.2b
Welded Flange Plate
Moment Connection
by ASD

GOAL: Design a welded flange plate beam-to-column moment connection.

GIVEN: A welded flange plate beam-to-column moment connection is shown in Figure 12.2b. The beam is a W18×50 and the column is a W14×90. Bolts are $^7/_8$-in., A325-N and the electrodes are E70. The shapes are A992 steel and the plates are A36. The ASD required strength is $M_a = 167$ ft-kips and $V_a = 30$ kips. Assume the moment will cause the top flange to be in tension and the bottom flange to be in compression.

SOLUTION

Step 1:   Obtain the beam and column properties from Manual Table 1-1.

$$\text{Beam} - \text{W18}\times 50 \quad d = 18.0 \text{ in.} \quad b_f = 7.50 \text{ in.}$$
$$t_w = 0.355 \text{ in.} \quad t_f = 0.570 \text{ in.}$$

$$\text{Column} - \text{W14}\times 90 \quad d = 14.0 \text{ in.} \quad b_f = 14.5 \text{ in.}$$
$$t_w = 0.440 \text{ in.} \quad t_f = 0.710 \text{ in.}$$

Step 2:   Determine the force to be carried in each flange plate.
Conservatively assume the moment arm is the depth of the beam. Thus

$$P_a = \frac{M_a}{d} = \frac{167(12)}{18.0} = 111 \text{ kips}$$

Step 3:   Determine the minimum plate area based on the limit state of yielding.

$$A_p = \frac{\Omega P_a}{F_y} = \frac{1.67(111)}{(36)} = 5.15 \text{ in.}^2$$

The top flange plate should be narrower than the beam flange to facilitate welding in the down position. Therefore, try a $^7/_8$- × 6.5-in. plate. $A_p = 5.69$ in.^{2}

Step 4:   Check the plate for tension rupture.
The shear lag factor, U, for a welded joint is given in Specification Table D3.1 Case 4. Here it is noted that the lowest value of U is 0.75. This value is used as a conservative approach at this time. Thus

$$P_n = UF_u A_n = 0.75(58)(5.69) = 248 \text{ kips}$$

and

$$\frac{P_n}{\Omega} = \frac{(248)}{2.00} = 124 > 111 \text{ kips}$$

Thus, the tension rupture limit state does not control regardless of the final weld length.

Step 5:   Select the fillet weld size based on weld rupture.
The minimum size weld for a $^7/_8$-in. plate attached to a 0.570-in. beam flange, based on Specification Table J2.4, is $^1/_4$ in. Therefore, determine the required length of a pair of $^1/_4$-in. fillet welds on the sides of the flange plate.

$$L = \frac{111}{2(0.928(4))} = 15.0 \text{ in.}$$

This length appears to be reasonable for this connection. Thus, the top plate is 6.5- × 17.0- × $^7/_8$-in., as shown in Figure 12.4.

Step 6:   Consider the block shear rupture of the top flange of the beam.
Because the plate is welded to the flange, the critical shear limit state is shear yielding. For the two blocks on each side of the web, as shown in Figure 12.4b, the

required areas are

$$A_{gv} = 16.5(0.570) = 9.41 \text{ in.}^2$$

$$A_{nt} = \frac{1}{2}(0.570) = 0.285 \text{ in.}^2$$

and the allowable strength is

$$\frac{R_n}{\Omega} = 2\left[\frac{(0.6(50)(9.41) + 1.0(65))(0.285)}{2.00}\right] = 300 > 111 \text{ kips}$$

Step 7: Determine the required compression flange plate.

This plate must be checked for local buckling and overall buckling. As a starting point, assume that yielding will be the controlling limit state. Thus, the same area will be required as for the tension plate. However, this plate should be wider than the beam flange so that the welds can again be placed in the downward position. Assume a plate width of 8.5 in. Thus

$$t_p = \frac{5.15}{8.5} = 0.606 \text{ in.}$$

Select a $\frac{5}{8}$-in. plate for further consideration.

Step 8: Check the compression plate for local buckling.

Local buckling of the compression plate is checked with the width/thickness limits from Specification Table B4-1. The width of plate between welds is treated as a stiffened plate and the width that projects beyond the weld is treated as an unstiffened plate.

For the stiffened plate, Case 14

$$\frac{b}{t} = \frac{7.5}{0.625} = 12.0 < 1.49\sqrt{\frac{E}{F_y}} = 42.3$$

For the unstiffened plate, Case 3

$$\frac{b}{t} = \frac{0.5}{0.625} = 0.80 < 0.56\sqrt{\frac{E}{F_y}} = 15.9$$

So the plate strength is not limited by local buckling.

Step 9: Determine the compressive strength of the plate.

The plate is assumed to have a length for compression buckling of 2.0 in. from the column flange to the end of the weld, as shown in Figure 12.4a. The effective length factor is taken as 0.65, the value recommended in the Commentary for a fixed-fixed column. Determine the slenderness ratio for this plate.

$$r = \sqrt{\frac{I}{A}} = \sqrt{\frac{h^2}{12}} = \sqrt{\frac{(0.625)^2}{12}} = 0.180$$

$$\frac{KL}{r} = \frac{0.65(2.0)}{0.180} = 7.22$$

For compression elements that are part of connections, Specification Section J4.4 indicates that, when the slenderness ratio is less than 25, $F_{cr} = F_y$. Thus, the selection of this plate for yielding, as was originally done, is correct and the 8.5- $\times$ $\frac{5}{8}$-in. plate is acceptable for the compression limit states.

Step 10: Determine the welds required to connect the flange plates to the column flange.

The force to be transferred is the same for both plates. A comparison of the plate width with the column flange width shows that they are compatible because each plate is narrower than the column flange width, $b_f = 14.5$ in. In addition, the force is perpendicular to the weld so the weld strength can be increased by 1.5. Thus, for fillet welds on both the top and bottom of the plate

$$D = \frac{111}{1.5(0.928)(2b_p)} = \frac{39.9}{b_p}$$

For the top flange plate

$$D = \frac{39.9}{6.5} = 6.14 \text{ sixteenths, therefore use a pair of } ^7/_{16}\text{-in. welds}$$

For the bottom flange plate

$$D = \frac{39.9}{8.5} = 4.69 \text{ sixteenths, therefore use a pair of } ^5/_{16}\text{-in. welds}$$

Step 11: Final design.
  The web connection design that was demonstrated in Example 12.1b must also be carried out here. The final geometry for the welded flange plate connection is shown in Figure 12.4.

12.3.3   Bolted Flange Plate Connection

The bolted flange plate connection is similar to the welded flange plate connection except that the attachment of the plate to the beam flange is through bolts. The addition of bolts to the beam tension flange means that a new limit state, the flexural strength of the beam based on rupture of the tension flange, must be assessed. The other limit states that result from the use of bolts have been described several times and are applicable again here.

  The bolted flange plate connection is an effective connection from the erection standpoint. The plates can be shop-welded to the column flange and the beam inserted between the plates and bolted in the field. To accommodate this field erection process, the top plate is usually set a bit high and a filler used once the beam is in place.

  The following example demonstrates the limit state checks associated with the transfer of the flange force, as was done for Example 12.2. The web connection will not be designed because no new limit states are to be considered.

EXAMPLE 12.3a
Bolted Flange Plate
Moment Connection
by LRFD

GOAL: Design a bolted flange plate beam-to-column moment connection.

GIVEN: A bolted flange plate beam-to-column moment connection is shown in Figure 12.2c. This connection is to be designed for the same conditions as those in Example 12.2a. The beam is a W18×50 and the column is a W14×90. Bolts are $^7/_8$-in., A325-N and the electrodes are E70. The shapes are A992 steel and the plate is A36. The required strength is $M_u = 250$ ft-kips and $V_u = 45.0$ kips. Assume the moment will cause the top flange to be in tension and the bottom flange to be in compression.

SOLUTION

Step 1: Determine the beam and column properties.
  The member dimensions are the same as those given for Example 12.2a. In addition, for the W18×50, from Manual Table 1-1, $S_x = 88.9$ in.[3]

Step 2: Check the reduced beam section for flexure.
  Although the connection has not yet been designed, it is known that at a section through the connection, there will be two bolt holes in the tension flange. This may

reduce the strength of the beam below the required strength. If that is the case, there will be no reason to continue with this connection design. Thus, the provisions of Specification Section F13 must be applied for the limit state of rupture of the tension flange.

Determine the gross and net areas of the tension flange.

$$A_{fg} = 7.5(0.570) = 4.28 \text{ in.}^2$$

$$A_{fn} = (7.5 - 2(7/8 + 1/8))(0.570) = 3.14 \text{ in.}^2$$

Check the yield stress to tensile strength ratio to determine a value for Y_t.

$$\frac{F_y}{F_t} = \frac{50}{65} = 0.76 < 0.8$$

Therefore, for all A992 shapes, $Y_t = 1.0$ and for this W18×50 beam with a pair of holes for $7/8$-in. bolts

$$F_y A_{fg} = 50(4.28) = 214 \text{ kips}$$

$$F_u A_{fn} = 65(3.14) = 204 \text{ kips}$$

Because $F_u A_{fn} < Y_t F_y A_{fg}$, the nominal moment strength is limited by Specification Equation F13-1 to

$$M_n = \frac{F_u A_{fn}}{A_{fg}} S_x = \frac{204}{4.28}(88.9)\left(\frac{1}{12}\right) = 353 \text{ ft-kips}$$

and

$$\phi M_n = 0.9(353) = 318 > 250 \text{ ft-kips}$$

So the flexural strength is adequate.

Step 3:   Check the flange plate for tension yield.

The flange plate will likely be similar to the one used in Example 12.2a. Try a $7\frac{1}{4}$- × $\frac{3}{4}$-in. plate

$$A_g = 7.25(0.750) = 5.44 \text{ in.}^2$$

$$\phi R_n = 0.9(36)(5.44) = 176 > 167 \text{ kips}$$

Step 4:   Check the plate for tension rupture.

$$A_n = (7.25 - 2(7/8 + 1/8))(0.750) = 3.94 \text{ in.}^2$$

$$\phi R_n = 0.75(58)(3.94) = 171 > 167 \text{ kips}$$

So the plate size is adequate based on tension.

Step 5:   Determine the number of bolts required based on the bolt shear rupture.

First consider the shear rupture of the bolts to determine the minimum number of bolts required.

For a $7/8$-in. bolt, from Manual Table 7-1, $\phi r_n = 21.6$ kips, therefore

$$\text{Required number of bolts} = \frac{167}{21.6} = 7.73$$

Thus, try an eight-bolt connection with bolt spacing of 3.0 in. and end distances of at least twice the bolt diameter so that the full bolt strength can be used.

Step 6:   Determine the bolt bearing strength on the plate.

$$R_n = 2.4(7/8)(0.750)(58) = 91.4 \text{ kips}$$

Thus, for the eight-bolt connection in the plate, the design strength is

$$\phi R_n = 0.75(8)(91.4) = 548 > 167 \text{ kips}$$

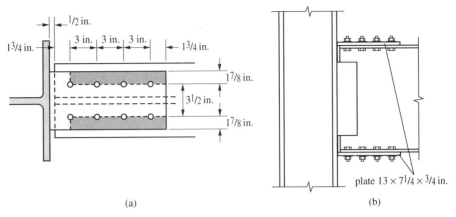

Figure 12.5 Connection for Example 12.3.

Step 7:   Determine the bolt bearing strength on the beam flange, again assuming that all bolts have sufficient clear distance to be controlled by bearing.

$$R_n = 2.4(7/8)(0.570)(65) = 77.8 \text{ kips}$$

Thus, for the eight-bolt connection in the beam flange, the design strength is

$$\phi R_n = 0.75(8)(77.8) = 467 > 167 \text{ kips}$$

The assumption of an end distance of $2d_b$ is not a problem because if only six bolts were considered, the limit state of bolt bearing would still not be critical.

Step 8:   Check the plate for block shear rupture.
      Check the plate for block shear using the geometry shown in Figure 12.5. Because there are two possible block shear failure patterns, one with the center portion failing in tension and the other with the two outside portions failing in tension, the worst case must be identified. The critical tension area for block shear will be the one associated with the least tension width. In this case it will be for the middle $3\frac{1}{2}$ in.-section and the critical net tension area is

$$A_{nt} = (3.5 - (7/8 + 1/8))(0.750) = 1.88 \text{ in.}^2$$

and the shear areas are

$$A_{gv} = 10.75(0.750) = 8.06 \text{ in.}^2$$
$$A_{nv} = (10.75 - 3.5(7/8 + 1/8))(0.750) = 5.44 \text{ in.}^2$$

Consider the shear yield and shear rupture and select the least strength, thus

$$0.6F_y A_{gv} = 0.6(36)(8.06) = 174 \text{ kips}$$

$$0.6F_u A_{nv} = 0.6(58)(5.44) = 189 \text{ kips}$$

Selecting the shear yield term and combining it with the tension rupture term gives a connection design block shear strength, with $U_{bs} = 1.0$, of

$$\phi R_n = 0.75(174 + 1.0(58)(1.88)) = 212 > 167 \text{ kips}$$

Step 9:   Check the beam flange for block shear.
      In this case, the beam web prevents a block shear failure in the middle portion so check the sum of the two outer portions.

$$A_{nt} = 2\left(2.00 - \frac{1}{2}(7/8 + 1/8)\right)(0.570) = 1.71 \text{ in.}^2$$

Step 12:   Determine the welds required to connect the flange plates to the column flange.
   The force to be transferred is the same for both plates. A comparison of the plate width with the column flange width shows that they are compatible because the plates are narrower than the column flange width, $b_f = 14.5$ in. In addition, the force is perpendicular to the weld so the weld strength can be increased by 1.5. Thus, for fillet welds on both the top and bottom of the plate

$$D = \frac{111}{1.5(0.928)(2b_p)} = \frac{39.9}{b_p}$$

For both flange plates

$$D = \frac{39.9}{7.25} = 5.50 \text{ sixteenths}$$

therefore, use a pair of $\frac{3}{8}$-in. welds which exceeds the minimum for this plate thickness.

Step 13:   Consider the web connection.
   The same web shear connection as was used in the welded flange plate connection could be used in this connection.

Step 14:   Final design.
   The flange plates of this connection are 13- $\times$ $7\frac{1}{4}$- $\times$ $\frac{3}{4}$-in. with eight $\frac{7}{8}$-in., A325N bolts, as shown in Figures 12.5a and b.

12.4   COLUMN STIFFENING

The connection designs illustrated in the previous examples treated the beam side of the connection. That is, they looked at only the connecting elements and their influence on the beam to which they were attached. They did not consider, however, the influence of the connection and transfer of forces to the supporting element. Normally, a fully restrained moment connection is made to the flange of a column. This is the most efficient use of the column because the strong axis is resisting the transferred moment.

As with each connecting element, the application of force to a supporting element requires a check of all applicable limit states. The typical moment connection, like those illustrated in Figures 12.2a through 12.2c, results in the transfer of a concentrated force to the column flange. The limit states for flanges and webs with concentrated forces that are applicable to the beam-column connection are defined in Specification Section J10 as

1. Flange local bending
2. Web local yielding
3. Web crippling
4. Web compression buckling
5. Web panel zone shear

Application of these limit states vary, depending on whether the applied force is tension or compression and whether the connection is on one side or both sides of the column. If the limit states are exceeded, either the column section should be changed or stiffeners and web doubler plates are required.

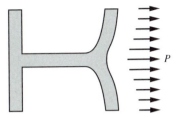

Figure 12.6 Flange Local Bending.

12.4.1 Flange Local Bending

Flange local bending (J10.1) is illustrated in Figure 12.6. This limit state is applicable only where a tensile force is applied to the column flange. The primary concern addressed through this limit state is the stress distribution in the weld if the flange deformation is excessive. Thus, the limit on the applied force is set in order to prevent excessive deformation. The nominal strength for flange local bending is

$$R_n = 6.25t_f^2 F_{yf}$$

and

$$\phi = 0.9 \, (\text{LRFD}) \qquad \Omega = 1.67 \, (\text{ASD})$$

If the force is applied over a small central portion of the column flange, less than 15% of the flange width, this limit state does not need to be checked because the force is applied close to the column web and very little flange deformation occurs.

If the force is applied close to the end of the column, the distribution of the force within the flange is limited by the proximity of the end of the column and the resulting deflection increases. Thus, if the force is applied closer than $10t_f$ to the end of the member, the nominal strength must be reduced by 50%. When this limit state is exceeded, a pair of half-depth transverse stiffeners are needed.

12.4.2 Web Local Yielding

Web local yielding (J10.2) is the same limit state that was considered for bearing of the web in a seated connection as discussed in both Chapter 11 and Section 6.14. Although previously discussed for compressive forces, this limit state is also applicable to tensile forces. Figure 12.7 illustrates the application of a concentrated force to the web of the

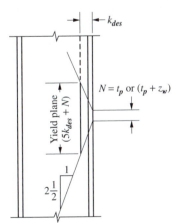

Figure 12.7 Distribution of Concentrated Forces on Column Web.

column. This force could be transfered through a directly welded beam flange or a beam flange plate. The bearing length N is taken as the thickness of the plate applying the force, or the plate thickness plus the weld width when attached with fillet welds. The force is distributed in both directions, provided the connection is at least the depth of the column from the column end. This distribution is on a slope of 2.5:1 over the depth given by k_{des} in Manual Table 1-1. In this case

$$R_n = (5k_{des} + N)F_y t_w$$

$$\phi = 1.0 \,(\text{LRFD}) \quad \Omega = 1.50 \,(\text{ASD})$$

If the connection is closer to the end of the column than the depth of the column, d, the distribution of the force can take place in only one direction and the factor 5 is replaced by 2.5. This is the relationship that was used for the seated connection. When this limit state is exceeded, a pair of half-depth stiffeners or a web doubler plate is needed.

12.4.3 Web Crippling

Web crippling (J10.3) applies only to compressive forces. It is the limit state that predicts the crumpling of the web beneath a compressive force. It is similar to local web yielding but occurs in more slender webs whereas local web yielding occurs in more stocky webs.

Web crippling strength depends on how close the force is applied with respect to the column end. For illustration here, it is assumed that the force is at least $d/2$ from the column end. When this limit state was considered for the seated connection, the force was assumed to be applied less than $d/2$ from the end. The nominal strength for this case is given by Specification Equation J10-4 as

$$R_n = 0.80t_w^2 \left[1 + 3\left(\frac{N}{d}\right)\left(\frac{t_w}{t_f}\right)^{1.5} \right] \sqrt{\frac{EF_y t_f}{t_w}}$$

$$\phi = 0.75 \,(\text{LRFD}) \quad \Omega = 2.00 \,(\text{ASD})$$

If the web crippling limit state is exceeded, a pair of half-depth stiffeners or a half-depth doubler plate are required.

12.4.4 Web Compression Buckling

Web compression buckling (J10.5) applies only when compressive forces are applied on opposite sides of the column, putting the web into compression. If the forces are close to the end of the column (less than $d/2$), the strength is reduced by 50%. The strength is given by Specification Equation J10-8 as

$$R_n = \frac{24t_w^3 \sqrt{EF_y}}{h}$$

$$\phi = 0.90 \,(\text{LRFD}) \quad \Omega = 1.67 \,(\text{ASD})$$

If the web compression buckling limit state is exceeded, a single full-depth stiffener, a pair of full-depth stiffeners, or a full-depth doubler plate are required.

12.4.5 Web Panel Zone Shear

Web panel zone shear (J10.6) within the boundaries of a fully rigid connection may be significant. The strength of the panel zone is based on shear yielding of the web unless a significant axial force also exists. In this case, shear/axial interaction is considered. When the effects of panel zone shear are not included in the structural analysis, the panel zone is expected to behave elastically. If the behavior of the panel zone is included in the structural analysis, the nonlinear behavior of the panel zone can be included and its strength increased accordingly. When panel-zone deformations are not included in the analysis, panel zone strength given by Specification Section J10.6 is

For $P_r \leq 0.4P_c$

$$R_n = 0.60F_y d_c t_w$$

For $P_r > 0.4P_c$

$$R_n = 0.60F_y d_c t_w \left(1.4 - \frac{P_r}{P_c}\right)$$

and

$$\phi = 0.90 \,(\text{LRFD}) \qquad \Omega = 1.67 \,(\text{ASD})$$

where P_r is the required strength and P_c is the yield strength, P_y for LRFD or $0.6P_y$ for ASD. The panel zone strength must be sufficient to resist the total shear in the panel zone, including the story shear carried by the column web. When this limit state is exceeded, a full-depth doubler plate is required.

In the discussion of each of these limit states, the concluding statement indicated that if the limit state was exceeded, a stiffener or doubler plate was required. Thus, this is a "go-no go" decision. It is possible that a stiffener may be required by a very small margin for only one of these limit states. Unfortunately, stiffeners are an expensive element to add to a connection, especially if they must be fitted between the column flanges as for a full-depth stiffener. In many cases, it is much more economical to have selected a column section that may be larger than required for the axial load but avoids the requirement of stiffeners. Stiffener requirements should not be left for the detailing stage of the design process, but addressed early in the design process so that these requirements are considered at a point in time when member sizes can still be revised.

If stiffeners cannot be avoided, they are designed to resist a force calculated as the applied force, either tension or compression, minus the resisting force as defined for each limit state. This net force is resisted by the cross section of the stiffeners, which are sized based on the provisions for tension or compression connecting elements in Specification Section J4. The Specification provides additional criteria for stiffeners and doubler plates in Section J10.8 and J10.9.

The arbitrary dimensional requirements for stiffeners are as follows:

1. The width of each stiffener plus half the column web thickness must be greater than one-third of the attached plate width, $b_s \geq (b_p/3 - t_w/2)$.
2. The thickness of the stiffener must be at least half the thickness of the attached plate and at least the plate width divided by 15, $t_s \geq t_p/2$, and $t_s \geq b_p/15$.
3. Transverse stiffeners must also extend at least one-half the depth of the column.

The strength requirements are also found in these sections. For stiffeners that resist tension forces, the provisions of Specification Chapter D must be satisfied. The weld between the loaded flange and stiffener must be sized to transfer the load that must be carried by the stiffener, and the weld to the web must transfer the difference between the forces on each end of the stiffener.

The strength of a compression stiffener must satisfy the requirements for compression connecting elements found in Specification Section J4.4. The stiffener may be designed to bear on the loaded flange or welded to transfer the force that the stiffener is required to resist. The weld to the web is designed to transfer the difference between the forces on the ends of the stiffener.

Doubler plates, when needed, must be designed for the forces they are required to resist according to the provisions for those forces. These include: for compression the provisions of Chapter E, for tension the provisions of Chapter D, and for shear the provisions of Chapter G. Additional limitations are:

1. The plate thickness and size must provide sufficient additional material to equal or exceed the strength requirements.

2. The welds of the doubler plate to the column web must develop the force transmitted to the doubler plate.

EXAMPLE 12.4a
Column Side Limit States by LRFD

GOAL: Check the column side limit states for a moment connection and design any needed stiffeners and doubler plates.

GIVEN: Consider the bolted flange plate connection of Example 12.3a. The flange plates are $7\frac{1}{4} \times \frac{3}{4}$ and resist a required force of $P_u = 167$ kips.

SOLUTION

Step 1: Determine the column flange strength based on flange local bending.

This limit state is applicable only for a tension force and the strength is

$$R_n = 6.25t_f^2 F_y = 6.25(0.710)^2(50) = 158 \text{ kips}$$

$$\phi R_n = 0.9(158) = 142 < 167 \text{ kips}$$

Because the strength is less than the applied force, half-depth stiffener plates are required.

Step 2: Determine the column web strength based on web local yielding.

This limit state applies to both tension and compression forces applied to the column web. The bearing length, N, is the sum of the plate thickness plus the $\frac{3}{8}$-in. fillet weld on each side of the plate, thus

$$N = 3/4 + 2(3/8) = 1.50 \text{ in.}$$

and from Manual Table 1-1, $k_{des} = 1.31$ so the web strength is

$$R_n = (5k_{des} + N)F_y t_w = (5(1.31) + 1.50)(50)(0.44) = 177 \text{ kips}$$

$$\phi R_n = 1.0(177) = 177 > 167 \text{ kips}$$

Therefore, this limit state is not exceeded and does not call for stiffeners.

Step 3: Determine the column web strength based on web crippling.

This limit state is applicable only for a compressive force applied to the column. The column web strength is

$$R_n = 0.8t_w^2 \left[1 + 3\left(\frac{N}{d}\right)\left(\frac{t_w}{t_f}\right)^{1.5}\right]\sqrt{\frac{EF_y t_f}{t_w}}$$

$$= 0.8(0.440)^2 \left[1 + 3\left(\frac{1.50}{14.0}\right)\left(\frac{0.440}{0.710}\right)^{1.5}\right]\sqrt{\frac{(29,000)(50)(0.710)}{(0.440)}} = 274 \text{ kips}$$

$$\phi R_n = 0.75(274) = 206 > 167 \text{ kips}$$

Therefore, no stiffeners are required for this limit state.

Step 4:   Determine the column web strength for web compression buckling.

This limit state does not need to be checked unless there are opposing compressive forces on opposite sides of the column. The connection described for this example did not mention any connection on the other side of the column. This limit state can be checked to establish any limits on future connections to this column. The value for h is not given explicitly in the Manual; however, h/t_w is given. Thus, $h = 25.9(0.440) = 11.4$ in. The column web strength is then

$$R_n = \frac{24t_w^3\sqrt{EF_y}}{h} = \frac{24(0.440)^3\sqrt{(29,000)(50)}}{11.4} = 216 \text{ kips}$$

$$\phi R_n = 0.9(216) = 194 > 167 \text{ kips}$$

Thus, this column web does not experience compression buckling if opposing forces less than 194 kips are applied on opposite sides of the column.

Step 5:   Determine the strength of the web for panel zone shear.

Based on yielding of the panel zone, without the interaction of any axial force in the column, the available panel zone shear strength is

$$R_n = 0.6F_y dt_w = 0.6(50)(14.0)(0.440) = 185 \text{ kips}$$

$$\phi R_n = 0.9(185) = 167 \text{ kips}$$

Because this is equal to the force applied by the connection, the panel zone cannot accommodate any additive story shear. For a typical exterior column connection, the story shear and the shear from the connection forces are not additive so this panel will not have a panel zone shear problem unless the column axial load is greater than $0.4P_y$.

Step 6:   Determine the force to be transferred by stiffeners.

The only column web limit state that calls for a stiffener in this example is that of flange local bending, which is an issue for the tension flange only. The force to be transferred through the stiffener is the difference between the applied force and that available through the web, thus

$$R_u = (167 - 142) = 25 \text{ kips}$$

This is clearly a small force to be transferred. Careful review of the limit state of flange local bending shows that if the column flange was 0.770 in. thick instead of 0.710 in. thick, no stiffener plates would be required. In this case, a W14×99 would have eliminated the stiffener problem.

Step 7:   Determine the required stiffener size.

Based on the dimensional requirements for a stiffener

The minimum width of each stiffener is

$$b_s \geq \left(\frac{b_p}{3} - \frac{t_w}{2}\right) = \left(\frac{7.25}{3} - \frac{0.440}{2}\right) = 2.20 \text{ in.}$$

The thickness of the stiffener must be at least

$$t_s \geq \frac{t_p}{2} = \frac{0.750}{2} = 0.375 \text{ in. or} \quad t_s \geq \frac{b_p}{15} = \frac{7.25}{15} = 0.483 \text{ in.}$$

Transverse stiffeners must also extend at least one-half the depth of the column. Therefore, try a 2.25- ×$\frac{1}{2}$-in. stiffener with a $\frac{3}{4}$-in. corner cut off, as shown in Figure 12.8.

For the tension stiffener, the design strength of one stiffener is

$$A_g = (2.25 - 0.750)(0.50) = 0.750 \text{ in.}^2$$

$$\phi R_n = 0.9(36)(0.750) = 24.3 \text{ kips}$$

Therefore, the pair of stiffeners provide $2(24.3) = 48.6$ kips, which is greater than the required strength of 25 kips.

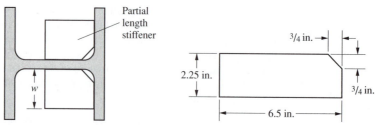

Figure 12.8 Column Stiffener for Example 12.4.

Step 8: Determine the required weld size.

The weld between the loaded flange and stiffener must be sized to transfer the 12.5 kips carried by each stiffener. Fillet welds will be used on the top and bottom of the stiffener. Thus

$$D = \frac{12.5}{1.5(1.392)(2)(1.50)} = 2.00 \text{ sixteenths}$$

and a minimum $\frac{3}{16}$-in. weld is required by Specification Table J2.4.

The weld to the web must transfer the difference between the forces on each end of the stiffener. Because this is a half-depth stiffener, the total force in the stiffener must be transferred to the column web, thus

$$D = \frac{12.5}{(1.392)(2)(6.50 - 0.750)} = 0.781 \text{ sixteenths}$$

and a minimum $\frac{3}{16}$-in. weld is required by Specification Table J2.4.

Step 9: Conclusion.

With the exception of the conditions covered in Steps 4 and 5, the $6\frac{1}{2}$- × $2\frac{1}{4}$- × $\frac{1}{2}$-in. stiffeners as shown in Figure 12.8 with $\frac{3}{16}$-in. fillet welds will be adequate.

EXAMPLE 12.4b
Column Side Limit
States by ASD

GOAL: Check the column side limit states for a moment connection and design any needed stiffeners and doubler plates.

GIVEN: Consider the bolted flange plate connection of Example 12.3b. The flange plates are $7\frac{1}{4} \times \frac{3}{4}$ and resist a required force of $P_a = 111$ kips.

SOLUTION

Step 1: Determine the column flange strength based on flange local bending.

This limit state is applicable only for a tension force and the strength is

$$R_n = 6.25t_f^2 F_y = 6.25(0.710)^2(50) = 158 \text{ kips}$$

$$\phi R_n = \frac{(158)}{1.67} = 94.6 < 111 \text{ kips}$$

Because the strength is less than the applied force, half-depth stiffener plates are required.

Step 2: Determine the column web strength based on web local yielding.

This limit state applies to both tension and compression forces applied to the column web. The bearing length, N, is the sum of the plate thickness plus the $\frac{3}{8}$-in. fillet weld on each side of the plate, thus

$$N = 3/4 + 2(3/8) = 1.50 \text{ in.}$$

and from Manual Table 1-1, $k_{des} = 1.31$ so the web strength is

$$R_n = (5k_{des} + N)F_y t_w = (5(1.31) + 1.50)(50)(0.440) = 177 \text{ kips}$$

$$\frac{R_n}{\Omega} = \frac{(177)}{1.5} = 118 > 111 \text{ kips}$$

Therefore, this limit state is not exceeded and does not call for stiffeners.

Step 3: Determine the column web strength based on web crippling.

This limit state is applicable only for a compressive force applied to the column. The column web strength is

$$R_n = 0.8 t_w^2 \left[1 + 3\left(\frac{N}{d}\right)\left(\frac{t_w}{t_f}\right)^{1.5} \right] \sqrt{\frac{E F_y t_f}{t_w}}$$

$$= 0.8(0.440)^2 \left[1 + 3\left(\frac{1.50}{14.0}\right)\left(\frac{0.440}{0.710}\right)^{1.5} \right] \sqrt{\frac{(29,000)(50)(0.710)}{(0.440)}} = 274 \text{ kips}$$

$$\frac{R_n}{\Omega} = \frac{(274)}{2.00} = 137 > 111 \text{ kips}$$

Therefore, no stiffeners are required for this limit state.

Step 4: Determine the column web strength for web compression buckling.

This limit state does not need to be checked unless there are opposing compressive forces on opposite sides of the column. The connection described for this example did not mention any connection on the other side of the column. This limit state can be checked to establish any limits on future connections to this column. The value for h is not given explicitly in the Manual; however, h/t_w is given. Thus, $h = 25.9(0.440) = 11.4$ in. The column web strength is then

$$R_n = \frac{24 t_w^3 \sqrt{E F_y}}{h} = \frac{24(0.440)^3 \sqrt{(29,000)(50)}}{11.4} = 216 \text{ kips}$$

$$\frac{R_n}{\Omega} = \frac{(216)}{1.67} = 129 > 111 \text{ kips}$$

Thus, this column web does not experience compression buckling if opposing forces less than 129 kips are applied at opposite sides of the column.

Step 5: Determine the strength of the web for panel zone shear.

Based on yielding of the panel zone, without the interaction of any axial force in the column, the available panel zone shear strength is

$$R_n = 0.6 F_y d t_w = 0.6(50)(14.0)(0.440) = 185 \text{ kips}$$

$$\frac{R_n}{\Omega} = \frac{(185)}{1.67} = 111 \text{ kips}$$

Because this is equal to the force applied by the connection, the panel zone cannot accommodate any additive story shear. For a typical exterior column connection, the story shear and the shear from the connection forces are not additive so this panel will not have a panel zone shear problem unless the column axial load is greater than $0.4 P_y$.

Step 6: Determine the force to be transferred by stiffeners.

The only column web limit state that calls for a stiffener in this example is that of flange local bending, which is an issue for the tension flange only. The force to be transferred through the stiffener is the difference between the applied force and that available through the web, thus

$$R_a = (111 - 94.6) = 16.4 \text{ kips}$$

This is clearly a small force to be transferred. Careful review of the limit state of flange local bending shows that if the column flange was 0.77 in. thick instead of 0.71 in. thick, no stiffener plates would be required. In this case, a W14×99 would have eliminated the stiffener problem.

Step 7: Determine the required stiffener size.

Based on the dimensional requirements for a stiffener
The minimum width of each stiffener is

$$b_s \geq \left(\frac{b_p}{3} - \frac{t_w}{2} \right) = \left(\frac{7.25}{3} - \frac{0.440}{2} \right) = 2.20 \text{ in.}$$

The thickness of the stiffener must be at least

$$t_s \geq \frac{t_p}{2} = \frac{0.750}{2} = 0.375 \text{ in. or} \quad t_s \geq \frac{b_p}{15} = \frac{7.25}{15} = 0.483 \text{ in.}$$

Transverse stiffeners must also extend at least one-half the depth of the column. Therefore, try a 2.25 × ½-in. stiffener with a ¾-in. corner cut off, as shown in Figure 12.8.

For the tension stiffener, the design strength of one stiffener is

$$A_g = (2.25 - 0.750)(0.50) = 0.750 \text{ in.}^2$$

$$\frac{R_n}{\Omega} = \frac{(36)(0.750)}{1.67} = 16.2 \text{ kips}$$

Therefore, the pair of stiffeners provide 2(16.2) = 32.4 kips, which is greater than the required strength of 16.4 kips.

Step 8: Determine the required weld size.

The weld between the loaded flange and stiffener must be sized to transfer the 8.20 kips carried by each stiffener. Fillet welds will be used on the top and bottom of the stiffener. Thus

$$D = \frac{8.20}{1.5(0.928)(2)(1.50)} = 1.96 \text{ sixteenths}$$

and a minimum ³⁄₁₆-in. weld is required by Specification Table J2.4.

The weld to the web must transfer the difference between the forces on each end of the stiffener. Because this is a half-depth stiffener, the total force in the stiffener must be transferred to the column web, thus

$$D = \frac{8.20}{(0.928)(2)(6.50 - 0.750)} = 0.768 \text{ sixteenths}$$

and a minimum ³⁄₁₆-in. weld is required by Specification Table J2.4.

Step 9: Conclusion.

With the exception of the conditions covered in Steps 4 and 5, the 6½- × 2¼ × ½-in. stiffener as shown in Figure 12.8 with ³⁄₁₆-in. fillet welds will be adequate.

12.5 PROBLEMS

1. Design a bolted flange-plate connection to connect a W21×57 beam to the flange of a W14×99 column. The connection must transfer a dead load moment of 36 ft-kips and a live load moment of 110 ft-kips, and a dead load shear of 6.7 kips and a live load shear of 20 kips. The plates are A36 steel and welded to the column with E70 electrodes. Use ³⁄₄-in., A325N

bolts. The beam and column are A992 steel. Design by (a) LRFD and (b) ASD.

2. For the design from Problem 1, determine the column stiffening requirements. If stiffeners or doubler plates are required, design the stiffeners and doublers by (a) LRFD and (b) ASD.

3. Design a welded flange-plate connection to connect a W24×103 beam to the flange of a W14×159 column. The connection must transfer a dead load moment of 167 ft-kips and a live load moment of 500 ft-kips, and a dead load shear of 12 kips and a live load shear of 35 kips. The beam and column are A992 steel and the plates are A36. Use E70 electrodes and $^3/_4$-in., A325N bolts. Design by (a) LRFD and (b) ASD.

4. For the design from Problem 3, determine the column stiffening requirements. If stiffeners or doubler plates are required, design the stiffeners and doublers by (a) LRFD and (b) ASD.

5. Design a bolted flange-plate connection to connect a W24×76 beam to the flange of a W14×120 column. The connection must transfer a dead load moment of 45 ft-kips and a live load moment of 135 ft-kips, and a dead load shear of 10 kips and a live load shear of 30 kips. The plates are A36 steel and welded to the column with E70 electrodes. Use $^3/_4$-in., A325N bolts. The beam and column are A992 steel. Design by (a) LRFD and (b) ASD.

6. For the design from Problem 5, determine the column stiffening requirements. If stiffeners or doubler plates are required, design the stiffeners and doublers by (a) LRFD and (b) ASD.

7. Design a welded flange-plate connection to connect a W24×117 beam to the flange of a W14×176 column. The connection must transfer a dead load moment of 150 ft-kips and a live load moment of 450 ft-kips, and a dead load shear of 10 kips and a live load shear of 30 kips. The beam and column are A992 steel and the plates are A36. Use E70 electrodes and $^3/_4$-in., A325N bolts. Design by (a) LRFD and (b) ASD.

8. For the design from Problem 7, determine the column stiffening requirements. If stiffeners or doubler plates are required, design the stiffeners and doublers by (a) LRFD and (b) ASD.

9. Design a direct-welded flange moment connection to connect a W24×76 beam to the flange of a W14×99 column. The connection must transfer a dead load moment of 80 ft-kips and a live load moment of 240 ft-kips, and a dead load shear of 15 kips and a live load shear of 45 kips. The beam and column are A992 steel and the web plate is A36. Use E70 electrodes and $^3/_4$-in., A325X bolts. Design by (a) LRFD and (b) ASD.

10. For the design from Problem 9, determine the column stiffening requirements. If stiffeners or doubler plates are required, design the stiffeners and doublers by (a) LRFD and (b) ASD.

Chapter 13

Safeco Field, Seattle.
Photo courtesy Michael Dickter/
Magnusson Klemencic Associates.

Steel Systems for Seismic Resistance

13.1 INTRODUCTION

For wind and gravity loads, structural analysis and design are normally performed by assuming that the structural response remains elastic. In seismic design, this assumption is too restrictive, particularly for applications that involve significant ground motion. That is, the structural response in a strong earthquake is naturally inelastic and an elastic analysis may unnecessarily overestimate the resulting forces and incorrectly underestimate the deformations.

This chapter provides an introduction to the design of steel building structures for seismic resistance. The requirements of the Specification are supplemented by the AISC Seismic Provisions for Structural Steel Buildings, ANSI/AISC 341-05, to provide appropriate guidance when designing for seismic loads. In this chapter, this standard is referred to as the Seismic Provisions.

To account for the inelasticity that is expected in the response of a structure to a seismic event, the approach used in the Seismic Provisions, the National Earthquake Hazard Reduction Program (NEHRP) Provisions, ASCE 7, and the *International Building Code* incorporates a seismic response modification factor, R, a drift amplification factor, C_d, and a system overstrength factor, Ω_o, which permit the use of an elastic analysis. These factors are incorporated as follows:

- R is used as a divisor when determining the seismic force for which the structure will be designed. Higher R values represent higher ductility levels in the structural system, thus reducing the resulting seismic forces in proportion to this ductility.
- C_d is used as a multiplier when determining the story drift. Lower C_d values represent higher levels of structural stiffness and therefore lower story drift.
- Ω_o is used as a multiplier in seismic load combinations. It increases the design loads to account for the level of overstrength present in a system so that the analysis reflects a more accurate prediction of the onset of inelastic behavior.

To determine the values of R, C_d, and Ω_o that are appropriate for a design, buildings are categorized based upon occupancy and use. In the NEHRP Provisions, buildings are assigned to one of three seismic use groups and then to a seismic design category based upon the seismic use group, the expected acceleration and soil characteristics, and the period of the building. Provisions in ASCE 7 and the *International Building Code* vary slightly but are similar.

Seismic design categories A, B, and C generally correspond to a classification of low to moderate seismicity. In these cases, the engineer can choose to use a basic steel structure with no special detailing, for which $R = 3$, $C_d = 3$, and $\Omega_o = 3$. Alternatively, the engineer can choose to use a system defined in the Seismic Provisions and take advantage of a higher R factor.

Seismic design categories D, E, and F generally correspond to a classification of high seismicity. In such cases, the engineer must use a structural system defined in the Seismic Provisions. The remainder of this chapter discusses the structural systems provided in the Seismic Provisions for resisting seismic forces—those in which R is taken greater than 3. The reader is encouraged to review the Seismic Provisions in detail.

13.2 EXPECTED BEHAVIOR

For gravity loads, wind loads, and seismic loads associated with smaller earthquakes, it is expected that the structural response will be elastic. However, for larger earthquakes, it is recognized that it may be impractical or impossible to prevent some inelastic behavior. For this reason, and because there is no guarantee that an actual earthquake will be less than that defined for design purposes in the building code, the Seismic Provisions are based upon a capacity design methodology. Accordingly, the provisions contained for each system are intended to result in a structure in which controlled inelastic deformations can occur during a strong earthquake to dissipate the energy imparted to the building by the ground motion. These inelastic deformations are forced to occur in a predictable manner and in specific elements and/or locations in the structural system. The remainder of the structure remains elastic as these deformations occur, protected in much the same way that a fuse protects the wiring in a circuit from overload.

Given this basic premise of the capacity design methodology, the fuse elements often establish the design requirements for the members and connections that surround them. This has varying implications for different types of systems.

- As illustrated in Figure 13.1, in a moment frame, the fuse element is typically a plastic hinge that forms in the girders just outside the girder-to-column connection. Accordingly, the girder-to-column connections, column panel zones, and columns must all be designed to develop the flexural strength of the girders connected to them.

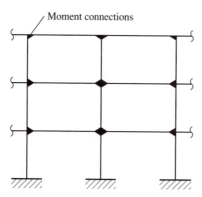

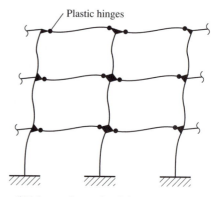

(a) Moment frame before seismic deformations

(b) Moment frame after deformations occur in large earthquake

Figure 13.1 Moment-Frame Systems.

- As illustrated in Figure 13.2, in a concentrically braced frame, the fuse element is usually a compression buckling/tension yielding mechanism formed in the bracing member itself. Accordingly, the brace-to-gusset connections, gussets, gusset connections, beams, and columns must all be designed to develop the tension yield strength and compression buckling strength of the braces that connect to them.

Regardless of the system chosen, fuse elements must deform in a predictable and controlled manner, and provide a ductility that exceeds the level of deformation anticipated. Thus, the systems are configured so that limit states with higher ductility, such as yielding, have control over limit states with lesser ductility, such as rupture.

The actual material properties, such as steel yield strength and strain hardening effects, can influence the behavior of the system. As discussed throughout this book, steel is specified by ASTM designation, which identifies the specified minimum yield strength, among other characteristics. The actual yield strength, however, is most likely higher than the specified value. Also, once strain hardening begins to take place, the effects of load reversals will tend to further elevate the apparent yield strength. The difference between the actual yield strength and specified minimum yield strength and strain hardening effects are important

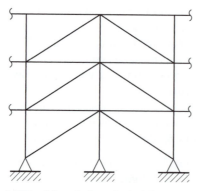

(a) Braced frame before seismic deformations

(b) Braced frame after deformations occur in large earthquake

Figure 13.2 Braced-Frame Systems.

in the capacity design methodology because they increase the strength required in the remainder of the structure to permit yielding in the fuse elements.

These effects are treated directly with multipliers in the Seismic Provisions. First, a multiplier R_y is given for each grade of steel. When applied to the specified minimum yield strength, F_y, the resulting quantity is the expected yield strength, R_yF_y. Second, an allowance is made for the effects of strain hardening, generally with a factor of 1.1. Thus, the Seismic Provisions use an elevated yield strength, generally equal to $1.1R_yF_y$, when determining the strength of fuse elements and the resulting design forces for connections and members surrounding the fuse elements.

13.3   MOMENT-FRAME SYSTEMS

The moment-frame systems given in the Seismic Provisions generally use flexural fuse elements, usually plastic hinges forming in the girders just outside the fully restrained (FR) girder-to-column moment connections. Three types of moment-frame systems are addressed in the Seismic Provisions: special moment frames (SMF) in Section 9, intermediate moment frames (IMF) in Section 10, and ordinary moment frames (OMF) in Section 11. SMF and IMF use connections that have demonstrated at least 0.03 radians and 0.01 radians inelastic rotation, respectively, in testing. Some typical seismic moment connections are shown in Figures 13.3 and 13.4. OMF use a prescriptive connection that provides for small inelastic demands. Assuming that the elastic drift of a moment frame is 0.01 radians and the inelastic drift is equal to the inelastic rotation at the connections, SMF, IMF and OMF provide for interstory drifts of 0.04, 0.02, and 0.01 radians, respectively.

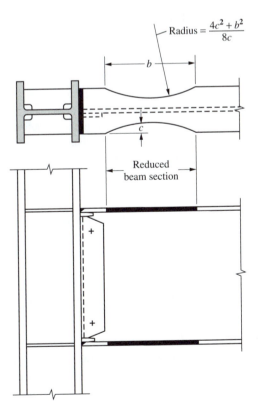

Figure 13.3 Typical Seismic-Reduced Beam Section (RBS) Moment Connection.

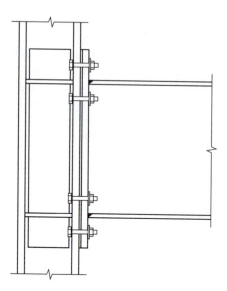

Figure 13.4 Typical Seismic End-Plate Moment Connection.

The values of R, C_d, and Ω_o provided in the NEHRP Provisions for each of these three systems are as follows:

System	R	C_d	Ω_o
SMF	8	3	$5\frac{1}{2}$
IMF	$4\frac{1}{2}$	3	4
OMF	$3\frac{1}{2}$	3	3

The use of SMF is not limited in any seismic design categories, whereas IMF and OMF usage is restricted based upon seismic design category, building height, and structural configuration.

13.3.1 Special Moment Frames (SMF)

SMF are configured to form fuses through plastic hinging in the beams, usually adjacent to the beam-to-column connection, to accommodate significant inelastic deformation during large seismic events. There may also be some inelastic deformation in the column panel zone. Several requirements are included in the Seismic Provisions to promote this behavior, as described in the ensuing sections.

Fuse Strength

With the plastic hinges forming in the beams, the fuse flexural strength is $1.1R_y M_p$. The girder-to-column connections, column panel zones, and columns must all be designed to allow the fuse to develop this flexural strength. Alternative approaches recognized in the Seismic Provisions include moment-frame systems with partially restrained (PR) connections and weak panel-zone systems, wherein the fuses would form through connection deformations and panel-zone shear deformations, respectively.

Beam-to-Column Connections

The moment connections used in SMF must have supporting tests demonstrating conformance with the ductility requirements, such as through the use of a connection listed in the AISC *Prequalified Connections for Special and Intermediate Steel Moment Frames for Seismic Applications* (AISC 358-05). Alternatively, the use of connections qualified by prior testing or project-specific testing is acceptable.

Panel Zone Requirements

Some inelastic deformation in the column panel zone is permitted, and is in many cases beneficial to the system performance. A panel zone consistent with tested assemblies is required, and requirements in the Seismic Provisions generally result in stiff panel zones with limited yielding. It may also be necessary to reinforce the column with a web doubler plate for shear and/or transverse stiffeners in the column at the beam flanges for the flange forces transferred to the column by the beam flanges.

Beam and Column Compactness

The compactness criteria in the AISC Specification are predicated based on a required ductility level of 3, whereas the expected member ductility demands for beams and columns in SMF can be on the order of 6 or 7. Accordingly, beams and columns in SMF must meet the more stringent width-thickness limits in the Seismic Provisions.

Prevention of Story Mechanisms

In SMF, a strong-column weak-beam relationship must be satisfied in proportioning the columns. This requirement is formulated as a check of the moment ratio between the beam(s) and column(s) at each moment-connected joint in the structure. However, this check is not intended to eliminate all column yielding. Rather, it is a simplified approach that results in a framing system with columns strong enough to force flexural yielding in beams at multiple levels of the frame. This prevents a story mechanism, as shown in Figure 13.5, from forming and achieves a higher level of energy dissipation. Some exceptions are permitted, as in the case for a one-story building, where it would not increase energy dissipation if the beams yielded instead of the columns.

Stability Bracing Requirements

Special stability bracing requirements apply in SMF because the bracing must be suitable to maintain the position of the braced elements well into the inelastic range. For beams, the permitted unbraced length is generally reduced, and bracing is required near the location

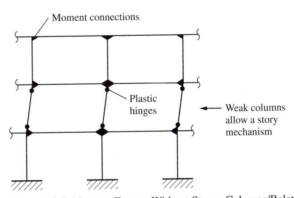

Figure 13.5 Moment Frames Without Strong Columns/Relationship.

where the plastic hinge is expected to form. The Seismic Provisions provide several options for bracing the beams at the beam-to-column connection and the column. Often, the configuration of the gravity framing and interconnection of the floor slab to the beam can be used to satisfy these requirements.

Protected Zones

The fuse regions in SMF—the plastic hinge regions in the beams—are expected to undergo significant inelastic deformations. Accordingly, attachments and other potential notch-effect-inducing conditions are prohibited in these areas.

13.3.2 Intermediate Moment Frames (IMF) and Ordinary Moment Frames (OMF)

IMF and OMF systems are similar in configuration to SMF, but do not provide as high a capacity to accommodate inelastic deformation during large seismic events. The Seismic Provisions requirements for SMF, IMF, and OMF emphasize that IMF and OMF are subject to lesser special requirements than SMF. In fact, IMF and OMF are often subject to no additional requirements beyond those in the AISC Specification.

IMF are based on the use of a tested connection design with a qualifying interstory drift angle of 0.02 radians. That is, IMF are subject to the same connection testing requirements as SMF, but with a lesser required interstory drift angle. OMF are based on a prescriptive design procedure and an expected interstory drift angle of 0.01 radians, which corresponds to a nominally elastic response.

13.4 BRACED-FRAME SYSTEMS

The braced-frame systems in the Seismic Provisions fall into two categories: concentric and eccentric. Concentrically braced frames generally use axial fuse elements—usually the braces themselves, which yield in tension and/or buckle in compression. Eccentrically braced frames generally use shear and/or flexural fuse elements—usually a segment, called a link, in the beams themselves between the braces.

Three types of braced-frame systems are addressed in the Seismic Provisions: special concentrically braced frames (SCBF) in Section 13, ordinary concentrically braced frames (OCBF) in Section 14, and eccentrically braced frames (EBF) in Section 15. The values of R, C_d, and Ω_o provided in the NEHRP Provisions for each of these three systems are as follows:

System	R	C_d	Ω_o
SCBF	6	2	5
OCBF	5	2	$4\frac{1}{2}$
EBF	8 or 7*	2	4

*$R = 8$ if beam-to-column connections away from EBF link is a moment connection; $R = 7$ otherwise.

All braced-frame systems have building height restrictions that vary based on the seismic design category.

13.4.1   Special Concentrically Braced Frames (SCBF)

SCBF are configured to form fuses through tension yielding and compression buckling of the braces between the end connections, to accommodate significant inelastic deformation during large seismic events. Several requirements are included in the Seismic Provisions to promote this behavior, as described in the ensuing sections.

Fuse Strength

The fuse axial strength in tension is $1.1R_yF_yA_g$, a quantity that is usually larger than the fuse axial strength in compression and, thus, controls the force requirements. The brace-to-gusset connections, gussets, gusset-to-beam and gusset-to-column connections, beams, and columns must all be designed to permit the brace to develop this full axial strength in tension.

Gusset Requirements

Most braces and gussets are detailed so that out-of-plane buckling occurs before in-plane buckling. When this is the case, weak-axis bending is induced in the gusset by the end rotations and the gusset must be detailed to accommodate these rotations. Accordingly, a free length of two times the plate thickness must be provided between the end of the brace and the bend line in the gusset plate, as illustrated in Figure 13.6. The bend line in the gusset is a line perpendicular to the brace axis that passes through the point on the gusset edge connection that is nearest to the brace end. Alternatively, the bracing connection can be detailed to force the deformation into the bracing member, with buckling occurring either in-plane or out-of-plane.

Brace Slenderness

The slenderness ratio, Kl/r, of the brace affects post-buckling cyclic performance of the system. Accordingly, a maximum brace slenderness ratio of 200 is permitted, and special provisions apply when the slenderness ratio exceeds $4.0\sqrt{E/F_y}$.

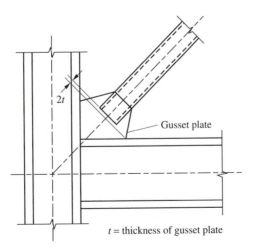

2t

Gusset plate

t = thickness of gusset plate

Figure 13.6 Typical Seismic Bracing Connection.

Brace Net Section Limitations

The net section of the brace must be large enough to allow tension yielding to control over tension rupture. Most end connections involve a net section that may require reinforcement to satisfy this requirement.

Distribution of Bracing

Braces must be used in a manner such that the lateral forces in all stories are resisted by a combination of tension yielding and compression buckling of the brace members. Although a 50-50 distribution is considered ideal, the provisions allow up to 70% of the lateral force to be resisted by tension or compression braces, unless it can be shown that the system response is essentially elastic. The mixing of tension and compression braces improves the buckling and post-buckling strength of the system and helps prevent accumulation of inelastic drifts in one direction.

Beam, Column, and Brace Compactness

The compactness criteria in the AISC Specification are predicated based on a required ductility level of 3, whereas the expected member ductility demands for beams and columns in SMF can be on the order of 6 or 7. Accordingly, beams and columns in SCBF must meet the more stringent width-thickness limits in the Seismic Provisions. The more stringent width-thickness limitations for braces also improve the fracture resistance and post-buckling cyclic performance of the braces.

Bracing Configurations

A variety of bracing configurations can be used. Some configurations require special considerations whereas others such as K-bracing, are not permitted. Figure 13.7 illustrates several bracing configurations.

In V-braced and inverted-V-braced frames, the expected yielding and buckling behavior of the braces creates an unbalanced vertical force because the tension brace remains effective as it yields but the compression brace is ineffective after buckling. This unbalanced force must be resisted by the intersecting beam, as well as its connections and supporting members. That is, the beam must be designed for the corresponding load redistribution in addition to the gravity loads. Alternatively, the bracing configuration can be altered to eliminate the potential for unbalanced loading. For example, the V and Inverted-V configurations can be

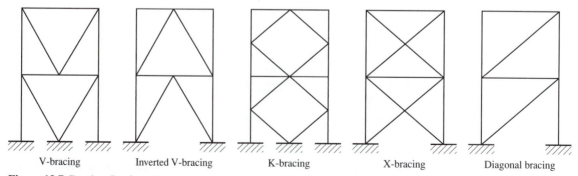

| V-bracing | Inverted V-bracing | K-bracing | X-bracing | Diagonal bracing |

Figure 13.7 Bracing Configurations.

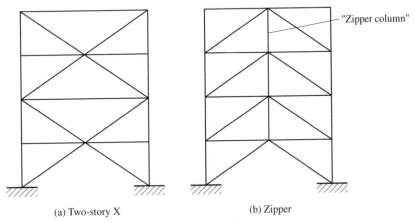

(a) Two-story X   (b) Zipper

Figure 13.8 Two-Story X and Zipper Configurations.

alternated to form a two-story X configuration. Another approach involves the addition of a zipper column. These bracing configurations are illustrated in Figure 13.8.

K-bracing (and knee bracing) is prohibited in SCBF because the configuration results in an unbalanced force in the columns.

X-bracing is allowed. However, the common tension-only design approach for wind forces is not permitted. Both diagonals of the X-bracing must be designed to resist the tension and compression forces that result from cyclic load reversals.

Single diagonal bracing is not permitted because all braces in this configuration are called upon to resist the same type force, either tension or compression, at the same time. In this case, for one direction of loading, 30% to 70% of the braces would not be in tension. Thus, the diagonal braced frame should be implemented as previously shown in Figure 13.2.

Stability Bracing Requirements

Special stability bracing requirements apply in SCBF because the bracing must be suitable to maintain the position of the braced elements well into the inelastic range. Often, the configuration of the gravity framing and interconnection of the floor slab to the beam can be used to satisfy these requirements.

Protected Zones

The fuse regions in SCBF—the braces and gussets–are expected to undergo significant inelastic deformations. Accordingly, attachments and other potential notch-effect-inducing conditions are prohibited in these areas.

13.4.2 Ordinary Concentrically Braced Frames (OCBF)

OCBF are similar in configuration to SCBF, but do not provide as high a capacity to accommodate inelastic deformation during large seismic events. The Seismic Provisions requirements for SCBF and OCBF emphasize that OCBF are subject to less special requirements than SCBF.

13.4.3   Eccentrically Braced Frames (EBF)

EBF are configured to form fuses through shear yielding, flexural yielding, or a combination of the two in the EBF link, in order to accommodate significant inelastic deformation during large seismic events. Unlike the behavior of SCBF and OCBF, the braces in EBF are intended to remain nominally elastic. The requirements in the Seismic Provisions to promote this behavior are described in the ensuing sections.

Fuse Strength

The fuse strength in shear is $1.25 R_y V_n$, where V_n is the nominal shear strength of the link, which is the lesser of the nominal plastic shear strength and the shear associated with flexural yielding of the link. The strain hardening multiplier used for EBF is 1.25. This is higher than the multiplier used in determining the strain hardening effects for other systems because EBF exhibit more strain hardening effects. The beam segments outside the link, gussets, gusset-to-beam and gusset-to-column connections, and columns must all be designed to develop the shear and/or flexural yielding mechanism in the links.

Link Location

EBF links are usually located as segments within the length of the beams, either between braces or between a brace and a beam-to-column connection. Alternatively, links can be provided as vertical elements between beams and V or inverted-V bracing. EBF configurations are shown in Figure 13.9.

When links are located as segments within the length of the beams, it is preferable to locate the links between the ends of the braces. When the links are located between braces and the beam-to-column connections, the beam-to-column connections require special consideration because the rotational demands are substantially higher than those at a beam-to-column connection in an SMF. In applications involving a significant flexural demand, a prequalified connection or a connection qualified by testing must be used. When the links used are short enough that shear yielding dominates, the need for qualification testing is eliminated if the connection is reinforced with haunches or other suitable reinforcement designed to preclude inelastic action in the reinforced zone adjacent to the column.

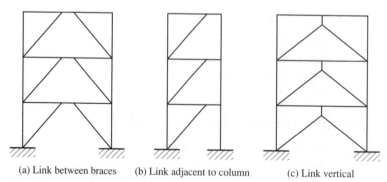

(a) Link between braces   (b) Link adjacent to column   (c) Link vertical

Figure 13.9 Configuration of EBF.

Link Rotations

Link rotations in EBF are limited to 0.02 radians for flexural links and 0.008 radians for shear links. For links that deform in combined shear and flexure, the rotation limit is determined by linear interpolation between these limits.

Link Stiffening and Bracing

To ensure that the required rotations can be achieved and yielding occurs without local buckling well into the inelastic range, links are stiffened as prescribed in Seismic Provisions Section 15.3. The use of web doubler plates to stiffen links is not permitted because this type of reinforcement does not deform consistently with the web deformations. Additionally, beam web penetrations within the link are not permitted and the link must be braced against out-of-plane displacement and twist at the ends of the link.

Braces and Beam Segments Outside of Links

Because the inelastic action in EBF is intended to occur primarily within the links, the braces and beam segments outside of the links must be designed to remain nominally elastic as the links deform. Limited yielding outside of the links is allowed, as long as the beam segments outside the links and braces have sufficient strength to develop the fully yielded and strain-hardened strength of the links. The braces and beam segments outside the links are normally designed as members subject to the combined effects of axial force and flexure.

13.5   OTHER FRAMING SYSTEMS

Several other systems are provided for in the Seismic Provisions, including special truss moment frames (STMF) in Section 12, buckling-restrained braced frames (BRBF) in Section 16, and special plate shear walls (SPSW) in Section 17. The values of R, C_d, and Ω_o provided in the NEHRP Provisions for each of these three systems are as follows:

System	R	C_d	Ω_o
STMF	7	3	$5\frac{1}{2}$
BRBF*	8 or 7	$2\frac{1}{2}$ or 2	5 or $5\frac{1}{2}$
SPSW	7	2	6

*The first number applies in each category if the beam-to-column connections are moment connections; the second number in each category applies otherwise.

Each of these systems has building height restrictions that vary based on the seismic design category.

Composite steel and reinforced concrete systems are also provided for in Part II of the Seismic Provisions.

13.5.1   Special Truss Moment Frames (STMF)

STMF are configured to form fuses through yielding in a special segment of the truss, to accommodate significant inelastic deformation during large seismic events. The special segment can be either a truss panel with diagonals or a Vierendeel truss panel. The remainder of the truss and framing in the system is designed to remain elastic as the special segment deforms. A schematic STMF is illustrated in Figure 13.10.

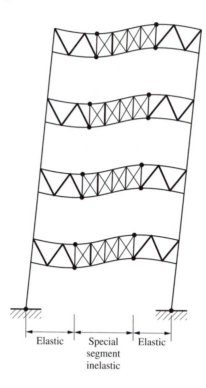

Figure 13.10 STMF Configuration.

When diagonals are used, the yielding of the special segment occurs by axial tension yielding and compression buckling of the diagonals. Diagonal web members used in the special segments of STMF systems are limited to flat bars only, and must meet a limiting width-thickness ratio of 2.5.

When a Vierendeel panel is used, yielding of the special segment occurs by flexural yielding of the chord members.

The size of the truss and size and location of the special segment are limited to correspond with the research on which the system is based. It is desirable to locate the STMF special segment near mid-span of the truss because shear due to gravity loads is generally lower in that region.

Other than the normal gravity loads carried by the frame, no major structural loading is permitted in the special segment.

13.5.2 Buckling-Restrained Braced Frames (BRBF)

BRBF are a type of concentrically braced frame system that has special bracing elements, as shown in Figure 13.11. These bracing elements provide essentially the same response in compression as they do in tension. The bracing elements are composed of a load-bearing core and a surrounding sleeve element that restrains the global buckling of the core, forcing yielding in compression rather than buckling. BRBF are configured to form fuses through tension yielding and compression yielding of these special bracing elements, to accommodate significant inelastic deformation during large seismic events.

The bracing elements must be qualified by testing to ensure that the braces used provide the necessary strength and deformation capacity. The required deformation capacity is amplified beyond what is required by an analysis in recognition that actual deformations can be larger than those predicted by analysis.

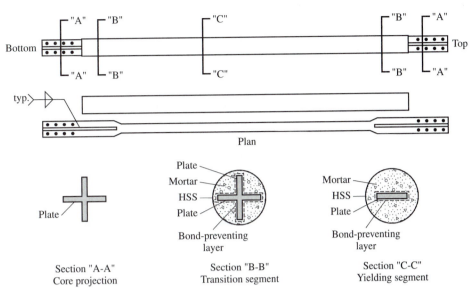

Figure 13.11 BRBF Bracing Element.

The steel core has a yielding segment that is designed with a cross-sectional area and length based on strength, stiffness, and strain demands. Because each bracing element is a manufactured item, the designer can specify an array of braces that promote distributed yielding throughout the frame.

The steel core projections beyond the yielding segment are designed to provide the transition from the core and connection to the remainder of the framing system. The projections are designed so that they remain nominally elastic like the rest of the frame as the yielding segment deforms.

13.5.3   Special Plate Shear Walls (SPSW)

SPSW have slender, unstiffened plate elements surrounded by and connected to horizontal and vertical boundary elements that are rigidly interconnected. A schematic SPSW system is illustrated in Figure 13.12. SPSW are configured to form fuses through plate yielding and buckling (tension field action), to accommodate significant inelastic deformation during large seismic events.

Although plastic hinging is anticipated at the ends of horizontal boundary elements, the boundary elements, like the rest of the framing system, are designed to remain essentially elastic as the plates deform. The tension-field action in SPSW is analogous to that in a plate girder, but the behavior and strength of SPSW differs from that of plate girders. Accordingly, the design requirements in the Seismic Provisions for SPSW differ from those in the AISC Specification for plate girders.

13.5.4   Composite Systems

A variety of composite structural systems are provided for in Part II of the Seismic Provisions. These systems include Composite Partially Restrained Moment Frames (C-PRMF), Composite Special Moment Frames (C-SMF), Composite Intermediate Moment Frames (C-IMF), and Composite Ordinary Moment Frames (C-OMF). The requirements of Part II of the Seismic Provisions are applied in addition to those of the Seismic Provisions and the Specification.

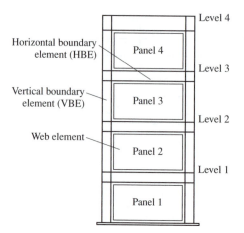

Horizontal boundary element (HBE)

Vertical boundary element (VBE)

Web element

Level 4

Level 3

Level 2

Level 1

Panel 4

Panel 3

Panel 2

Panel 1

Figure 13.12 SPSW System.

13.6   OTHER GENERAL REQUIREMENTS

13.6.1   Bolted and Welded Connections

Connections in the seismic force resisting system must be configured such that a ductile limit state in the fuse controls—the deformations occur in the fuse elements before failure occurs in the connections. This generally means that connections in the seismic force resisting system are much larger than they would be if designed for gravity, wind, and low-seismic applications. There are additional special requirements for the use of bolts and welds in the Seismic Provisions.

Bolted joints in shear are designed as pretensioned bearing joints with faying surfaces prepared as for Class A or better slip-critical connections. These are not slip-critical connections—they are bearing joints with some slip resistance. Because slip cannot and need not be prevented in large ground motions, the intent is to control slip in lesser ground motions and pretension the bolts because large ground motions can cause full reversal of design load.

Hole type usage is restricted to standard holes and short-slotted holes perpendicular to the loading direction, unless another hole type is shown acceptable by testing. One exception provided is that oversized holes are permitted in brace diagonals within certain limits.

For design purposes, bolt bearing checks are required to be made at the *deformation considered* level, to prevent excessive deformations of bolted joints due to bearing on the connected material, primarily to minimize damage in lesser ground motions.

In welded connections, filler metal with a minimum specified Charpy V-notch toughness of 20 ft-lbs at 0°F is required in all welds involved in the seismic load path, except for *demand critical* welded joints, which have more stringent notch toughness requirements.

It is prohibited to share a common force between bolts and welds because seismic deformation demands generally exceed the deformation compatibility required for loads to be shared between welds and bolts.

13.6.2   Protected Zones

The fuse elements in the various systems covered in the Seismic Provisions may undergo significant inelastic deformations when subjected to large ground motions. Accordingly, construction operations that might cause discontinuities must be restricted from these areas. Thus, the Seismic Provisions designate protected zones in each system that must be kept free of sharp transitions, penetrations, notches, and so forth. Discontinuities that are inadvertently created in these zones must generally be repaired.

13.6.3   Local Buckling

The yielding of fuse elements requires member ductility of 6 or 7, which is more than the normal ductility of 3 used in the development of the compactness criteria in the Specification. Thus, in the Seismic Provisions, more stringent seismic compactness criteria are provided in Table I-8-1.

13.6.4   Column Requirements

Special requirements for columns and column splices in the seismic force resisting system are stipulated in the Seismic Provisions. Minimum design forces are specified to preclude column and column splice failure in compression or tension. This approach does not necessarily preclude yielding of the column, and some guidance is provided in the Commentary for cases in which yielding of the column might be of concern.

Column splices must be located away from the beam-to-column connections, generally within the middle third of the story height in which the splice occurs, to reduce the effects of flexure. Additionally, if partial-joint-penetration groove welds are used to make column splices, a 100% increase in required strength is specified and the use of notch-tough filler metal is required.

There are also requirements for columns that are not a part of the seismic load resisting system, because these columns are still active in distributing the seismic shears between the floors.

13.6.5   Column Bases

To increase frame stiffness, column bases are normally treated similarly to beam-to-column moment connections, accounting for the inherent differences, such as the increased flexibility due to deformations in longer anchor rods, compressibility of the grout and concrete, and foundation rocking effects.

13.7   CONCLUSIONS

This introduction to the design of steel structures for seismic force resistance is intended to provide a starting point for further study. The detailed provisions are found in the Seismic Provisions for Steel Buildings, ANSI/AISC 341-05, and additional guidance is found in the AISC Seismic Design Manual. The interested student is encouraged to study these two documents for a more in-depth treatment of seismic design of steel structures.

13.8   PROBLEMS

1. What is the major difference between the analysis and design of a structure for wind and gravity loads versus seismic loads?

2. Explain the use of the R, C_d, and Ω_o factors. What do these factors account for?

3. How are the R, C_d, and Ω_o factors determined for a particular analysis?

4. What is the purpose of fuse elements in seismic design? Provide some examples of structural fuse elements.

5. What type of fuse elements are typically used in moment frame systems?

6. What are the three types of moment frames considered in the Seismic Provisions? What are the respective values for R, C_d, and Ω_o for each of these systems?

7. For Special Moment Frames, what type of relationship should exist between the column and beams to prevent a story mechanism?

8. Which type of moment frame has a ductility requirement for connections of an inter-story drift angle of 0.02 radians?

9. Name the two categories of braced frames provided for in the Seismic Provisions. What type of fuse element is used by each of these?

10. List three types of braced frame systems addressed in the Seismic Provisions and their corresponding values for R, C_d, and Ω_o.

11. How do SCBF and OCBF allow for inelastic deformations in structures?

12. List some examples of CBF configurations.

13. How do EBF differ from CBF in their performance during large seismic events?

14. Where are the fuse elements located for Eccentrically Braced Frames?

15. List some other seismic force resisting systems mentioned in the Seismic Provisions and indicate the corresponding fuse element for each of these.

16. How does the size of connections in seismic force resisting systems differ from connections designed for gravity and wind systems? Why?

17. Is it permitted to share a common force between bolts and welds in seismic design? Why or why not?

18. For seismic design, where should column splices be located and why?

Index